D0085007

WITHDRAWN-UNL

SEISMIC DESIGN OF BUILDINGS AND BRIDGES

SEISMIC DESIGN OF BUILDINGS AND BRIDGES

FOR CIVIL AND STRUCTURAL ENGINEERS

ALAN WILLIAMS, Ph.D., S.E., C.ENG.

Registered Structural Engineer, California; Chartered Engineer, United Kingdom;
Senior Engineer, Department of Transportation, State of California

Engineering Press, Inc **San Jose, California**

© Copyright 1995, Engineering Press

All rights reserved. Except as permitted under the United States Copyright Act of 1976, no part of this publication may be reproduced or distributed in any form or by any means, or stored in a data base or retrieval system, without the prior written permission of the copyright owner.

This book describes various code requirements to help educate engineers in the details of seismic design. It is not a complete enumeration of all code rules, equations or restrictions, hence anyone undertaking structural design must refer to the published codes. Engineering Press and Alan Williams will not be responsible for any direct, incidental or consequential damages resulting from the use of this book.

ISBN 0-910554-04-8

Library of Congress Cataloging-in-Publication Data

Williams, Alan.
Seismic design of buildings and bridges : for civil and structural engineers /Alan Williams.
p. cm.
Includes bibliographical references and index.
ISBN 0-910554-04-8 (hardbound)
1. Earthquake resistant design. 2. Structural design.
3. Bridges--Design and construction. I. Title.
TA658.44.W55 1995
624.1'762--dc20 94-42928
CIP

Printed in the United States of America
4 3 2 1

Engineering Press P.O. Box 1 San Jose CA 95103-0001

CONTENTS

SEISMIC DESIGN OF BUILDINGS AND BRIDGES

Studying for your Professional License?
Over 125 Review Materials....
Books
Videos
Audio Cassettes
... Much More

Request your free Catalog today.

Engineering Press
P.O. Box 1
San Jose, CA 95103-0001
(800) 800-1651 (408) 258-4503
FAX (800) 700-1651 (408) 259-5510

INTRODUCTION

OBJECTIVE OF THE BOOK

The purpose of this textbook is to assist candidates in preparing for examinations requiring a knowledge of seismic principles for the design of buildings and bridges. It is intended to serve as a comprehensive guide and reference for self study of the subject, with the emphasis placed on those analytical and design methods which lead to the quickest and simplest solution of any particular problem. The text is illustrated with the solution of more than one hundred problems, several taken from recent examination papers. Some of the material in this book has appeared in two previous publications[1,2], and the opportunity has been taken to update the text to conform to the requirements of the 1994 edition of the Uniform Building Code[3]. The format of the new Code has been completely revised and many new technical provisions, dealing with seismic design, have been introduced.

This text provides comprehensive clarification and interpretation of the applicable Building Codes and Standard Specifications and extensive reference publications are cited to reflect current design procedures. In addition, an appendix contains calculator programs, written for the HP–28S calculator, to facilitate the solution of frequently occurring calculation procedures. These are easily applied to give a rapid solution to the problem while also presenting all the intermediate stages in the calculation. With minor modifications, the programs are also suitable for use on the Hewlett Packard series 48 calculators.

CALIFORNIA STATE BOARD PROFESSIONAL ENGINEER EXAMINATION

In addition to passing the eight–hour Professional Engineer examination of the National Council of Examiners for Engineering and Surveying (NCEES), the California State Board of Registration requires the candidate to obtain a passing score in two special supplemental two–hour thirty–minute examinations. These examinations cover the topics of seismic principles and engineering surveying principles. The national examination addresses the full scope of civil engineering practice while the Special Civil Engineer examinations assess the candidate's level of competence in the two areas of practice that are of particular concern to the California profession. The scope of the seismic examination is comparable to the seismic design requirements of Part A of the California State Board Structural Engineer examination. The examination currently takes the form of a multiple choice examination and past examination papers are not publicly available. However, typical questions, considered representative of the examination questions, are presented in an appendix together with their solutions. Chapters one, two, and three of this text are of particular relevance to the Special Seismic examination.

CALIFORNIA STATE BOARD STRUCTURAL ENGINEERING EXAMINATION

Registration as a Structural Engineer is required by all Civil Engineers in California who wish to design school and hospital structures and buildings exceeding 160 feet in height. Authority to use the title "Structural Engineer" is conferred in California by the California State Board of Registration for Professional Engineers and Land Surveyors. The title is awarded to those applicants who satisfy the following requirements:

- Possession of a valid registration as a Civil Engineer in California.
- Completion of three years of qualifying experience in responsible charge of structural engineering work.
- Provide satisfactory references from four existing Structural Engineers.
- Obtain a passing score in the 16–hour written structural engineer examination.

In addition to California, the examination is utilized by the States of Nevada, Washington, Hawaii, and Idaho.

The written examination consists of four separate four–hour papers, parts A, B, C and D. Part A examines general structural principles and seismic design. Part B examines structural steel design principles. Part C examines reinforced and prestressed concrete design principles. Part D examines timber design with one question on masonry design. Each paper is allotted a total score of twenty five points and generally contains four to six problems with the maximum number of points awarded to each problem indicated. Each paper contains one multiple–choice question with the remainder being in essay type format. All questions are compulsory in Part A and Part D. In Part B and Part C, the candidate may choose two alternative problems covering the design of bridges or bridge elements. In allocating the time to be spent on each problem, the candidate should allow 9.6 minutes for each point awarded to the problem.

All four parts of the examination are open–book, allowing the candidate to use textbooks, handbooks and bound reference materials. A battery–operated, nonprinting, silent calculator is allowed but loose materials and writing pads are prohibited. To complete all problems within the allotted time, the candidate must work quickly and decisively without spending time searching for solution procedures. Copies of past examination papers are obtainable from the California State Board of Registration[4].

NATIONAL COUNCIL STRUCTURAL ENGINEERING EXAMINATION

The special structural engineering examination of the National Council of Examiners for Engineering and Surveying (NCEES) is utilized by many state licensing boards to provide the qualifying examination for candidates seeking registration as a Structural Engineer. The examination questions are intended to assess the candidate's knowledge and understanding of basic structural engineering analysis and design principles and to determine the candidate's familiarity with design office procedures, standard references and current codes [3,5,6,7,8,9,10,11,12]. Removal of examination booklets from the examination room is prohibited and past examination papers are not publicly available. However, a handbook[13], published by NCEES, is available which illustrates representative problems which have appeared in past examinations. The content of the NCEES Special Structural I examination is similar to the examination administered by the California state licensing board.

ORGANIZATION OF THE TEXT

The text is organized into eight separate sections. Chapter 1 covers general seismic principles. Chapters two and three cover the determination of lateral forces for buildings based on the 1994 edition of the Uniform Building Code. These chapters contain appropriate questions from the Structural Engineer examination papers published by the California State Board of Registration. Chapters 4, 5, 6 and 7 deal with steel, concrete, wood and masonry problems, respectively, and chapter 8 covers the seismic design of bridges. Representative problems, selected from California Structural Engineer examinations, illustrate the subject matter presented in these chapters.

The notation adopted in the text is that used in the Uniform Building Code[3] and in the Standard Specifications for Highway Bridges[6], as appropriate, and conforms to current usage. Occasional duplication of the meaning of a particular symbol has been unavoidable, but the symbols are explained in the text as they occur and ambiguities should not arise.

The author wishes to express his gratitude to the California State Board of Registration for Professional Engineers and Land Surveyors for their kind permission to reproduce material from past examination papers. Thanks are also due to the Building Seismic Safety Council for permission to use the illustration[15] on the dust cover.

In a book of this type, errors may be expected to occur and the author will appreciate being informed of any errors that may be found in the text. Any suggestions or comments may be sent to the author at the publisher's address.

The following abbreviations are used in the text to denote commonly occurring reference sources:

UBC[3]
AISC[5]
AASHTO[6]
AASHTOSD[7]
ACI[8]
AITC[9]
NDS[10]
SEAOC[14]

REFERENCES

1. Williams, A. *Structural Engineer Registration.* Engineering Press Inc., San Jose, 1992 (P.O. Box 1, San Jose, CA 95103. Tel: 800–800–1651)

2. Newnan, D. Editor. *Civil Engineering License Review, Twelfth Edition.* Engineering Press Inc., San Jose, 1995 (P.O. Box 1, San Jose, CA 95103. Tel: 800–800–1651)

3. International Conference of Building Officials. *Uniform Building Code - 1994.* Whittier, 1994 (5360 Workman Mill Road, Whittier, CA 90601. Tel: 310–692–4226)

4. Board of Registration for Professional Engineers and Land Surveyors. *Examination Papers.* Sacramento (P.O. Box 659005, Sacramento, CA 95865. Tel: 916–920–7466)

5. American Institute of Steel Construction. *Manual of Steel Construction, Ninth Edition.* Chicago, 1989 (400 Michigan Avenue, Chicago, IL 60611. Tel: 312–670–2400)

6. American Association of State Highway and Transportation Officials. *Standard Specifications for Highway Bridges, Fifteenth Edition.* Washington, 1989 (AASHTO, 444 North Capitol Street, N.W., Washington, DC 20001. Tel: 202–624–5800)

7. American Association of State Highway and Transportation Officials. *Standard Specifications for Highway Bridges,Fifteenth Edition:Division I-A:Seismic Design.* Washington, 1989 (AASHTO, 444 North Capitol Street, N.W., Washington, DC 20001. Tel: 202–624–5800)

8. American Concrete Institute. *Building Code Requirements and Commentary for Reinforced Concrete (ACI 318-89).* Detroit, 1989 (P.O. Box 19150, Redford Station, Detroit, MI 48219. Tel: 313–532–2600)

9. American Institute of Timber Construction. *Timber Construction Manual, Fourth Edition*. John Wiley & Sons, New York, 1994.

10. American Forest and Paper Association. *National Design Specification for Wood Construction, Eleventh Edition (ANSI\NFoPA NDS-1991)*. Washington, 1991 (1111, 19th Street, NW, Washington, DC 20036. Tel: 202-463-2700)

11. American Forest and Paper Association. *Commentary on the National Design Specification for Wood Construction, Eleventh Edition (ANSI\NFoPA NDS-1991)*. Washington, 1991 (1111, 19th Street, NW, Washington, DC 20036. Tel: 202-463-2700)

12. American Concrete Institute. *Building Code Requirements for Masonry Structures (ACI 530-88)*. Detroit, 1988 (P.O. Box 19150, Redford Station, Detroit, MI 48219. Tel: 313-532-2600)

13. National Council of Examiners for Engineering and Surveying. *Handbook for Structural Engineers*. Clemson, 1991 (P.O. Box 1686, Clemson, SC 29633-1686. Tel: 803-654-6824)

14. Structural Engineers Association of California. *Recommended Lateral Force Requirements and Commentary*. Sacramento, 1990 (SEAOC, P.O. Box 19440, Sacramento, CA 95819-0440. Tel: 916-427-3647)

15. Building Seismic Safety Council. *NEHRP recommended provisions for the development of seismic regulations for new buildings: Part 2, Commentary*. Washington, DC, 1991.

This page left blank intentionally.

1

General seismic principles

1.1 EARTHQUAKE PHENOMENA

1.1.1 Basic seismology

Most earthquakes are produced by the sudden rupture or slip of a geological fault[1,2] at the intersection of two tectonic plates. Along the west coast of the United States the boundaries of two large tectonic plates, the Pacific Plate and the North American Plate, are located. The major fault occurring in this region is the six hundred mile long San Andreas fault which is a nearly vertical right-lateral, strike-slip fault and is readily identified where it intersects the surface of the earth. The Loma Prieta earthquake of October 1989 occurred on the San Andreas fault[3,4], as shown in Figure 1-1, and the direction of motion[3,5] is shown in Figure 1-2.

The sudden release of energy at the focus or hypocenter of the earthquake causes seismic waves to propagate through the earth's crust and produces vibrations on the earth's surface. The amplitude of the vibrations diminishes with distance from the epicenter, the point on the earth's surface immediately above the hypocenter, and may last for a few seconds or for more than one minute. Typically, a California earthquake is of short duration with the Loma Prieta earthquake lasting only eight seconds. Two principal types of seismic waves are generated, body waves which travel from the hypocenter directly through the earth's lithosphere and surface waves which travel from the epicenter along the surface of the earth. Body waves consist of the primary, or P wave, which is a compression wave and the secondary, or S wave, which is a transverse shear wave[6]. Surface waves consist of the Love wave which produces a sideways motion and the Rayleigh wave which produces a rotary wave-like motion[6]. Body waves have a higher frequency range and attenuate more rapidly than surface waves. Hence, structures with longer natural periods, such as high–rise buildings and bridges, are more at risk some distance from the epicenter than low–rise buildings which have a short natural period.

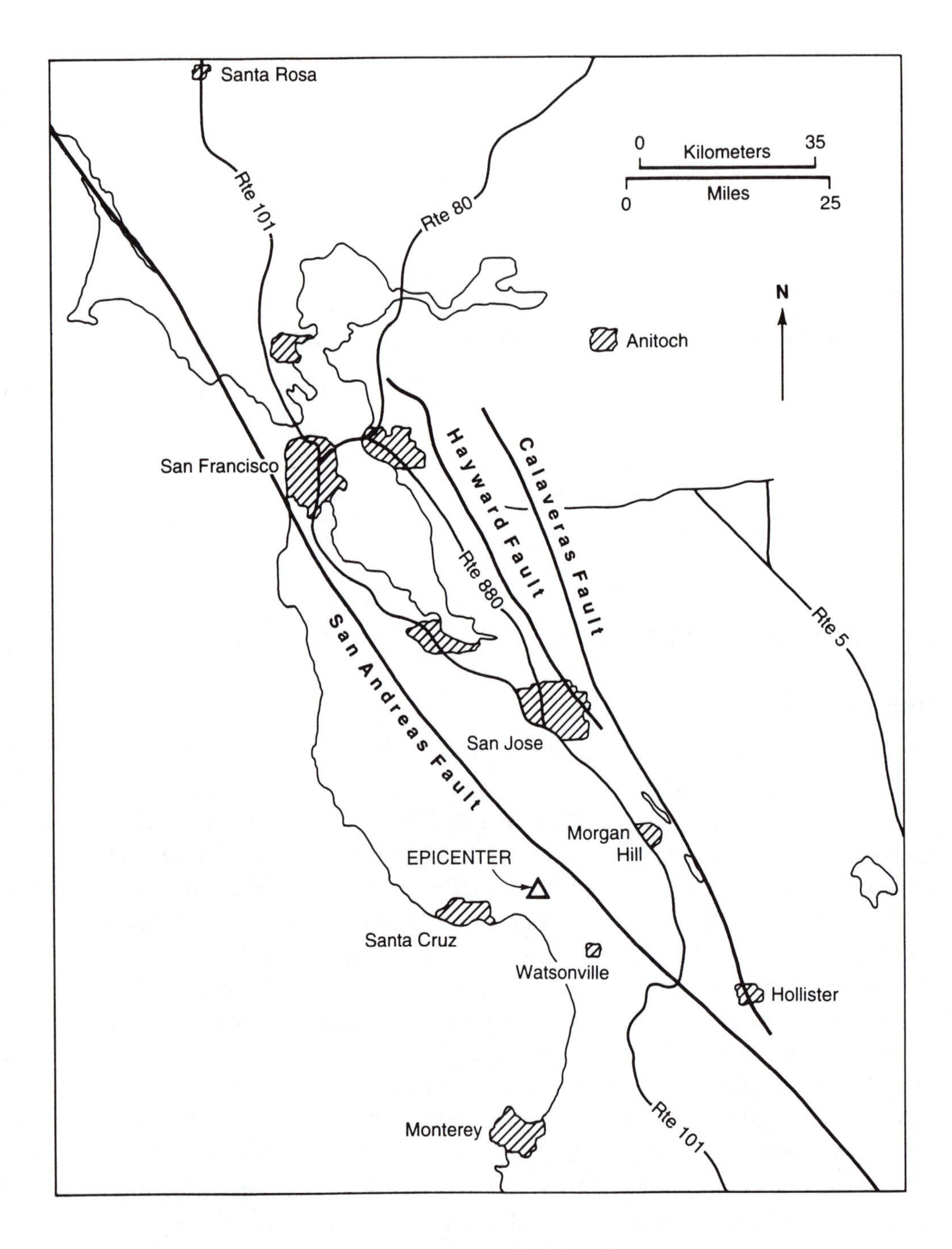

Figure 1–1. Location of Loma Prieta earthquake[4]

1.1.2 Measurement of earthquake magnitude and intensity

The Richter magnitude scale is a logarithmic based scale which utilizes the amplitude of seismic vibrations, recorded on a standard seismograph, to determine the strength of an earthquake. A unit increment on the scale represents a ten-fold increase in amplitude and an increase in energy release of 31.6 times. Earthquakes of Richter magnitude 6, 7 and 8 are categorized respectively as moderate, major and great earthquakes. Richter magnitude 8.5 is estimated to be the maximum that may be anticipated in California. The Loma Prieta earthquake had a magnitude of 7.1 and was the result of a rupture along a 25 mile long segment of the San Andreas fault between Los Gatos and Watsonville[3]. The San Francisco earthquake of April 1906 had a magnitude of 8.25 and was the result of a rupture along a 270 mile long segment of the San Andreas fault.

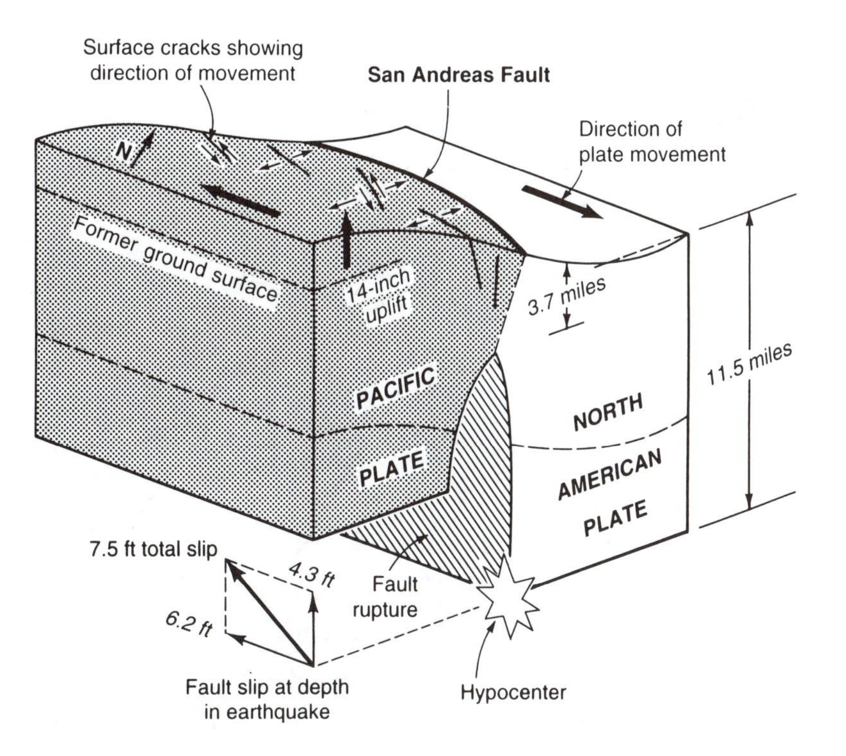

Figure 1–2. Motion on the San Andreas fault during the Loma Prieta earthquake[5]

Earthquake intensity is measured on the modified Mercalli index which is based on the observed effects of an earthquake at a specific site and a qualitative assessment of the damage caused and is an indication of the severity of ground shaking at that site. Modified Mercalli intensity values range from a value of I to a value of XII. Index value VII is classified as strong shaking causing damage to older masonry structures, chimneys and furniture. Index value VIII is classified as very strong shaking causing collapse of unreinforced masonry structures, towers and monuments. The soil conditions at a particular location affect the observed impact on the surrounding area and this is illustrated by the isoseismal map[3,5] for the Loma Prieta earthquake shown in Figure 1–3. The seismic damage caused at a particular site is influenced by the magnitude, duration and frequency of the ground vibration, distance from the epicenter, geological conditions between the epicenter and the site, soil properties at the site, and the building type and characteristics.

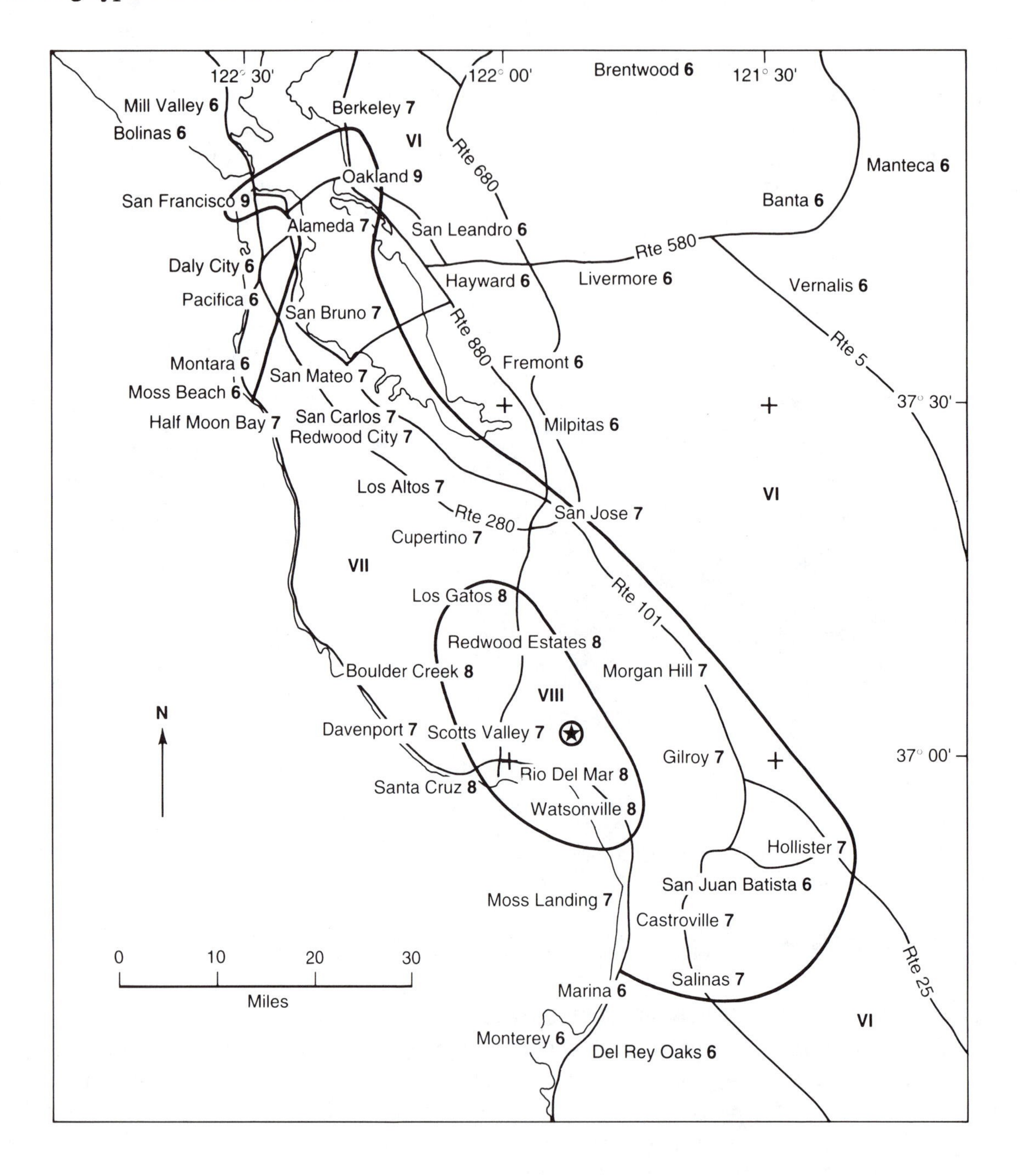

Figure 1–3. Isoseismal map of Mercalli intensities for the Loma Prieta earthquake[5]

1.1.3 Seismic effects

Structural damage during an earthquake is caused by the response of the structure to the ground motion input at its base. The dynamic forces produced in the structure are due to the inertia of its vibrating elements. The magnitude of the effective peak acceleration reached by the ground vibration directly effects the magnitude of the dynamic forces observed in the structure. Accelerograms for the Loma Prieta earthquake[3] obtained at rock sites situated five to twenty kilometers from the epicenter are shown in Figure 1-4. Figure 1-5 shows similar accelerograms[3] for rock sites located seventy-six to eighty kilometers from the epicenter.

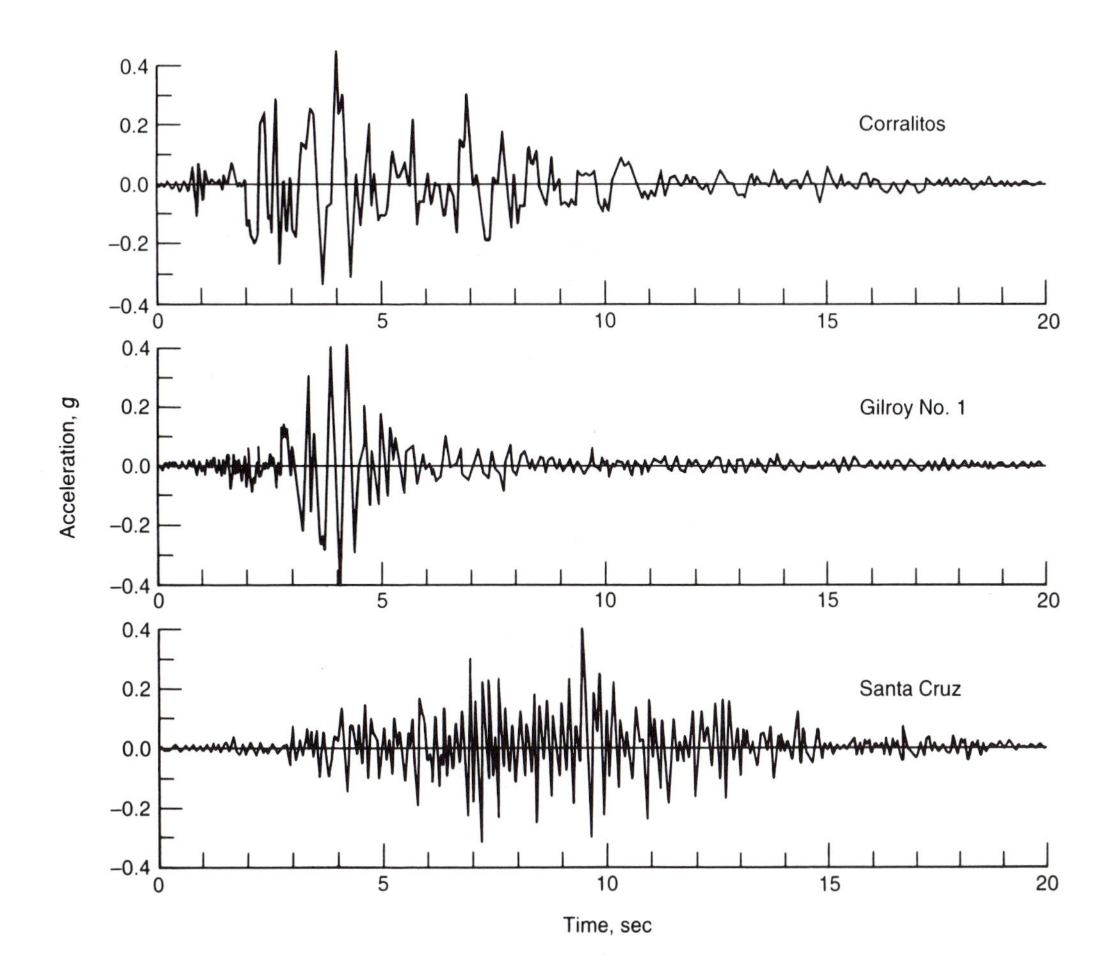

Figure 1-4 Accelerograms of motions recorded at rock sites within 20 kilometers of the source[3]

The response of the structure exceeds the ground motion and this dynamic magnification depends on the duration and frequency content of the ground vibration, the soil properties at the site, distance from the epicenter and the dynamic characteristics of the structure. The average spectral response for four different soil types for single degree of freedom systems with five percent damping have been determined for the Loma Prieta earthquake[3] and are shown in Figure 1-6. The spectral shapes[3] obtained by normalizing the response curves, with respect to the peak ground acceleration, are shown in Figure 1–7.

Soil liquefaction is another effect produced by earthquakes. A saturated uniform, fine grained sand or silt, when subjected to repeated vibration, experiences an increase in pore water pressure due to a redistribution of its particles, with a consequent reduction in shear strength. This produces a quicksand type condition, with a loss of bearing capacity, causing settlement and collapse of structures. In the Loma Prieta earthquake, liquefaction occurred close to the epicenter and in a number of susceptible areas in San Francisco and Oakland. A number of methods are available in order to prevent liquefaction. Drainage may be installed in order to lower the ground water table and remove the pore water. However, the resulting settlement may effect adjacent structures. Preconsolidation of the soil may also be achieved by vibroflotation techniques and this, also, may effect adjacent structures.

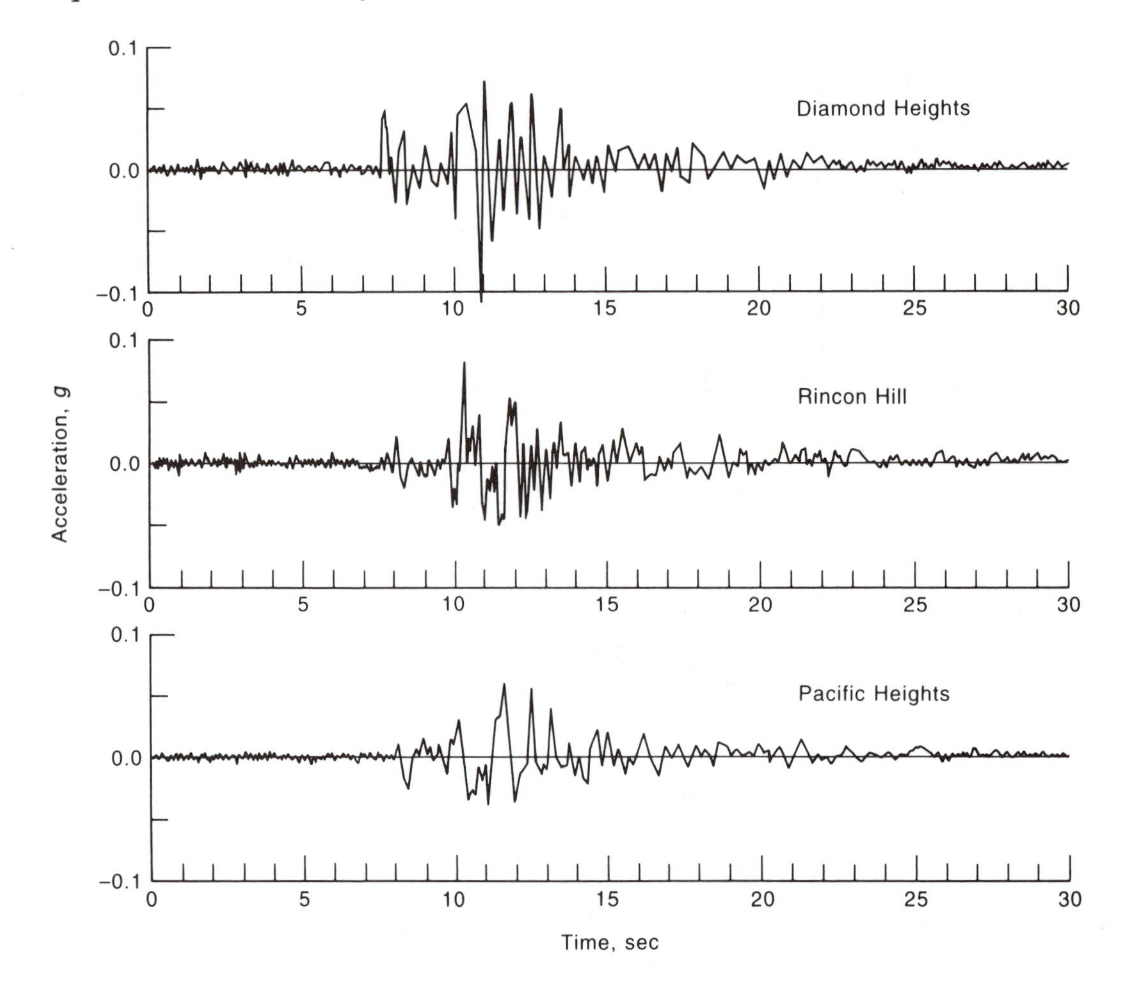

Figure 1–5 Accelerograms of motions recorded at rock sites in San Fransisco[3]

Placing a porous overburden over the site will produce preconsolidation and result in increased pore pressures being required before liquefaction can occur. In order to increase the shear strength of the soil, soil grouting or chemical injection may be employed. Alternatively, all deleterious soil may be removed and replaced with sound material, or pile foundations may be employed with the piles penetrating the unsatisfactory layer to found at a stable level.

When the epicenter of an earthquake is located on a sea bed, a destructive tidal wave, or tsunami, may be produced. The tidal wave may reach a height of up to fifty feet. Destructive tsunamis occurred in Hawaii in 1946 and in Alaska in 1964.

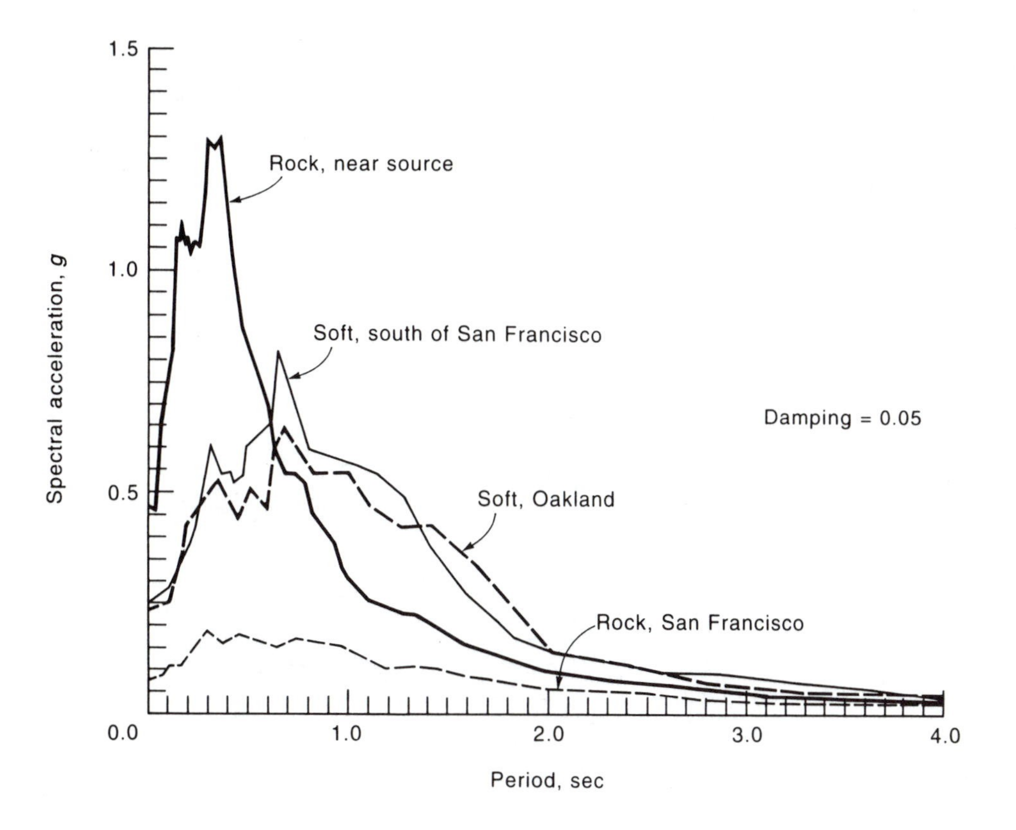

Figure 1–6 Average spectral shapes for motions recorded at various sites[3]

1.2 STRUCTURAL DYNAMICS

1.2.1 Undamped free vibrations

The single bay, single story frame shown in Figure 1–8 may be considered as having all of its mass concentrated, or lumped, at the infinitely rigid roof structure. If axial deformation in the columns is neglected, the frame exhibits only one degree of dynamic freedom, the lateral or sway displacement indicated.

A dynamic model of the structure consists of a single column with stiffness k supporting a mass of magnitude m to give the inverted pendulum, or lollipop structure shown. If the mass of this simple oscillator is subjected to an initial displacement and released, with no external forces acting, free vibrations occur about the static position.

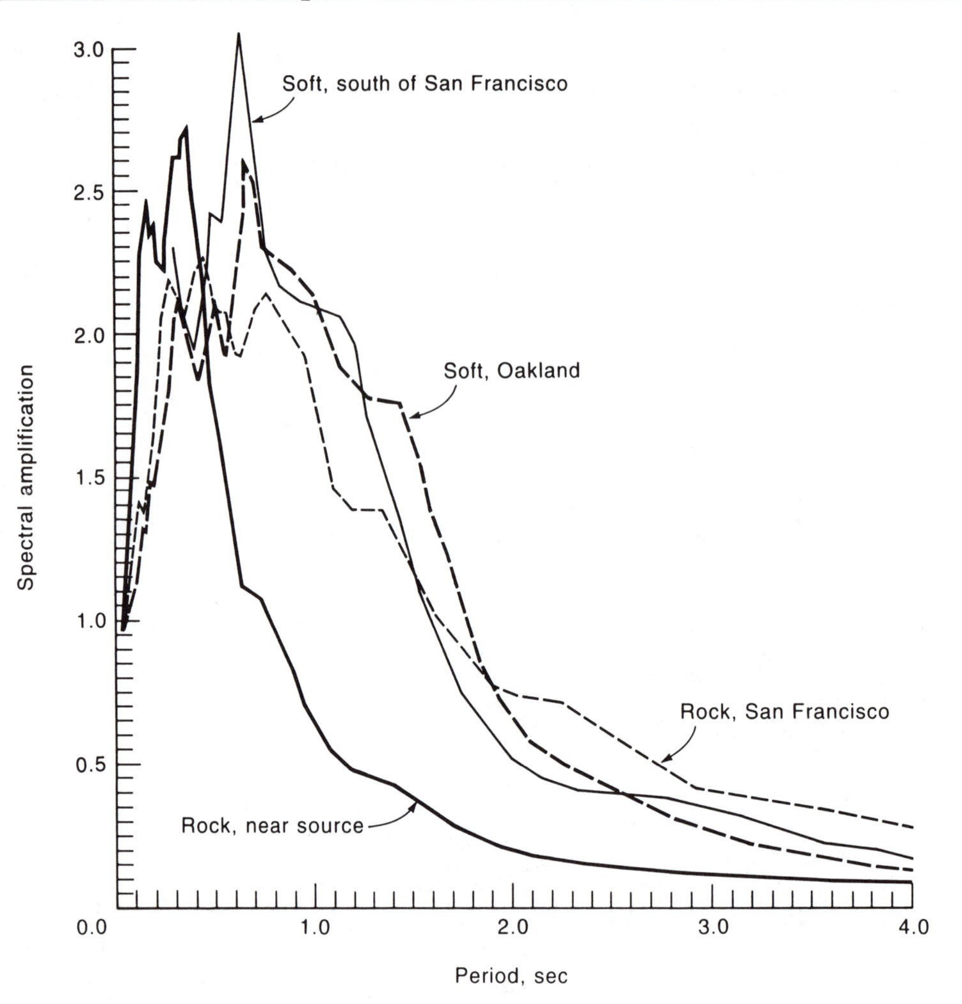

Figure 1–7 Average normalized spectral shapes for motions recorded at various sites[3]

The force required to produce a displacement x is

$$P = kx$$

Applying Newton's second law of motion to derive the inertial force produced and denoting the acceleration of the mass by $\ddot{x}$ gives the expression

$$P = -m\ddot{x}$$

Hence, the time dependent equation of dynamic equilibrium is

$$0 = -m\ddot{x} + kx$$

This is a second order, homogeneous, linear differential equation and has the solution

$$x = (\dot{x}_0/\omega)\sin\omega t + x_0\cos\omega t$$

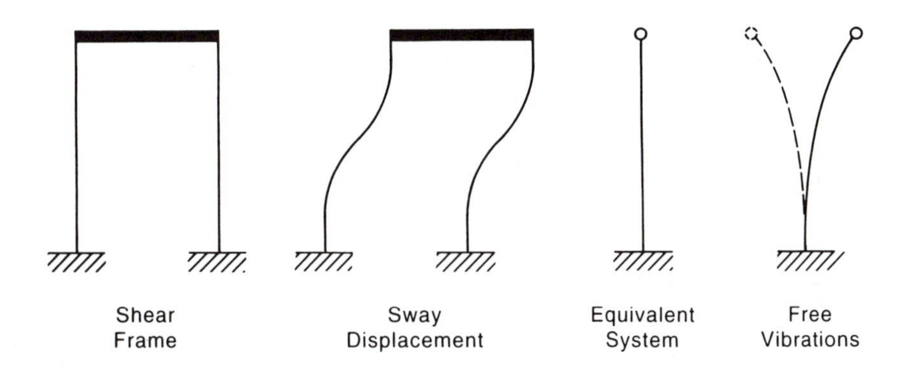

Figure 1–8 Simple sway oscillator

This expression is a representation of simple harmonic motion, as shown in Figure 1–9, with

	x	= displacement at time t
	x_0	= displacement at time $t = 0$
	$\ddot{x}$	= velocity at time $t = 0$
	ω	= circular natural frequency or angular velocity
		$= (k/m)^{1/2}$
		$= 2\pi/T$
		$= 2\pi f$
where,	T	= natural period
	f	= natural frequency of vibration

The form of these expressions indicates that the natural period increases as the mass of the system increases and the stiffness decreases.

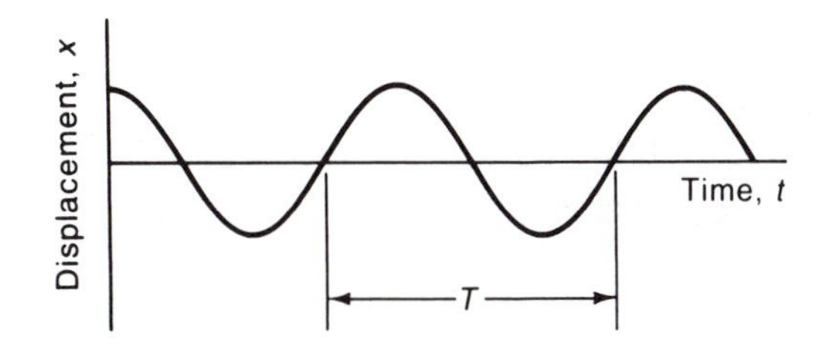

Figure 1–9 Undamped free vibrations

Example 1–1 (Determination of dynamic properties)
The space grid roof structure of the single story building shown in Figure 1–10 may be considered infinitely rigid and has a dead load of twenty pounds per square foot. The side sheeting has a dead load, including columns and side rails, of ten pounds per square foot. All columns are size W10 × 30 and may be considered axially inextensible. If damping may be neglected, determine the dynamic properties of the structure in the east–west direction.

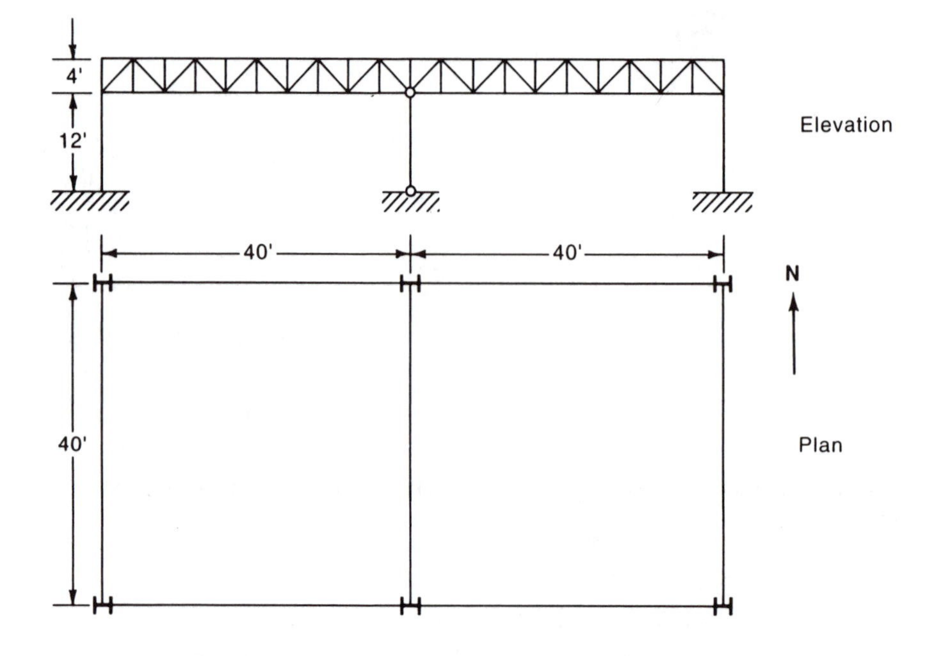

Figure 1–10 Building for Example 1–1

Solution

The weight tributary to the roof is given by

$$
\begin{aligned}
w &= 0.02 \times 40 \times 80 + 2 \times 0.01(6 + 4)(40 + 80) \\
&= 88 \text{ kips}
\end{aligned}
$$

The total stiffness of the two east columns is

$$
\begin{aligned}
k_E &= 12EI/l^3 \\
&= 12 \times 29{,}000(2 \times 170)/(12 \times 12)^3 \\
&= 39.63 \text{ kips per inch}
\end{aligned}
$$

The total stiffness of the two center columns is

$$k_C = 0$$

The total stiffness of the two west columns is

$$
\begin{aligned}
k_W &= 3EI/l^3 \\
&= 9.91 \text{ kips per inch}
\end{aligned}
$$

The total stiffness in the east-west direction is

$$
\begin{aligned}
k &= k_E + k_C + k_W \\
&= 39.63 + 9.91 \\
&= 49.54 \text{ kips per inch}
\end{aligned}
$$

The circular natural frequency is given by

$$\begin{aligned} \omega &= (k/m)^{1/2} \\ &= (kg/w)^{1\backslash 2} \\ &= (49.54 \times 386.4/88)^{1/2} \\ &= 14.75 \text{ radians per second} \end{aligned}$$

The natural frequency is given by

$$\begin{aligned} f &= \omega/2\pi \\ &= 2.35 \text{ hertz} \end{aligned}$$

The natural period is given by

$$\begin{aligned} T &= 1/f \\ &= 0.43 \text{ seconds} \end{aligned}$$

1.2.2 Damped free vibrations

In practice, the internal frictional resistance in an oscillatory system will cause any induced vibration to die out. The frictional resistance, or damping, dissipates the energy of the system by transforming it into heat. By assuming that the resistance is equivalent to a viscous damping force, which is proportional to the velocity of motion of the system, the dynamic motion may be readily analyzed[7]. Applying a damping coefficient, c, to the simple oscillator of Figure 1–8 provides the differential equation

$$0 = m\ddot{x} + c\dot{x} + kx$$

Substituting the function $x = A\exp(qt)$ in this expression yields the auxiliary equation

$$0 = mq^2 + cq + k$$

which has the roots

$$q_1, q_2 = -c/2m \pm [(c/2m)^2 - \omega^2]^{1/2}$$

Three complementary functions are possible for the system depending on whether the roots of the auxiliary equation are equal, real and distinct, or complex. These three conditions are termed critical damping, overdamping and underdamping.

1.2.3 Critical damping

The roots of the auxiliary equation are equal if

$$0 = (c/2m)^2 - \omega^2$$

Hence $c = 2m\omega = c_c$

where c_c = critical damping coefficient.

The complementary function reduces to

$$x = (A + Bt)\exp(c_c t/2m)$$

where the constants A and B may be determined from the initial conditions. The critical damping coefficient is the minimum value of the damping coefficient sufficient to bring the system to rest exponentially without oscillation.

1.2.4 Overdamping

The roots of the auxiliary equation are real and distinct if

$$(c/2m)^2 > \omega^2$$
$$c > 2m\omega$$
$$c > c_c$$

The complementary function reduces to

$$x = A\exp(q_1 t) + B\exp(q_2 t)$$

The motion of the overdamped system is similar to that of the critically damped system.

1.2.5 Underdamping

For the general condition of an underdamped system with a damping coefficient less than c_c, the roots of the auxiliary equation are complex and may be written in the form

$$q_1, q_2 = -c/2m \pm i\omega_D$$

where $i^2 = -1$

The complementary function reduces to

$$x = (A\sin\omega_D t + B\cos\omega_D t)e^{-ct/2m}$$

which is the equation of oscillatory motion in which successive amplitudes form a descending geometrical progression whose common ratio is $\exp(-c\pi/2m\omega_D)$. The damped circular frequency of the system is given by

$$\begin{aligned}\omega_D &= [\omega^2 - (c/2m)^2]^{1/2}\\ &= \omega[1 - (c/2m\omega)^2]^{1/2}\\ &= \omega(1 - (c/c_c)^2]^{1/2}\\ &= \omega(1 - \xi^2)^{1/2}\end{aligned}$$

where ξ is defined as the damping ratio and is a measure of the damping capacity of a system. The damped natural period of the system is given by

$$\begin{aligned}T_D &= 2\pi/\omega_D\\ &= 2\pi/\omega(1 - \xi^2)^{1/2}\\ &= T/(1 - \xi^2)^{1/2}\end{aligned}$$

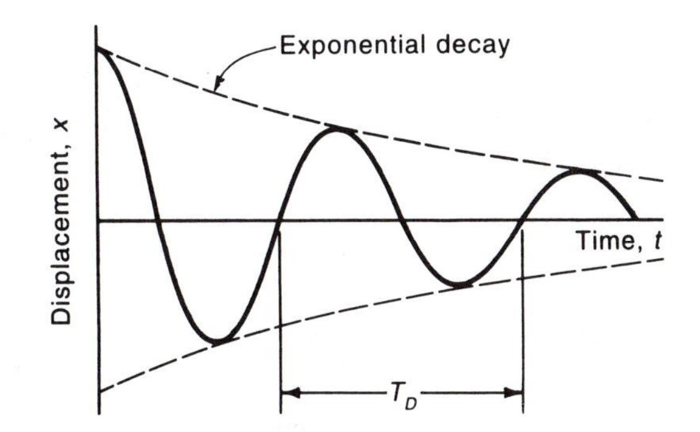

Figure 1–11 Damped free vibrations

Since the amount of viscous damping in buildings is less than 15% of the critical value, the damped circular frequency and the undamped circular frequency are almost identical and the undamped value is customarily used for the damped value in practical applications. Similarly, the amplitude peaks may be assumed to lie on the exponential curve x = exp(−ct/2m), as shown in Figure 1–11, and the ratio between any two successive amplitude peaks, a time interval T_D apart, is constant. The logarithmic decrement is defined as the natural logarithm of this ratio and is given by

$$\begin{aligned}\delta &= cT_D/2m \\ &= \xi\omega T_D \\ &= 2\pi\xi/(1-\xi^2)^{1/2} \\ &= 2\pi\xi\end{aligned}$$

As it is not possible to analytically determine either c or ξ, this latter expression provides a practical means of experimentally obtaining the damping characteristics of a vibrating system. Typical values of the damping coefficient range from 2 percent of the critical damping coefficient for welded steel structures, 5 percent for concrete framed structures, 10 percent for masonry shear walls and 15 percent for wood structures.

Example 1–2 (Determination of damping characteristics)
The frame shown in Figure 1–12 has an infinitely rigid beam and axially inextensible columns.

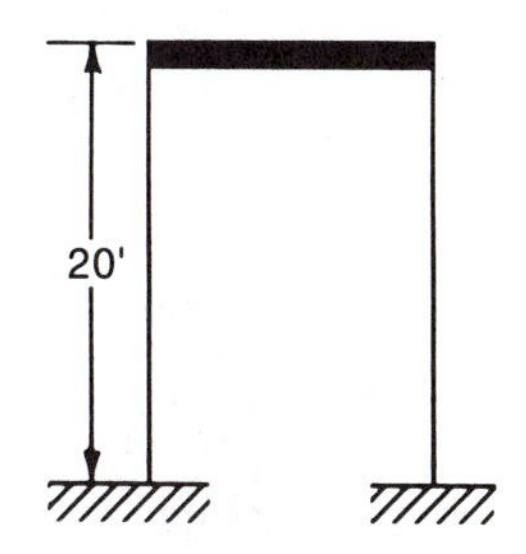

Figure 1–12 Frame for Example 1–2

The frame weight of thirty kips may be considered concentrated at the level of the beam and each column has a flexural rigidity of 1.69×10^6 kips–inches. Sway oscillations are induced in the frame and it is determined that the amplitude of vibration reduces to 25 percent of its initial value in ten full cycles of oscillation. Determine the damping characteristics.

Solution

The mass of the frame is given by

$$m = 30/32.2$$
$$= 0.932 \text{ kips–second}^2 \text{ per foot}$$

The stiffness of the frame in the sway mode is given by

$$k = 2 \times 12EI/l^3$$
$$= 2 \times 12 \times 1.69 \times 10^6 \times 12/(240)^3$$
$$= 35.208 \text{ kips per foot}$$

The circular natural frequency of the frame is

$$\omega = (k/m)^{1/2}$$
$$= 6.147 \text{ radians per second}$$

The logarithmic decrement is given by

$$10\delta = \ln(100/25)$$
$$\delta = 0.1386$$

The damping ratio is given by

$$\xi = \delta/2\pi$$
$$= 0.02206$$

The damped circular frequency is

$$\omega_D = \omega(1 - \xi^2)^{1/2}$$
$$= 6.1455 \text{ radians per second}$$

The critical damping coefficient is given by

$$c_c = 2m\omega$$
$$= 11.458 \text{ kips–seconds per foot}$$

The damping coefficient is

$$c = \xi c_c$$
$$= 0.2528 \text{ kips–seconds per foot}$$

The damped natural period of the system is given by

$$T_D = 2\pi/\omega_D$$
$$= 1.0225 \text{ seconds}$$

The natural period is

$$T = 2\pi/\omega$$
$$= 1.0223 \text{ seconds.}$$

1.2.6 Multiple-degree-of-freedom systems

The multistory structure shown in Figure 1–13 may be idealized as a multistory shear building by assuming that the mass is lumped at the floor and roof diaphragms, the diaphragms are infinitely rigid, and the columns are axially inextensible but laterally flexible. The dynamic response of the system is represented by the lateral displacements of the lumped masses with the number of degrees of dynamic freedom, or modes of vibration, n, being equal to the number of masses. The resultant vibration of the system is given by the superposition of the vibrations of each lumped mass. Each individual mode of vibration has its own period and may be represented by a single-degree-of-freedom system of the same period, and each mode shape, or eigenvector, remains of constant relative shape regardless of the amplitude of the displacement. A reference amplitude of a given mode shape may be assigned unit value to give the normal mode shape. The actual amplitudes must be obtained from the initial conditions. Figure 1–13 shows the first five modes of the five story shear building. The mode of vibration with the longest period (lowest frequency) is termed the first fundamental mode. Modes with shorter periods (higher frequencies) are termed higher modes or harmonics.

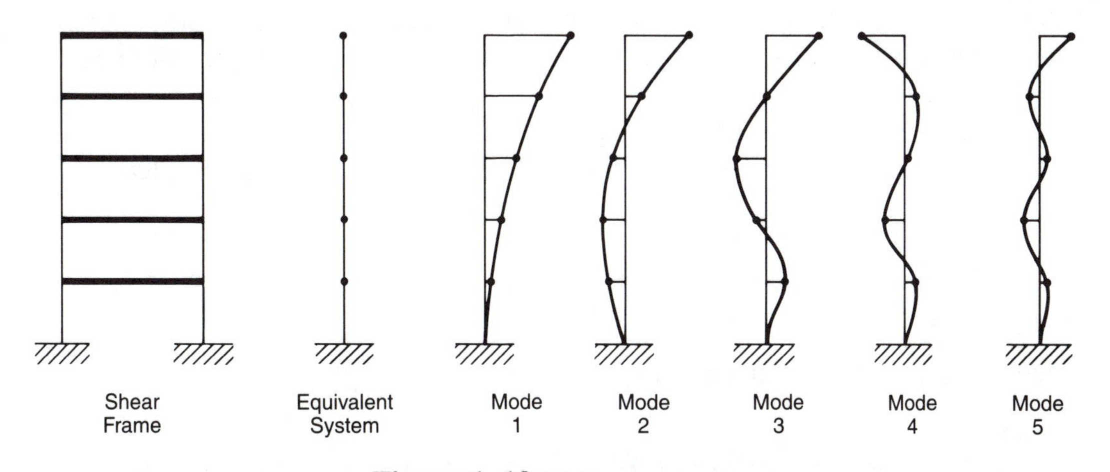

Figure 1–13 Multistory structure

A modal analysis procedure may be utilized to determine the dynamic response of a multiple-degree-of-freedom structure[7,8]. The maximum response for the separate modes is obtained by modeling each as an individual single-degree-of-freedom oscillator. As the maximum values cannot all occur simultaneously, these values are combined statistically in order to obtain the total response. The square-root-of-the-sum-of-the squares method is acceptable[8] for two-dimensional structures when the periods of any higher mode to any lower mode is 0.75 or less and the damping ratio does not exceed five percent.

Matrix methods

Since each degree of dynamic freedom provides one equation of dynamic equilibrium, the resultant vibration of the system consists of n such equations and may be expressed in matrix form, for undamped free vibrations, as

$$\{0\} = [M]\{\ddot{x}\} + [K]\{x\}$$

For simple harmonic motion, this reduces to

$$\{0\} = ([K\} - \omega^2[M])\{x\}$$

This expression is a representation of the eigenvalue equation with

[K] = stiffness matrix of the system
[M] = mass matrix, the diagonal matrix of lumped masses
{x} = eigenvector or mode shape associated with the eigenvalue ω

The eigenvalue equation has a nontrivial solution only if the determinant of the coefficient matrix is zero. Thus, the frequency determinant is

$$|\,[K] - \omega^2[M]\,| = 0$$

Expansion of this determinant yields the characteristic polynomial of degree n in (ω^2), the roots of which provide the eigenvalues. Back substituting the eigenvalues in the eigenvalue equation yields the eigenvectors for each mode. From the eigenvalues, the corresponding natural periods are obtained and the spectral accelerations determined from the appropriate response curve.

The participation factors, for a multiple-degree-of-freedom system, are defined in matrix notation by[7]

$$\{P\} = [\Phi\}^T[M\}\{1\}/[\Phi]^T[M][\Phi]$$

where {P} = column vector of partition factors for all modes considered
[Φ] = mode shape matrix or eigenvectors
{1} = column vector of ones
[M] = mass matrix, the diagonal matrix of lumped masses
$[\Phi]^T[M][\Phi]$ = modal mass matrix

For a specific system, the participation factors have the property[9] that

$$\Sigma P_j\phi_{1j} = 1.0$$

where P_j = the participation factor associated with the specific mode j
ϕ_{1j} = the component, for the first node of the system, of the eigenvector associated with the specific mode j

The matrix of maximum node displacements is defined in matrix notation by

$$[x] = [\Phi][P][S_d]$$
$$= [\Phi][P][S_v][\omega]^{-1}$$
$$= [\Phi][P][S_a][\omega^2]^{-1}$$

where [P] = diagonal matrix of participation factors
$[S_d]$ = diagonal matrix of spectral displacements
$[S_v]$ = diagonal matrix of spectral velocities

$[S_a]$ = diagonal matrix of spectral accelerations
$[\omega]$ = diagonal matrix of modal frequencies
$[\omega^2]$ = diagonal matrix of squared modal frequencies

Similarly the matrix of peak acceleration response is defined in matrix notation by

$$
\begin{aligned}
[\ddot{x}] &= [\Phi][P][S_a] \\
&= [\Phi][P][S_v][\omega] \\
&= [\Phi][P][S_d][\omega^2]
\end{aligned}
$$

The matrix of lateral forces at each node of the system is given by Newton's second law of motion as

$$
\begin{aligned}
[F] &= [M][\ddot{x}] \\
&= [M][\Phi][P][S_a] \\
&= [M][\Phi][P][S_v][\omega] \\
&= [M][\Phi][P][S_d][\omega^2] \\
&= [K][\Phi][P][S_d] \\
&= [K][\Phi][P][S_v][\omega]^{-1} \\
&= [K][\Phi][P][S_d][\omega^2]^{-1} \\
&= [K][x\ ]
\end{aligned}
$$

The column vector of total base shear forces is given by

$$
\begin{aligned}
\{V\} &= [F]^T\{1\} \\
&= [P][S_a][\Phi]^T[M]\{1\}
\end{aligned}
$$

The eigenvectors may be normalized so as to produce the orthogonal property[7,9]

$$[\Phi]^T[M][\Phi] = [I]$$

where $[I]$ = identity matrix
$[\Phi]^T[M][\Phi]$ = modal mass matrix

The components of the normalized modal matrix are given by

$$\phi_{ij} = u_{ij}/(\Sigma m_{ii}u_{ij}^2)^{1/2}$$

where ϕ_{ij} = the component, for node i, of the normalized mode shape associated with the specific mode j
m_{ii} = the mass concentrated at node i
u_{ij} = the component, for node i, of the eigenvector associated with mode j

By using normalized eigenvectors, the column vector of participation factors reduces to

$$\{P\} = [\Phi]^T[M]\{1\}$$

Example 1-3 (Two story shear building)
A two story shear building with the properties shown in Figure 1-14 and with a damping ratio of five percent is located on a rock site near the source of the Loma Prieta earthquake. Determine the lateral forces and displacements at each level using the spectral ordinates of Figure 1-6.

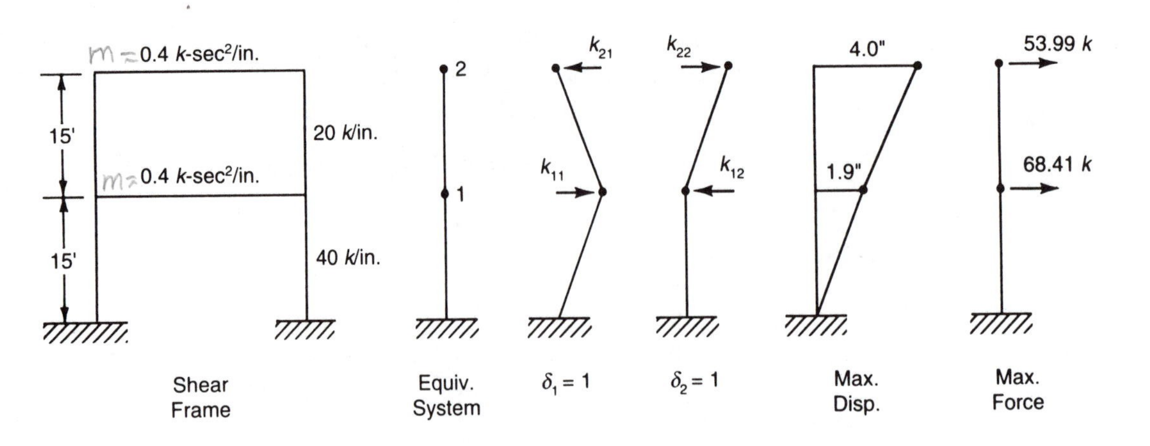

Figure 1–14 Building for Example 1–3

Solution

Unit shear displacement is imposed on each node in turn and the coefficient k_{ij} of the stiffness matrix is obtained as the force produced at node i by a unit displacement at node j. The stiffness matrix is, then,

$$[K] = \begin{bmatrix} k_{11} & k_{12} \\ k_{21} & k_{22} \end{bmatrix}$$

$$= \begin{bmatrix} (40+20) & -20 \\ -20 & 20 \end{bmatrix}$$

$$= \begin{bmatrix} 60 & -20 \\ -20 & 20 \end{bmatrix}$$

The diagonal mass matrix is

$$[M] = \begin{bmatrix} 0.4 & 0.0 \\ 0.0 & 0.4 \end{bmatrix}$$

The eigenvalue equation is

$$0 = \begin{bmatrix} 60 & -20 \\ -20 & 20 \end{bmatrix} - \omega^2 \begin{bmatrix} 0.4 & 0.0 \\ 0.0 & 0.4 \end{bmatrix} \begin{bmatrix} x_1 \\ x_2 \end{bmatrix}$$

The frequency determinant is

$$T = \begin{vmatrix} (60 - 0.4\omega^2) & -20 \\ -20 & (20 - 0.4\omega^2) \end{vmatrix}$$

$$= 0.16\omega^4 - 32\omega^2 + 800$$

Equating this polynomial in ω to zero provides the circular natural frequencies for the two modes of vibration

$$\omega_1 = 5.41 \text{ radians per second}$$
$$\omega_2 = 13.07 \text{ radians per second}$$

The corresponding natural periods are

$$T_1 = 1.16 \text{ seconds}$$
$$T_2 = 0.48 \text{ seconds}$$

From the response curves of Figure 1-6, the spectral accelerations are

$$S_{a1} = 0.25g = 96.6 \text{ inches per second per second}$$
$$S_{a2} = 0.85g = 328.4 \text{ inches per second per second}$$

Substituting these values in the eigenvalue equation, and setting the first components of the mode shape factors to unity, provides the eigenvectors or matrix of relative modal shapes

$$[u] = \begin{bmatrix} 1.0 & 1.0 \\ 2.414 & -0.414 \end{bmatrix}$$

The normalized modal matrix has the orthogonality property[7]

$$[I] = [\Phi]^T[M][\Phi]$$

where $[I]$ = the identity matrix

The components of the normalized modal matrix are given by

$$\phi_{ij} = u_{ij}/(\Sigma m_{ii}u_{ij}^2)^{1/2}$$

where ϕ_{ij} = the component for node i of the normalized mode shape associated with the specific mode j

For mode 1:

$$(\Sigma m_{ii}u_{i1}^2)^{1/2} = [0.4(1 + 2.414^2)]^{1/2} = 1.653$$

For mode 2:

$$(\Sigma m_{ii}u_{i2}^2)^{1/2} = [0.4(1 + 1.414^2)]^{1/2} = 0.685$$

The normalized modal matrix is, then,

$$[\Phi] = \begin{bmatrix} 0.605 & 1.461 \\ 1.461 & -0.605 \end{bmatrix}$$

The column vector of participation factors[8] is defined by

$$\{P\} = [\Phi]^T[M]\{1\}/[\Phi]^T[M][\Phi]$$

where $\{1\}$ = column vector of ones

and $[\Phi]^T[M][\Phi]$ = modal mass matrix = $[I]$ for a normalized modal matrix

Hence, the column vector of participation factors reduces to

$$\{P\} = [\Phi]^T[M]\{1\} = \begin{bmatrix} 0.826 \\ 0.342 \end{bmatrix}$$

Assuming the structure remains elastic, the matrix of actual nodal displacements is given by

$$[x] = [\Phi][P][S_a]/[\omega^2]$$

where
$[P]$ = diagonal matrix of participation factors
$[S_a]$ = diagonal matrix of spectral accelerations
$[\omega^2]$ = diagonal matrix of squared modal frequencies

Executing the necessary matrix operations gives

$$[x] = \begin{bmatrix} 1.648 & 0.961 \\ 3.980 & -0.398 \end{bmatrix}$$

The resultant maximum displacement at each node is obtained from the square-root-of-the-sum-of-the squares[8] of the relevant row vector and is given by the column vector

$$[x_c] = \begin{bmatrix} 1.9 \\ 4.0 \end{bmatrix}$$

The matrix of lateral forces at each node is given by

$$[F] = [K][x]$$
$$= \begin{bmatrix} 19.29 & 65.63 \\ 46.64 & -27.19 \end{bmatrix}$$

Using the square-root-of-the-sum-of-the squares method, the resultant maximum lateral force at each node is given by

$$[F_c] = \begin{bmatrix} 68.41 \\ 53.99 \end{bmatrix}$$

The row vector of base shears is

$$[V] = ([F]^T\{1\})^T$$
$$= [65.93 \quad 38.44]$$

Taking the square-root-of-the-sum-of-the squares, the resultant base shear is

$$V_c = (V_{11}^2 + V_{12}^2)^{1/2}$$
$$= 76.32 \text{ kips}$$

Example 1.4 (Structural Engineering Examination 1982, Section A, weight 4.0 points)

GIVEN: Figure 1-15 represents a three-story building. The effective dead loads are shown on each floor.

CRITERIA: The following dynamic properties of the plane frame are given:

Eigenvectors

$$[\Phi] = \begin{bmatrix} 1.680 & -1.208 & -0.714 \\ 1.220 & 0.704 & 1.697 \\ 0.572 & 1.385 & -0.984 \end{bmatrix}$$

$$[\Phi]^T[M][\Phi] = [I]$$

Eigenvalues

$$\{\omega\} = \begin{bmatrix} 8.77 \\ 25.18 \\ 48.13 \end{bmatrix} \text{ radians per second}$$

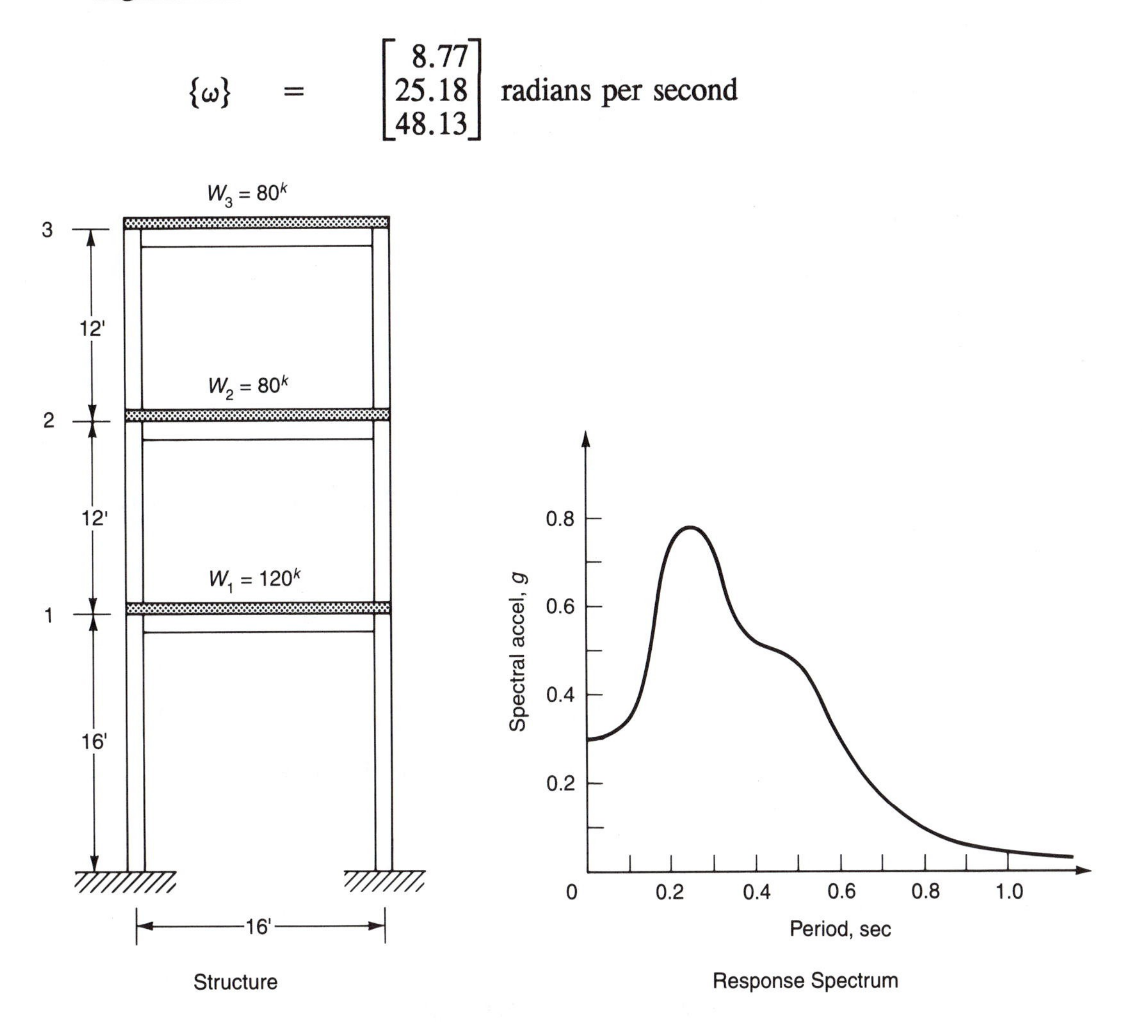

Figure 1–15 Details for Example 1–4

REQUIRED:
a) Compute the participation factors
b) How can you check that your participation factors are correct?
c) Calculate the displacements of each floor based on the spectra given.
d) Calculate the interstory drift for each floor.

SOLUTION

a. and b. PARTICIPATION FACTORS

The participation factors for a multiple-degree-of-freedom system, are defined in matrix notation by[7]

$$\{P\} = [\Phi]^T[M]\{1\}/[\Phi]^T[M][\Phi]$$

where $\{P\}$ = column vector of partition factors for all modes considered
$[\Phi]$ = mode shape matrix or eigenvectors
$\{1\}$ = column vector of ones
$[M]$ = diagonal matrix of lumped masses or mass matrix
and $[\Phi]^T[M][\Phi]$ = modal mass matrix
= [I] as given in the problem statement
= identity matrix

Hence the matrix of eigenvectors is normalized and the column vector of participation factors is given by

$$\{P\} = [\Phi]^T[M]\{1\}$$

The mass matrix is

$$[M] = \begin{bmatrix} 80 & 0 & 0 \\ 0 & 80 & 0 \\ 0 & 0 & 120 \end{bmatrix} 1/g$$

Executing the necessary matrix operations gives the column vector of participation factors for the three modes as

$$\{P\} = \begin{bmatrix} 301.44 \\ 125.88 \\ -39.44 \end{bmatrix} 1/g$$

Summing the product of the partition factors and the eigenvectors for the first node gives

$$\begin{aligned} \Sigma P_j\phi_{1j} &= (301.44 \times 1.608 - 125.88 \times 1.208 + 39.44 \times 0.714)/386 \\ &= 382.52/386 \\ &\approx 1.0 \end{aligned}$$

Hence the participation factors are correct[9]

c. FLOOR DISPLACEMENTS

The natural periods for each of the three modes is obtained from the given eigenvalues using the expression

$$T_n = 2\pi/\omega_n$$

and

$$\begin{aligned} T_1 &= 2\pi/8.77 = 0.717 \text{ seconds} \\ T_2 &= 2\pi/25.18 = 0.250 \\ T_3 &= 2\pi/48.13 = 0.131 \end{aligned}$$

The spectral accelerations for each of the three modes are obtained from the response curve given and are

$$S_{a1} = 0.17g$$
$$S_{a2} = 0.80g$$
$$S_{a3} = 0.40g$$

The matrix of actual node displacements is defined in matrix notation by

$$[x] = [\Phi][P][S_d]$$
$$= [\Phi][P][S_a][\omega^2]^{-1}$$

where
$[P]$ = diagonal matrix of participation factors
$[S_d]$ = diagonal matrix of spectral displacements
$[S_a]$ = diagonal matrix of spectral accelerations
$[\omega^2]$ = diagonal matrix of squared modal frequencies

Executing the necessary matrix operations gives the matrix of node displacements for the three modes as

$$[x] = \begin{bmatrix} 1.126 & -0.192 & 0.0049 \\ 0.813 & 0.112 & -0.0115 \\ 0.381 & 0.220 & 0.0067 \end{bmatrix}$$

The total displacements at each node are obtained[8] as the absolute value of the square-root-of-the-sum-of-the-squares for each row vector and are given by the column vector

$$[x_c] = \begin{bmatrix} 1.142 \\ 0.821 \\ 0.440 \end{bmatrix}$$

d. INTERSTORY DRIFT

The interstory drifts for the three modes are obtained by subtracting the row vector of node displacements for each node from the row vector of node displacements for the node above. The matrix of story drifts is then

$$[\Delta] = \begin{bmatrix} 0.313 & -0.304 & 0.0164 \\ 0.432 & -0.108 & -0.0182 \\ 0.381 & 0.220 & 0.0067 \end{bmatrix}$$

The interstory drifts for the combined displacements are similarly obtained as

$$[\Delta_c] = \begin{bmatrix} 0.321 \\ 0.381 \\ 0.440 \end{bmatrix}$$

Numerical Methods

Numerical methods[8] may be used to facilitate the modal analysis procedure. For a given mode of vibration, the participation factor is defined by

$$P = \Sigma M_i\phi_i/M$$

where M_i = mass at floor level i

ϕ_i = mode shape component for node point i for the given mode

M = modal mass

$= \Sigma M_i\phi_i^2$

and the summation extends over all the nodes in the structure.

The effective mass is defined by

$$\begin{aligned} M^E &= (\Sigma M_i\phi_i)^2/\Sigma M_i\phi_i^2 \\ &= P\Sigma M_i\phi_i \\ &= (\Sigma M_i\phi_i)^2/M \\ &= P^2M \end{aligned}$$

Similarly, the effective weight is defined by

$$W^E = (\Sigma W_i\phi_i)^2/\Sigma W_i\phi_i^2$$

where W_i = weight at floor level i

The peak acceleration at a node is defined by

$$\ddot{x} = \phi_i PS_a$$

where S_a = spectral acceleration for the given mode

The maximum displacement at a node is defined by

$$x_i = \phi_i PS_d$$

where S_d = spectral displacement for the given node

The lateral force at a node is given by Newton's Law as

$$\begin{aligned} F_i &= M_i\ddot{x} \\ &= M_i\phi_i PS_a \\ &= M_i\omega^2 x_i \end{aligned}$$

The total base shear is given by

$$\begin{aligned} V &= \Sigma F_i \\ &= PS_a\Sigma M_i\phi_i \\ &= P^2MS_a \\ &= M^ES_a \\ &= W^ES_a/g \end{aligned}$$

where g = acceleration due to gravity

= 386 inches per second^2

The lateral force at a node may also be determined by distributing the base shear over the node points as

$$\begin{aligned} F_i &= M_i\phi_i PS_a \\ &= (M_i\phi_i/PM)V \\ &= (M_i\phi_i/\Sigma M_i\phi_i)V \\ &= (W_i\phi_i/\Sigma W_i\phi_i)V \end{aligned}$$

For normalized eigenvectors, these expressions reduce to

$$\begin{aligned} M &= \text{modal mass} \\ &= \Sigma M_i \phi_i^2 \\ &= 1.0 \\ P &= \text{participation factor} \\ &= \Sigma M_i \phi_i \\ M^E &= \text{effective mass} \\ &= (\Sigma M_i \phi_i)^2 \\ W^E &= \text{effective weight} \\ &= (\Sigma W_i \phi_i)^2/g \end{aligned}$$

Example 1–5 (Structural Engineering Examination 1977, Section A, weight 5.0 points)

GIVEN: Figure 1–16 represents a three story building with plan dimensions of 100 feet × 100 feet. The effective dead load on each floor is shown on the Figure.

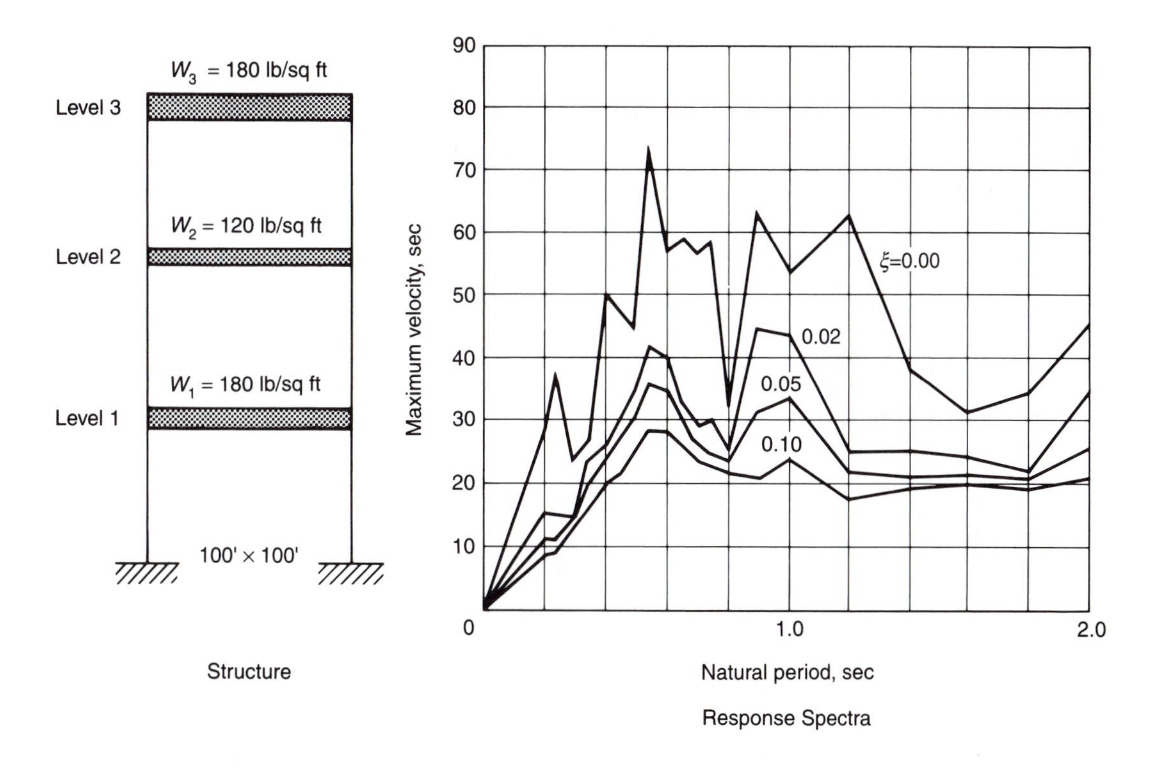

Figure 1–16 Details for Example 1–5

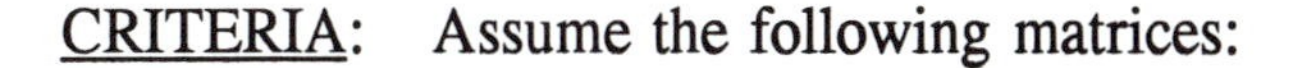

CRITERIA: Assume the following matrices:

Mode shape matrix

$$[\Phi] = \begin{bmatrix} 2.860 & -0.657 & 0.387 \\ 1.950 & 0.725 & -1.610 \\ 1.000 & 1.000 & 1.000 \end{bmatrix}$$

Modal frequency matrix

$$[\omega] = \begin{bmatrix} 15.5 \\ 38.5 \\ 61.7 \end{bmatrix} \text{ radians per second}$$

REQUIRED:
1. Determine the base shear for each mode by using the design response spectrum with $\xi = 0.05$ (ξ = damping ratio).
2. Determine the lateral load at each level for each mode.
3. What is the most probable base shear?

SOLUTION

DEAD LOAD

The dead loads tributary to each level are

$$\begin{aligned} \text{Level 1} &= 0.18 \times 100 \times 100 = 1800 \text{ kips} \\ \text{Level 2} &= 0.12 \times 100 \times 100 = 1200 \text{ kips} \\ \text{Level 3} &= 0.18 \times 100 \times 100 = 1800 \text{ kips} \end{aligned}$$

FIRST MODE

The natural period of the first mode is given by

$$\begin{aligned} T_1 &= 2\pi/\omega_1 \\ &= 2\pi/15.1 \\ &= 0.416 \text{ seconds} \end{aligned}$$

For a damping coefficient of five percent, the corresponding spectral velocity is obtained from the response spectra as

$$S_v = 25 \text{ inches per second}$$

The spectral acceleration is given by

$$\begin{aligned} S_a &= \omega_1 S_v \\ &= 15.1 \times 25 \\ &= 377.5 \text{ inches per sec}^2 \end{aligned}$$

From Table 1-1 the effective weight is derived as

$$\begin{aligned} W^E &= (\Sigma W_i \phi_1)^2 / \Sigma W_i \phi_1^2 \\ &= (9288)^2/21086 \\ &= 4091 \text{ kips} \end{aligned}$$

The base shear is given by

$$V = W^E S_a/g$$
$$= 4091 \times 377.5/386$$
$$= 4001 \text{ kips}$$

Level	W_i	ϕ_i	$W_i\phi_i$	$W_i\phi_i^2$	F_i
1	1800	2.86	5148	14723	2218
2	1200	1.95	2340	4563	1008
3	1800	1.00	1800	1800	775
Total	–	–	9288	21086	4001

Table 1–1 First mode: determination of effective weight

The lateral force at each level is given by

$$F_i = V(W_i\phi_i/\Sigma W_i\phi_i)$$
$$= 4001 W_i\phi_i/9288$$
$$= 0.431 W_i\phi_i$$

and the values are given in Table 1–1.

SECOND MODE

The natural period of the second mode is given by

$$T_2 = 2\pi/\omega_2$$
$$= 2\pi/38.5$$
$$= 0.163 \text{ seconds}$$

For a damping coefficient of five percent, the corresponding spectral velocity is obtained from the response spectra as

$$S_v = 9 \text{ inches per second.}$$

The spectral acceleration is given by

$$S_a = \omega_2 S_v$$
$$= 38.5 \times 9$$
$$= 346.5 \text{ inches per sec}^2$$

From Table 1–2 the effective weight is derived as

$$W^E = (\Sigma W_i\phi_i)^2/\Sigma W_i\phi_i^2$$
$$= (1488)^2/3208$$
$$= 690 \text{ kips}$$

The base shear is given by

$$
\begin{aligned}
V &= W^E S_a/g \\
&= 690 \times 346.5/386 \\
&= 620 \text{ kips}
\end{aligned}
$$

Level	W_i	ϕ_i	$W_i\phi_i$	$W_i\phi_i^2$	F_i
1	1800	-0.657	-1182	777	-492
2	1200	0.725	870	631	362
3	1800	1.000	1800	1800	750
Total	-	-	1488	3208	620

Table 1-2 Second mode: determination of effective weight

The lateral force at each level is given by

$$
\begin{aligned}
F_i &= V(W_i\phi_i/\Sigma W_i\phi_i) \\
&= 620W_i\phi_i/1488 \\
&= 0.416W_i\phi_i
\end{aligned}
$$

and the values are given in Table 1-2.

THIRD MODE

The natural period of the third mode is given by

$$
\begin{aligned}
T_3 &= 2\pi/\omega_3 \\
&= 2\pi/61.7 \\
&= 0.102
\end{aligned}
$$

For a damping coefficient of five percent, the corresponding spectral velocity is obtained from the response spectra as

$$S_v = 5 \text{ inches per second}$$

The spectral acceleration is given by

$$
\begin{aligned}
S_a &= \omega_3 S_v \\
&= 61.7 \times 5 \\
&= 308.5 \text{ inches per sec}^2
\end{aligned}
$$

From Table 1-3 the effective weight is derived as

$$
\begin{aligned}
W^E &= (\Sigma W_i\phi_i)^2/\Sigma W_i\phi_i^2 \\
&= (565)^2/5181 \\
&= 62 \text{ kips}
\end{aligned}
$$

Level	W_i	ϕ_i	$W_i\phi_i$	$W_i\phi_i^2$	F_i
1	1800	0.387	697	270	61
2	1200	-1.610	-1932	3111	-169
3	1800	1.000	1800	1800	158
Total	–	–	565	5181	50

Table 1–3 Third mode: determination of effective weight

The base shear is given by

$$\begin{aligned} V &= W^E S_a/g \\ &= 62 \times 308.5/386 \\ &= 50 \text{ kips} \end{aligned}$$

The lateral force at each level is given by

$$\begin{aligned} F_i &= V(W_i\phi_i/\Sigma W_i\phi_i) \\ &= 50W_i\phi_i/565 \\ &= 0.088W_i\phi_i \end{aligned}$$

and the values are given in the Table.

3. MOST PROBABLE BASE SHEAR
The ratios of the natural periods for the three modes are given by

$$\begin{aligned} T_3/T_2 &= 0.102/0.163 \\ &= 0.63 \\ &< 0.75 \\ T_2/T_1 &= 0.163/0.416 \\ &= 0.39 \\ &< 0.75 \end{aligned}$$

Hence, the most probable base shear may be determined from the square-root-of-the-sum-of-the-squares method[8] and is given by

$$\begin{aligned} V_c &= (4001^2 + 620^2 + 50^2)^{1/2} \\ &= 4049 \text{ kips} \end{aligned}$$

Iterative Methods

For a low rise building, not exceeding five stories, modal analysis may be limited to the fundamental mode. The structural system may be modeled as a shear building with rigid floor slabs. Lateral displacements of the nodes are then the result of column flexure with no rotations occurring at the joints. The stiffness of a particular story is given by

	k_i	$= 12(E\Sigma I/h^3)_i$
where	E	= modulus of elasticity of the columns
	h	= the height of story i
	ΣI	= total moment of inertia of all columns in story i

In addition, the story mass is assumed concentrated at the floor slab as shown in Figure 1–17.

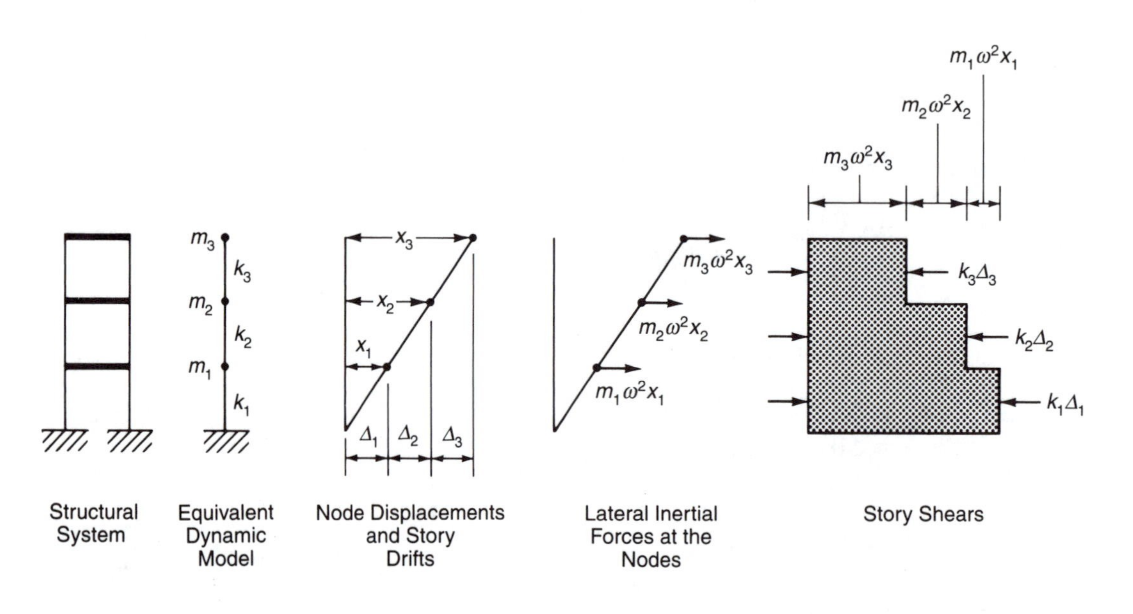

Figure 1–17 Modal analysis of a shear building

Using these assumptions, iterative techniques have been developed based on the methods proposed by Rayleigh[9], Stodola[10], and Holzer[10]. The method presented here is an adaptation[11] of the Holzer method. In the dynamic model described, when a node attains its maximum lateral displacement x_i, the velocity is zero and the inertial force at the node is given by

	F_I	$= m_i\ddot{x}_i$
		$= m_i\omega^2x_i$
where	m_i	= mass concentrated at floor level i
	$\ddot{x}_i$	= acceleration of the mass
	x_i	= maximum displacement of the mass
	ω	= circular natural frequency

The total shear force at any story equals the product of the story stiffness and the drift of that story. The increment in shear force at a node is produced by the inertial force at that level. The increment in shear force is given by

F_s	$= k_i\Delta_i - k_{i+1}\Delta_{i+1}$
k_i	= stiffness of story i
Δ_i	= drift of story i
	$= x_i - x_{i-1}$
$k_I\Delta_i$	= total shear force at story i

Equating the inertial force and increment in shear force gives

$$F_I = F_s$$
$$m_i\omega^2x_i = k_i\Delta_i - k_{i+1}\Delta_{i+1}$$

A solution may be obtained by assuming an initial mode shape with unit displacement at the top level. From this is calculated the inertial force or incremental shear force, in terms of the circular natural frequency, at each level. Summing the incremental shear force from the top story downward provides the total shear force at each story. Dividing these values by the appropriate story stiffness gives the drift of each story. Summing the drift values from the bottom story upward provides the lateral displacement at each level. Dividing these displacements values by the displacement at the top of the structure gives a revised mode shape. This revised mode shape may be used as a new initial mode shape and the iteration process repeated until agreement is obtained between the revised and initial mode shapes.

Example 1–6 (Structural Engineering Examination 1974, Section A, weight 5.0 points)

GIVEN: A two-story steel frame structure is shown in Figure 1–18. Consider the structure to have seven percent damping. Each story is known to deflect 0.1 inch under a 10 kips story shear.

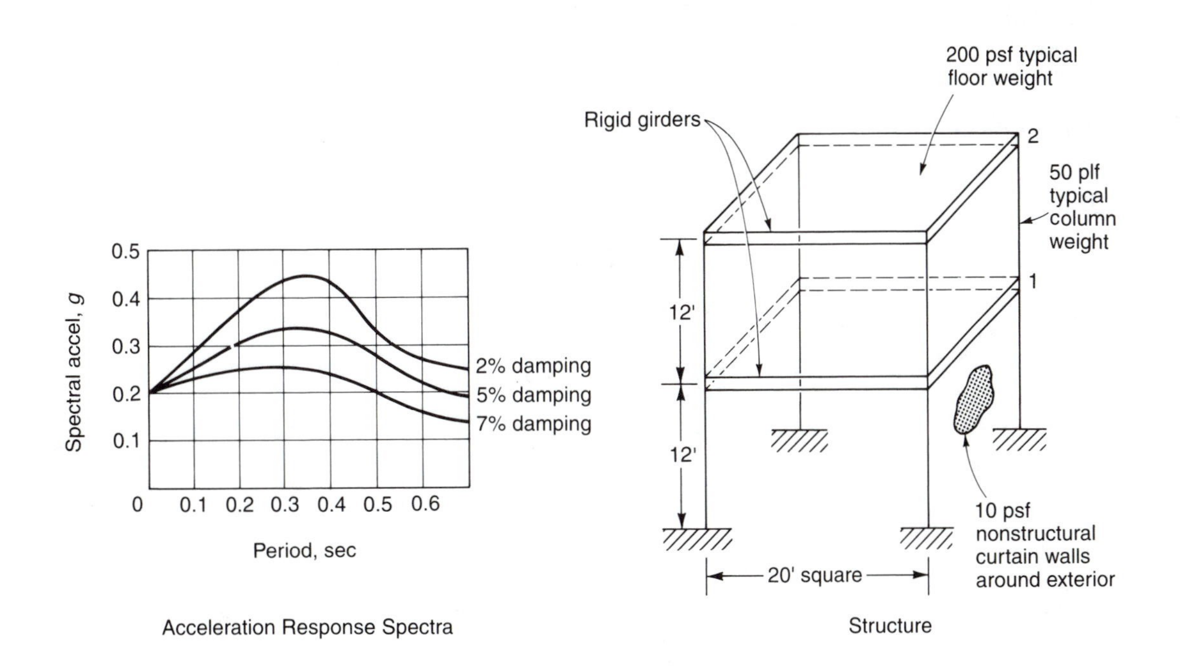

Figure 1–18 Details for Example 1–6

REQUIRED: 1. Determine the mathematical model for dynamic analysis and summarize this in a sketch indicating story masses and stiffness.

2. Sketch the approximate first mode shape, and calculate the fundamental period of vibration of the mathematical model. Do not use UBC equations to calculate period.
3. For the first mode assuming $T_1 = 0.5$ second, and that mode shape is $\phi_2 = 1.0$ and $\phi_1 = 0.67$, calculate the first mode story forces.
4. What is the peak ground acceleration?

SOLUTION

DEAD LOAD

The dead load tributary to the roof level is obtained by summing the contributions from the roof self weight, the columns and the walls. Thus,

Roof	$= 0.2 \times 20 \times 20$	=	80.0 kips
Four walls	$= 0.01 \times 20 \times (12/2) \times 4$	=	4.8
Four columns	$= 0.05 \times (12/2) \times 4$	=	1.2
Total	$= W_2$	=	86.0

The dead load tributary to the second floor is

Roof	$= 0.2 \times 20 \times 20$	=	80.0 kips
Four walls	$= 0.01 \times 20 \times 12 \times 4$	=	9.6
Four columns	$= 0.05 \times 12 \times 4$	=	2.4
Total	$= W_1$	=	92.0

STORY STIFFNESS

The stiffness of each story is the shear force required to produce a unit displacement of that story and is given by

$$k_1 = k_2 = \text{story shear/displacement} = 10/0.1 = 100 \text{ kips per inch}$$

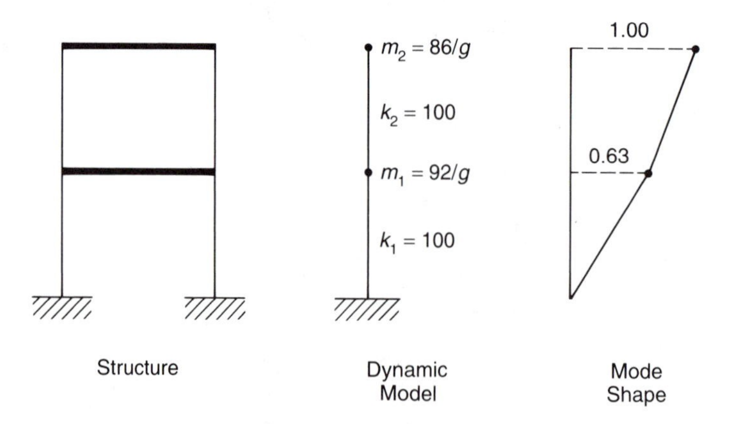

Figure 1–19 Dynamic model and mode shape for Example 1–6

1. DYNAMIC MODEL

The two-story frame may be considered a shear structure with all mass concentrated at the rigid floor slabs and having one degree of freedom, a horizontal translation, at each floor slab. The dynamic model is shown in Figure 1-19.

2. MODE SHAPE AND PERIOD

An iteration technique may be used to determine the mode shape. The procedure is illustrated in Table 1-4 with the initial mode shape defined by

$$x_2 = 1.00$$
$$x_1 = 0.67$$

Floor Level i	Story	Floor Mass m_i	Story Stiff k_i	Initial Mode x_i	Incrm Shear $m_i\omega^2x_i$	Story Shear $k_i\Delta_i$	Story Drift Δ_i	Floor Disp $\Sigma\Delta_i$	Revised Mode x_i
2		$86/g$		1.00	$86\omega^2/g$			$2.34\omega^2/g$	1.00
	2		100			$86\omega^2/g$	$0.86\omega^2/g$		
1		$92/g$		0.67	$62\omega^2/g$			$1.48\omega^2/g$	0.63
	1		100			$148\omega^2/g$	$1.48\omega^2/g$		

Table 1-4 First iteration

The iterative procedure is now repeated in Table 1-5 with the initial mode shape defined by

$$x_2 = 1.00$$
$$x_1 = 0.63$$

Floor Level i	Story	Floor Mass m_i	Story Stiff k_i	Initial Mode x_i	Incrm Shear $m_i\omega^2x_i$	Story Shear $k_i\Delta_i$	Story Drift Δ_i	Floor Disp $\Sigma\Delta_i$	Revised Mode x_i
2		$86/g$		1.00	$86\omega^2/g$			$2.30\omega^2/g$	1.00
	2		100			$86\omega^2/g$	$0.86\omega^2/g$		
1		$92/g$		0.63	$58\omega^2/g$			$1.44\omega^2/g$	0.63
	1		100			$144\omega^2/g$	$1.44\omega^2/g$		

Table 1-5 Second iteration

The revised mode shape is identical to the initial shape and another iteration is unnecessary. The final mode shape is shown in Figure 1-19.

The natural circular frequency of the first mode is obtained by equating the final value of the largest displacement component to its initial value. Thus,

$$2.30\omega^2/g = 1.00$$
$$\omega = (386/2.30)^{1/2}$$
$$= 12.96 \text{ radians per second.}$$

The fundamental period is given by

$$T = 2\pi/\omega$$
$$= 0.49 \text{ seconds}$$

3. LATERAL FORCES

The natural period for the first mode is given as

$$T = 0.5 \text{ seconds}$$

and, for a damping coefficient of seven percent, the corresponding spectral acceleration is obtained from the response spectra as

$$S_a = 0.2g$$

From Table 1-6 the participation factor is obtained as

$$P = \Sigma W_i\phi_i/\Sigma W_i\phi_i^2$$

where W_i = weight at floor level i

ϕ_i = mode shape component for node point i for the first mode

Then $P = 148/127$
$= 1.16$

The lateral force at a node is given by

$$F_i = W_i\phi_i P S_a/g$$
$$= W_i\phi_i \times 1.16 \times 0.2g/g$$
$$= 0.232W_i\phi_i$$

and the force at each level and the shear at each story V_i is shown in Table 1-6.

Level	W_i	ϕ_i	$W_i\phi_i$	$W_i\phi_i^2$	F_i	V_i
Roof	86	1.00	86	86	20.0	–
2nd Floor	92	0.67	62	41	14.4	20.0
1st Floor	–	–	–	–	–	34.4
Total	–	–	148	127	34.4	–

Table 1-6 Determination of participation factor

4. PEAK GROUND ACCELERATION

The peak ground acceleration occurs at time T = 0 and is obtained from the response spectra as

$$S_a = 0.2g$$

1.3 RESPONSE SPECTRA

1.3.1 Dynamic response of structures

The member forces produced in a structure by gravity loads are static forces which are time independent. Seismic forces are produced in a structure by a variable ground vibration which causes a time dependent response in the structure. The response generated depends on the magnitude, duration, and harmonic content of the exciting ground motion, the dynamic properties of the structure and the characteristics of the soil deposits at the site. A response curve is a graph of the maximum, or spectral, response of a range of single-degree-of-freedom oscillators to a specified ground motion, plotted against the frequency or period of the oscillators. The damping ratio of the oscillator may be varied and the response recorded may be displacement, velocity or acceleration. The relationship between these functions[9] may be determined by noting that when a node attains its maximum spectral displacement S_d the spectral velocity S_v is zero and the maximum inertial force equals the spring force. Hence

$$mS_a = kS_d$$

and since

$$\omega^2 = k/m$$

$$S_a = \omega^2 S_d$$

The spectral velocity is defined as

$$S_v = \omega S_d$$

The relationships between the functions are

$$S_a = \omega S_v = \omega^2 S_d$$

where
S_a = spectral acceleration
S_v = spectral velocity
S_d = spectral displacement.

Because of this interrelationship, all three spectra may be plotted on the same graph using tripartite axes and logarithmic scales.

1.3.2 Normalized design spectra

The response of a simple oscillator to variable seismic ground motions may be determined by the Duhamel-integral technique[7,12]. This results in the response curve for a ground motion specific to a particular accelerogram. For design purposes, the response curve must be representative of the characteristics of all seismic events which may be experienced at a particular location. In determining the response of a system, the effective peak ground acceleration (EPA)[8,13] is used as this has a more significant effect on the system than the peak ground acceleration (PGA). The selected spectra, which should include both near and distant earthquakes, are averaged and normalized with respect to the peak ground acceleration. The average curve is then smoothed to eliminate irregularities which could cause large variations in response for a slight change in period. Since soil conditions at a site substantially effect spectral shapes, separate response curves are required for each representative soil type. The normalized design spectra presented in the Uniform Building Code[14] are shown in Figure 1-20. These spectra are applicable to systems with a five percent damping ratio and individual curves are provided for three different soil types.

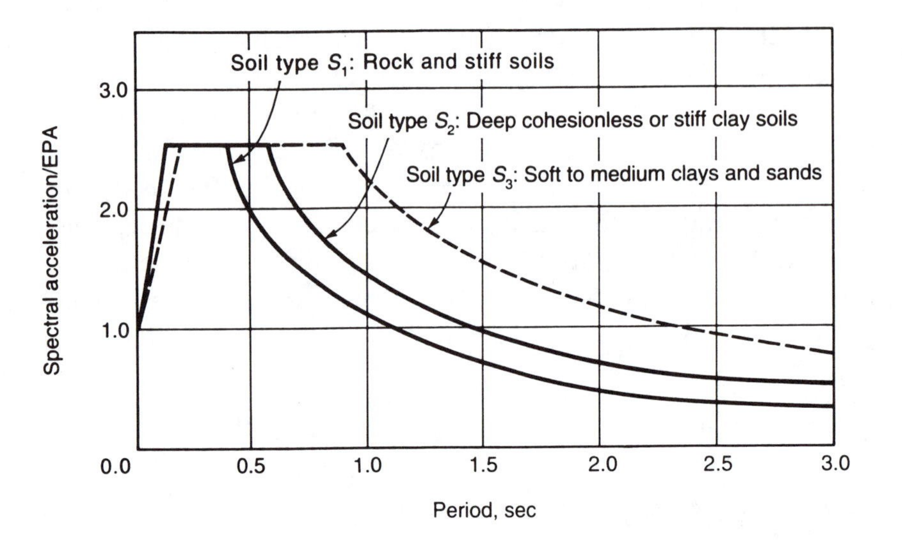

Figure 1-20 Normalized response spectra[14]

The vertical response of a structure is considered equal to two thirds of the corresponding horizontal response[14].

References

1. Harris, J.R. *Overview of seismic codes.* BSCES/ASCE Structural group lecture series, Boston, 1991.

2. Smith, S.W. Introduction to seismological concepts related to earthquake hazards in the Pacific Northwest. *Societal implications: selected readings.* Building Seismic Safety Council. Washington, D.C., 1985.

3. Thiel C.C. Editor. *Competing against time.* Report by the Governor's Board of Inquiry on the 1989 Loma Prieta earthquake. State of California, Office of Planning and Research, Sacramento, CA, 1990.

4. Shakal, A.M. et al. *CSMPI strong motion records from the Santa Cruz mountains (Loma Prieta) earthquake of October 17, 1989.* California Department of Conversation, Division of Mines and Geology, Sacramento, CA, 1989.

5. Plafker, G. and Galloway, J.P. *Lessons learned from Loma Prieta, California, earthquake, of October 17, 1989.* U.S. Geological Survey, Washington, D.C., 1989.

6. Building Seismic Safety Council. *Improving the seismic safety of new buildings.* Washington, D.C., 1986.

7. Paz, M. *Structural dynamics.* Van Nostrand Reinhold, New York, 1991.

8. Structural Engineering Association of California. *Recommended lateral force requirements and commentary.* Sacramento, CA, 1990.

9. Corps. of Engineers, *Seismic design guidelines for essential buildings.* NAVFAC, Technical Manual P-335.1, Washington, 1986.

10. Clough, R.W. and Penzien, J. *Dynamics of structures.* McGraw-Hill, New York, 1975.

11. Ramasco, R. An iterative method for the calculation of natural vibration frequencies in plane frames. *Il Cemento,* Vol. 3, No. 4, 1968, pp. 66-69.

12. Federal Highway Administration. *Seismic design and retrofit manual for highway bridges.* Washington, D.C., 1987.

13. Building Seismic Safety Council. *NEHRP recommended provisions for the development of seismic regulations for new buildings: Part 2, Commentary,* Washington, D.C., 1991.

14. International Conference of Building Officials. *Uniform building code - 1994.* Whittier, CA, 1994.

$$\omega = (k/m)^{1/2} = (kg/W)^{1/2} = 2\pi/T = 2\pi f \text{ rad/sec.}$$

$$g = 386.4; \quad \omega = 0.32\left(\frac{k}{W}\right)^{1/2}$$

This page left blank intentionally.

2

Static lateral force procedure for buildings

2.1 DETERMINATION OF LATERAL FORCES

2.1.1 Seismic zone factor

The seismic zone factor Z, given in UBC Table 16–I, is the Code estimate of the applicable site dependent effective peak ground acceleration expressed as a function of the gravity constant g. The values of Z range from 0.075 to 0.40 with the USA being divided into five different seismic zones in UBC Figure 16–2. The zone factor corresponds to ground motion values with a recurrence interval of 475 years which gives a ten percent probability of being exceeded in a fifty year period. These values are based on historical records and geological data and are also adjusted in order to provide consistent design criteria within local jurisdictions. The zone factor is used to scale the normalized design spectra given in UBC Figure 16–3 in order to provide the response spectrum envelopes required for the dynamic lateral force procedure specified in UBC Section 1629.

2.1.2 Seismic force coefficient

The seismic force coefficient C, given in UBC Section 1628, is the Code estimate of the dynamic amplified response of a structure to the input ground motion. The force coefficient is given by UBC Formula (28–2) as

$$C = 1.25S/T^{2/3}$$

where S = site coefficient, for a specific soil type, from UBC Table 16–J

T = fundamental period of vibration.

The form of this expression indicates that the force coefficient increases as the site coefficient increases and the natural period reduces.

The site coefficient allows for the effect soil conditions have on structural response. The Code specifies four different soil types and requires the use of soil type S_3 when insufficient geotechnical data is available to accurately establish the soil type. Soil type S_4 allows for the large amplifications which occur on sites with very soft clay deposits. UBC Figure 16–3 indicates that:

- for soil type S_1, the maximum response occurs over a period of 0.15 to 0.40 seconds
- for soil type S_2, the maximum response occurs over a period of 0.15 to 0.60 seconds
- for soil type S_3, the maximum response occurs over a period of 0.20 to 0.90 seconds

Two methods are provided in the UBC for determining the natural period of vibration, method A and method B. Method A utilizes UBC Formula (28–3) which is

$$T = C_t(h_n)^{3/4}$$

where h_n = height in feet of the roof above the base, not including the height of penthouses or parapets

C_t = 0.035 for steel moment-resisting frames
= 0.030 for reinforced concrete moment-resisting frames and eccentric braced steel frames
= 0.020 for all other buildings.

The form of this expression indicates that the natural period increases as the height of the structure increases and is greater for steel frames than for concrete frames.

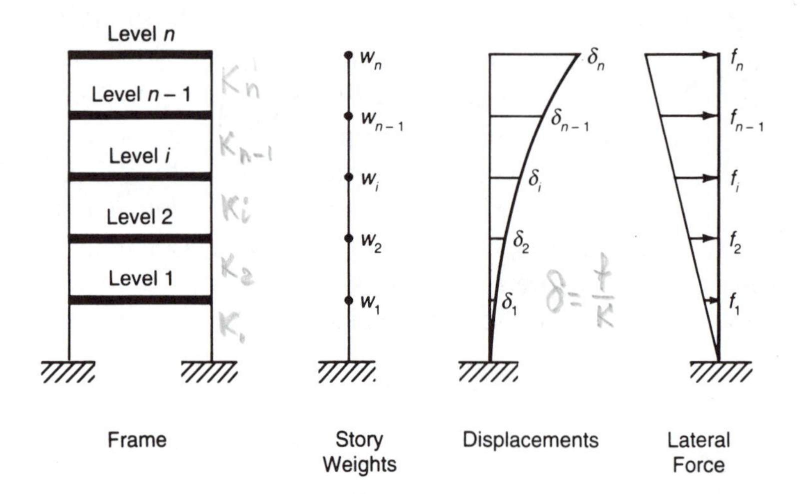

Figure 2–1 Rayleigh procedure

Method B utilizes the Rayleigh procedure as given in UBC Formula (28–5) which is

$$T = 2\pi(\Sigma w_i \delta_i^2 / g \Sigma f_i \delta_i)^{1/2}$$

where δ_i = elastic deflection at level i

f_i = lateral force at level i

w_i = seismic dead load located at level i

g = acceleration due to gravity

The lateral forces f_i represent any lateral force distribution increasing approximately uniformly with height as shown in Figure 2-1. This distribution, in the form of an inverted triangle, corresponds to the distribution of base shear which is assumed in the Code and is equivalent to the inertial forces produced in a frame with uniform mass distribution, equal story heights, and with acceleration increasing uniformly with height. If the contribution of the non structural elements to the stiffness of the structure is underestimated, the calculated deflections and natural periods are overestimated, giving a value for the force coefficient which is too low. To reduce the effects of this error, UBC Section 1628 specifies that the value of the natural period determined by method B may not exceed the value determined by method A by more than thirty percent in seismic zone 4 nor forty percent in seismic zones 1, 2, and 3 .

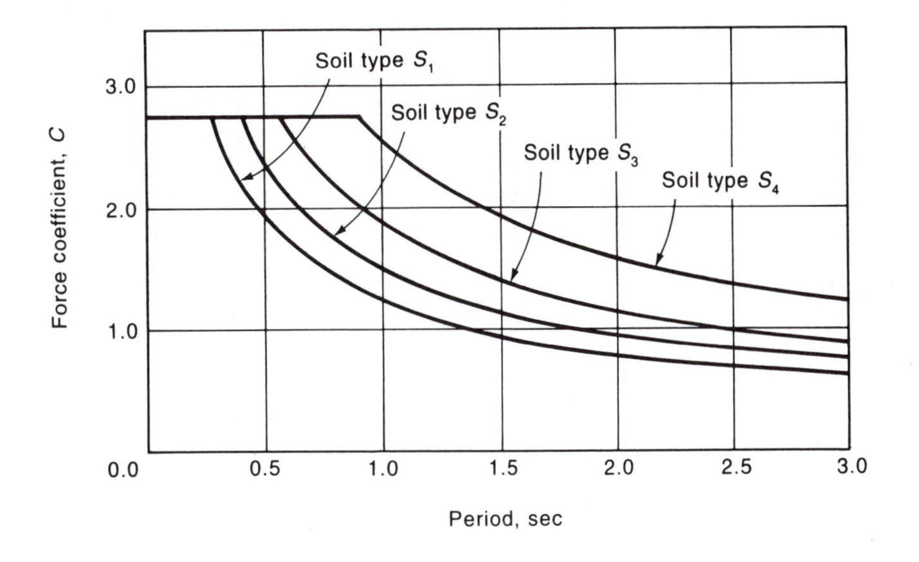

Figure 2-2 Seismic force coefficient

The value of the force coefficient need not exceed 2.75 and the value of the force coefficient C as given by UBC Formula (28-2), represents a normalized response spectrum. The site coefficient S adjusts the shape to allow for the appropriate soil characteristics, and the use of the factor $T^{2/3}$ rather than T adjusts the spectra to provide an equivalent normalized multi-mode response envelope to include an allowance for higher mode effects[1]. These equivalent multi-mode normalized response spectra are shown in Figure 2-2. The product ZC provides the response spectrum envelopes, as a fraction of g, required for the static design method, and represents the base shear induced in a linear elastic structure subjected to the maximum expected seismic ground vibration.

Example 2.1 (Determination of seismic force coefficient)
A three story, steel, moment-resisting frame with the properties shown in Figure 2-3 and with a damping ratio of five percent is located on a site in zone 3 with an undetermined soil profile. Calculate the natural period of vibration T using UBC method A and method B and determine the corresponding value of the seismic force coefficient C.

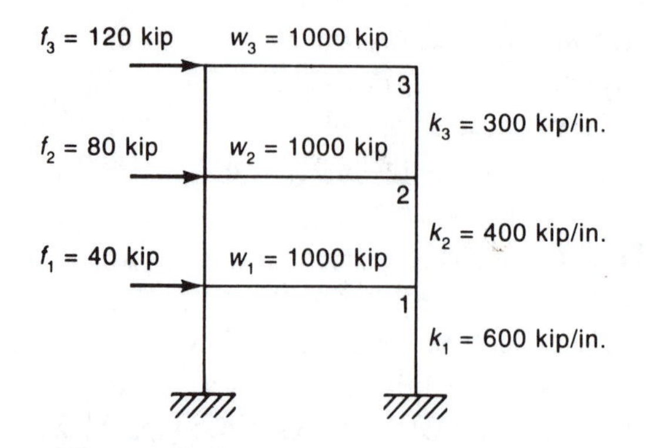

Figure 2-3 Frame for Example 2-1

Solution

Using method A, the natural period is given by UBC Formula (28-3) as

$$T_A = C_t(h_n)^{3/4}$$

where C_t = 0.035 for a steel moment-resisting frame
h_n = roof height = 36 feet.

Then the natural period is

$$T_A = 0.035(36)^{3/4}$$
$$= 0.51 \text{ seconds}$$

The seismic force coefficient is given by UBC Formula (28-2) as

$$C_A = 1.25S/(T_A)^{2/3}$$

where S = 1.5 from UBC Table 16-J for a site with unknown soil properties.

Then the force coefficient is

$$C_A = 1.25 \times 1.5/(0.51)^{2/3}$$
$$= 2.92$$
$$> 2.75$$

Hence use the upper limit of

$$C_A = 2.75$$

Using method B, and applying the force system indicated, the displacements at each level are given by

$$\delta_1 = (f_3 + f_2 + f_1)/k_1$$
$$= (120 + 80 + 40)/600$$
$$= 0.4 \text{ inches}$$
$$\delta_2 = (f_3 + f_2)/k_2 + \delta_1$$
$$= (120 + 80)/400 + 0.4$$
$$= 0.9 \text{ inches}$$
$$\delta_3 = f_3/k_3 + \delta_2$$
$$= 120/300 + 0.9$$
$$= 1.3 \text{ inches}$$

The natural period is given by UBC Formula (28–5) as

$$T_B = 2\pi(\Sigma w_i\delta_i^2/g\Sigma f_i\delta_i)^{1/2}$$

The relevant values are given in Table 2–1.

Level	w_i	f_i	δ_i	$w_i\delta_i^2$	$f_i\delta_i$
3	1000	120	1.3	1690	156
2	1000	80	0.9	810	72
1	1000	40	0.4	160	16
Total	–	–	–	2660	244

Table 2–1 Rayleigh procedure

Thus

$$\begin{aligned} T_B &= 0.32(2660/244)^{1/2} \\ &= 1.06 \text{ seconds} \end{aligned}$$

The natural period obtained by method B, in accordance with UBC Section 1628.2.2, is limited to $1.4T_A$ for structures in zone 3.

$$\begin{aligned} 1.4T_A &= 1.4 \times 0.51 \\ &= 0.71 \text{ seconds} \\ &< 1.06 \text{ seconds} \end{aligned}$$

Hence use the maximum value of

$$T_B = 0.71 \text{ seconds}$$

The force coefficient is given by UBC Formula (28–2) as

$$\begin{aligned} C_B &= 1.25 \times 1.5/(0.71)^{2/3} \\ &= 2.35 \text{ ... governs} \end{aligned}$$

2.1.3 Ductility

Ductility is a measure of the ability of a structural system to deform beyond its elastic load carrying capacity without collapse. This allows a redundant structure to absorb energy while successive plastic hinges are formed. For applied static loading, collapse of the structure occurs when a sufficient number of hinges have formed to produce a mechanism. In the case of cyclic seismic loading, the structure undergoes successive loading and unloading and the force–displacement relationship follows a sequence of hysteresis loops. For an idealized elastic–plastic system, this is illustrated in Figure 2–4, where the enclosed area is a measure of the hysteretic energy dissipation.

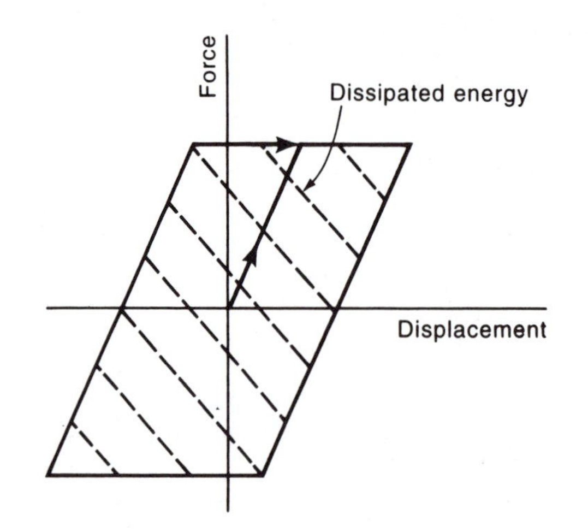

Figure 2–4 Hysteretic energy dissipation

2.1.4 Response modification factor

As it is uneconomical to design a structure to remain within its elastic range for a major earthquake, advantage is taken of the non–linear energy absorbing capacity of the system to allow limited structural damage without impairing the vertical load carrying capacity and endangering life safety. In addition, as yielding occurs, the natural period and the damping ratio increase thus reducing the seismic force developed in the structure.

The structure response modification factor R_W given in UBC Table 16–N is the ratio of the seismic base shear, which would develop in a linearly elastic system, to the prescribed design base shear and is a measure of the ability of the system to absorb energy and sustain cyclic inelastic deformations without collapse. In addition to compensating for the energy dissipation capability, lateral force system redundancy and increase in natural period and damping ratio, the response modification factor allows for the applicable design load factors and safety factors, the provision of secondary lateral support systems and the observed performance of specific materials and structural systems in past earthquakes.

2.1.5 Classification of structural systems

UBC Section 1627.6 details four general categories of building types and these are illustrated in Figure 2–5. Fundamental aspects of the determination of the response modification factor are that the Code detailing provisions for each type of construction material must be strictly adhered to and the necessary inspection and observation during construction is performed as specified in UBC Section 108.

In a bearing wall system, shear walls or braced frames provide support for all or most of the gravity loads and for resisting all lateral loads. In general, a bearing wall system has comparably lower values for R_W since the system lacks redundancy and the lateral support members also carry gravity loads and failure of the lateral support members will result in collapse of the gravity load carrying capacity. In seismic zones 3 and 4, the concrete and masonry shear walls are required to be specially detailed to satisfy UBC Sections 1921 and 2106. Steel braced frames are required to be specially detailed to satisfy UBC Section 2211.

A building frame system has separate systems to provide support for lateral forces and gravity loads. A frame provides support for essentially all gravity loads with independent shear walls or braced frames resisting all lateral forces. The gravity load supporting frame does not require special ductile detailing, but it is required to satisfy the deformation compatibility requirements of UBC Section 1631.2.4 and this imposes a practical limitation on the height of a building frame system.

Moment-resisting frames are specially detailed to provide good ductility and support for both lateral and gravity loads by flexural action. In seismic zones 3 and 4, the moment-resisting frames are required to be specially detailed to satisfy UBC Sections 1921 or 2211.

In general, a dual system has a comparably higher value for R_W since a secondary lateral support system is available to assist the primary nonbearing lateral support system. Nonbearing walls or bracing supply the primary lateral support system with a moment-resisting frame providing primary support for gravity loads and acting as a backup lateral force system. The moment-resisting frame must be designed to independently resist at least twenty-five percent of the base shear and, in addition, the two systems shall be designed to resist the total base shear in proportion to their relative rigidities.

Restrictions on building heights and on the use of the different building types in specific seismic zones are also imposed in UBC Table 16-N.

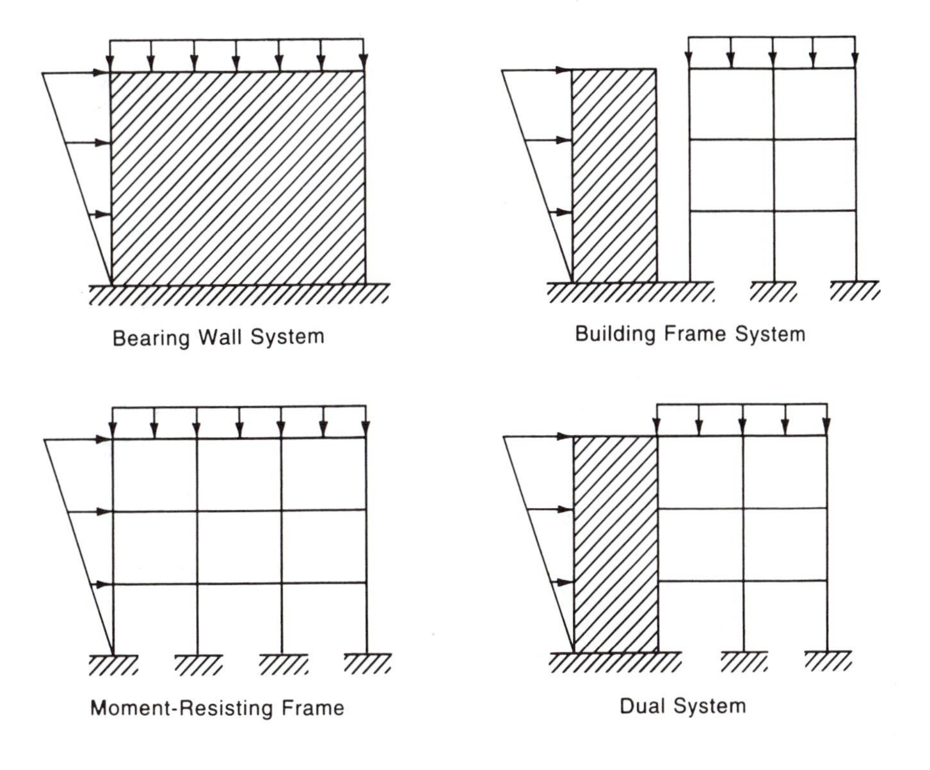

Figure 2-5 Structural systems

2.1.6 Combinations of structural systems

For those cases where different structural systems are employed over the height of the structure, the entire system must be designed, in accordance with UBC Section 1628.3.2, using the lowest R_W value of any story. This requirement is to prevent a concentration of inelastic behavior in the lower stories of a structure and may be relaxed when the dead load above the particular story is less than ten percent of the total structure dead load. This effectively permits, without penalty, the construction of a braced frame penthouse on a moment-resisting frame.

For those designs where a bearing wall system is employed along only one axis of a structure, UBC Section 1628.3.3 requires that the same value for R_W be used in the orthogonal direction. This requirement is to limit the nonelastic deformations perpendicular to the bearing wall system which may, otherwise, impair its gravity load capacity.

2.1.7 Building performance criteria

Normal building structures designed in accordance with the UBC Code may be expected to resist an upper level earthquake with a recurrence interval of 475 years without collapse and without endangering life safety[1,2]. It is anticipated that structural and nonstructural damage will occur which will necessitate the shutdown of the facility until repairs can be effected. In some circumstances, this will be an unacceptable situation and a design is required which will ensure the survival of operational capacity.

2.1.8 Occupancy categories and importance factors

Essential facilities are defined in UBC Table 16–K as hospitals, fire and police stations, emergency response centers and buildings housing equipment for these facilities. Hazardous facilities are defined as structures housing materials which will endanger the safety of the public if released. In order to ensure that essential and hazardous facilities remain functional after an upper level earthquake, an importance factor I of 1.25 is assigned to these facilities. This has the effect of increasing the prescribed design base shear by twenty–five percent, and raises the seismic level at which inelastic behavior occurs and the level at which the operation of essential facilities is compromised. In the case of some business facilities, the loss of operational capacity, following an earthquake, may constitute an unacceptable impact on business competitiveness. In addition, downtime costs caused by an earthquake may exceed the costs necessary to increase the seismic safety of the structure. In these circumstances, the building owner may find that improving seismic performance is economically justifiable. An assessment of the seismic risk may be determined by the earthquake risk management process[3], thus enabling an informed decision to be made before design is undertaken.

2.1.9 Seismic dead load

The seismic dead load W as specified in UBC Section 1628.1, is the total weight of the building and that part of the service load which may be expected to be attached to the building. This consists of

- Twenty-five percent of the floor live load for storage and warehouse occupancies.
- A minimum allowance of ten pounds per square foot for moveable partitions.
- Snow loads exceeding thirty pounds per square foot which may be reduced by seventy-five percent depending on the roof configuration and anticipated ice buildup.
- The total weight of permanent equipment and fittings.

Roof and floor live loads, except as noted above, are not included in the value of W.

2.1.10 Design base shear

The prescribed design base shear is given by UBC Formula (28-1) as

$$V = (ZIC/R_W)W$$

This formula is based on the assumption that the structure will undergo several cycles of inelastic deformation and dissipate energy without collapse. Forces and displacements in the structure are derived assuming linear elastic behavior[1]. The actual forces and displacements produced in the structure are assumed to be greater than these values by a factor $(3R_W/8)$ as specified for critical elements in UBC Sections 1627.9.1, 1628.7.2, 2211.5.1, 2211.11, 2211.7.1.2 for force determination, and Sections 1631.2.4, 1631.2.4.2, 1631.2.11, 2211.10.4 for deformation determination. The correct values, however, are somewhat lower for forces and higher for deformations[1,4]. The idealized force-displacement relationship is shown in Figure 2-6.

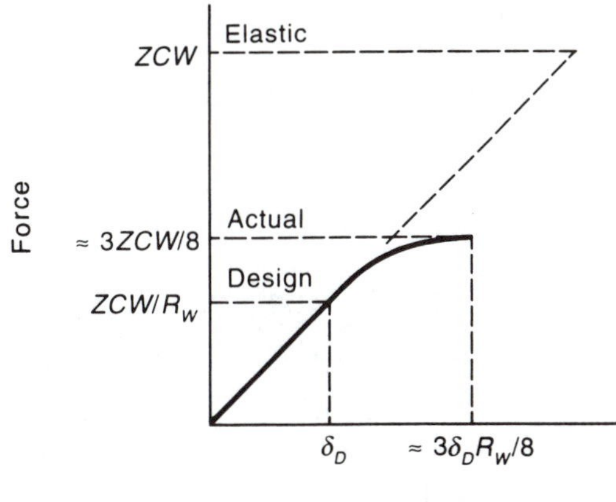

Figure 2-6 Assumed force-displacement curve

The Figure indicates that the maximum base shear developed in a fully elastic structure is given by

$$V_E = ZCW$$

This is reduced by the response modification factor to produce the design value of the base shear

$$V_D = (ZC/R_W)W$$

If the calculated displacement, for this design value, is δ_D and the displacement amplification factor is $3R_W/8$ the actual displacement is assumed, by the UBC Code, to be

$$\delta_A = \delta_D(3R_W/8)$$

Studies[4] indicate, however, that a more accurate value is

$$\delta_A' = \delta_D R_W$$

Example 2–2 (Seismic base shear)
The floor plan of a single story commercial building located in seismic zone 3 is shown in Figure 2–7. The 14 feet high masonry shear walls are load bearing and have a weight of 70 pounds per square foot. The weight of the roof is 50 pounds per square foot and all other weights may be neglected. Determine the seismic base shear.

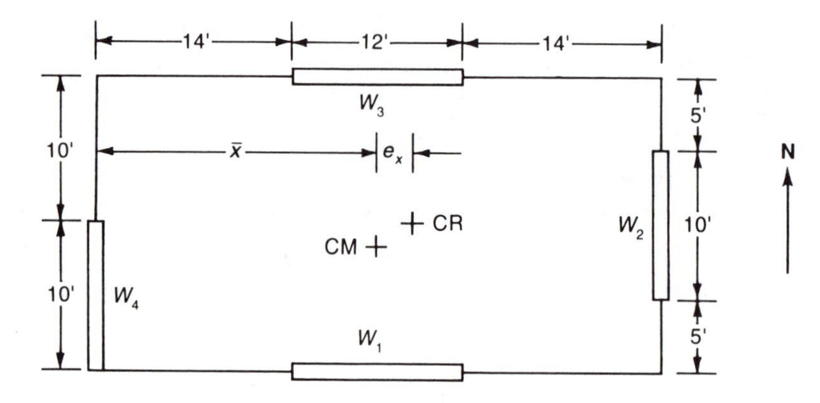

Figure 2–7 Building for Example 2–2

Solution
The relevant dead loads are given by

Roof	$= W_R$	$= 0.05 \times 40 \times 20$	$= 40$ kips
North wall	$= W_3$	$= 0.07 \times 12 \times 14$	$= 11.76$ kips
South wall	$= W_1$	$= 11.76$ kips	
East wall	$= W_2$	$= 0.07 \times 10 \times 14$	$= 9.80$ kips
West wall	$= W$ $= 9.80$ kips		

Total seismic dead load is, then

$$W = W_R + W_1 + W_2 + W_3 + W_4$$
$$= 83.12 \text{ kips}$$

The seismic base shear is given by UBC Formula (16–1) as

$$V = (ZIC/R_W)W$$

where Z = 0.3 for zone 3 from UBC Table 16–I
I = 1.0 for a standard occupancy structure as defined in UBC Table 16–K
C = 2.75, the maximum value specified by UBC Section 1628.2.1

R_W = 6 from UBC Table 16-N for a bearing wall system
W = 83.12 kips, as calculated

Then the seismic base shear is

$$V = (0.3 \times 1 \times 2.75/6)W$$
$$= 0.1375\ W$$
$$= 11.43 \text{ kips}$$

2.2 MULTISTORY STRUCTURES

2.2.1 Vertical distribution of seismic forces

The distribution of base shear over the height of a building is obtained as the superposition of all the modes of vibration of the multiple-degree-of-freedom system. The magnitude of the lateral force at a particular node depends on the mass of that node, the distribution of stiffness over the height of the structure, and the nodal displacements in a given mode and is given by[5,6]

$$F_x = V' w_x \phi_x / \Sigma\, w_i \phi_i$$

where V' = modal base shear
w_i = seismic dead load located at level i
ϕ_i = mode shape component at level i for the given mode.

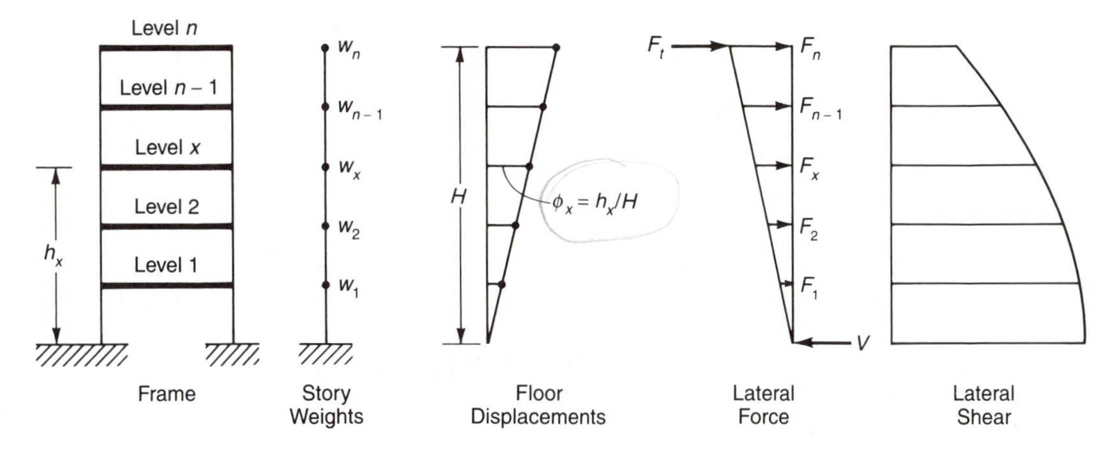

Figure 2–8 Vertical force distribution

For a structure with a uniform distribution of mass over its height and assuming a linear mode shape, as shown in Figure 2–8, this reduces to the expression

$$F_x = V_1 w_x h_x / \Sigma\, w_i h_i$$

where h_i = height above the base to level i.

If only the fundamental mode shape is considered, V_1 represents the design base shear for the fundamental mode and the nodal force distribution is linear. In order to account for higher mode effects in long period buildings, when T exceeds 0.7 seconds, an additional force F_t is then added at the top of the structure with

	F_t	$= 0.07TV$
where	V	$= F_t + V_1$
		$= F_t + \Sigma F_x$
		= total design base shear including the additional increment to account for higher mode effects.

The design lateral force at level x is then given by UBC Formula (28-8) as

$$F_x = (V-F_t)w_x h_x/\Sigma w_i h_i$$

Example 2-3 (Vertical distribution of base shear)
The two story bearing wall structure shown in Figure 2-9 has a roof and second floor weighing 20 pounds per square foot and walls weighing 100 pounds per square foot. The seismic base shear may be assumed to be given by $V = 0.183W$. Determine the vertical force distribution.

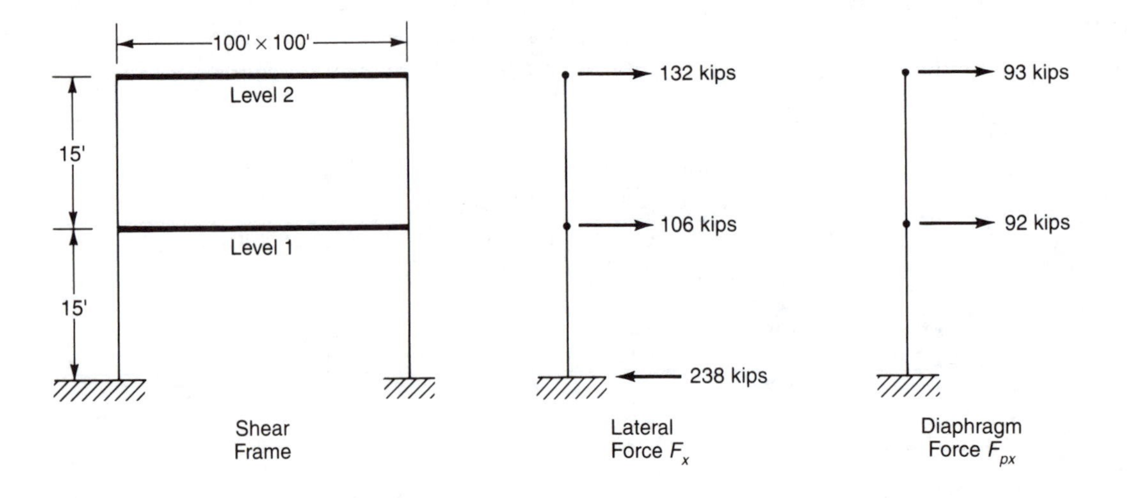

Figure 2-9 Building for Example 2-3

Solution
Using Method A, the natural period is given by UBC Formula (28-3) as

$$T = C_t(h_n)^{3/4} \text{ where,}$$
$$C_t = 0.02 \text{ for a bearing wall system}$$
$$h_n = \text{roof height} = 30 \text{ feet.}$$

Then the natural period is

$$T = 0.02(30)^{3/4}$$
$$= 0.26 \text{ seconds}$$
$$< 0.7 \text{ seconds}$$

Hence $F_t = 0$ and UBC Formula (28-8) reduces to

$$F_x = Vw_x h_x/\Sigma w_i h_i$$

The seismic dead load located at level 2 is given by

$$\begin{aligned} \text{Roof} &= 0.02 \times 100 \times 100 = 200 \text{ kips} \\ \text{Walls} &= 4 \times 0.10 \times 100 \times 15/2 = 300 \text{ kips} \\ w_2 &= 200 + 300 \\ &= 500 \text{ kips} \end{aligned}$$

The seismic dead load located at level 1 is given by

$$\begin{aligned} \text{Second floor} &= 0.02 \times 100 \times 100 = 200 \text{ kips} \\ \text{Walls} &= 4 \times 0.10 \times 100 \times 15 = 600 \text{ kips} \\ w_1 &= 200 + 600 \\ &= 800 \text{ kips} \end{aligned}$$

The relevant values are given in Table 2-2

Level	w_x	h_x	$w_x h_x$	F_x
2	500	30	15,000	132
1	800	15	12,000	106
Total	1300	-	27,000	238

Table 2-2 Vertical force distribution

$$\begin{aligned} V &= 0.183W \\ &= 0.183 \times 1300 \\ &= 238 \text{ kips} \\ F_x &= Vw_x h_x/\Sigma\, w_i h_i \\ &= 238 w_x h_x/27{,}000 \\ &= 0.00882 w_x h_x \end{aligned}$$

The values of F_x are given in Table 2-2 and shown on Figure 2-9.

2.2.2 Overturning

In accordance with UBC Section 1628.7.1, buildings shall be designed to resist seismic overturning effects with only 85 percent of the dead loads participating as specified in UBC Section 1631.1. In addition, where vertical discontinuities are present in the lateral force resisting elements, UBC Section 1628.7.2 requires that the supporting columns be designed for $(3R_w/8)$ times the prescribed forces to resist the seismic loads in the inelastic structure due to a major earthquake.

In accordance with UBC Section 1809.4, when determining stresses at the foundation-soil interface, the force F_t may be omitted from the calculation of overturning effects for regular, structures. This is justified by the fact that F_t represents the higher mode lateral forces and the forces in all stories do not attain their maxima simultaneously. In addition, a momentary uplifting of one edge of the foundation causes a reduction in the seismic forces. This provision, however, may not be applied to the design of the foundation itself. Although the actual seismic forces produced in the structure may be ($3R_w/8$) times the design forces, the above factors ensure that overturning is not critical.

Example 5–7 (Calculation of overturning moment)
Determine the factor of safety against overturning for the structure detailed in Example 2–3.

Solution
The overturning moment is given by

$$\begin{aligned} M_O &= \Sigma F_x h_x \\ &= 134 \times 30 + 108 \times 15 \\ &= 5640 \text{ kip feet} \end{aligned}$$

The restoring moment, allowing for the reduction in dead load specified in UBC Section 1631.1, is given by

$$\begin{aligned} M_R &= 0.85WB/2 \\ &= 0.85 \times 1300 \times 100/2 \\ &= 55{,}250 \text{ kip feet} \end{aligned}$$

The factor of safety against overturning is

$$\begin{aligned} M_R/M_O &= 55{,}250/5640 \\ &= 9.8 \end{aligned}$$

2.2.3 Story drift

Story drift is the lateral displacement of one level of a multistory structure relative to the level below. For buildings with a natural period less than 0.7 seconds, UBC Section 1628.8.2 limits drift to a minimum of $0.04/R_W$ or 0.005 times the story height. For buildings with a natural period of 0.7 seconds or greater, the more stringent limitation is imposed of $0.03/R_W$ or 0.004 times story height. These limitations are to ensure a minimum level of stiffness so as to control inelastic deformation and possible instability. Particularly in taller buildings, large deformations with heavy vertical loads may lead to significant secondary moments and instability.

Since the value of the natural period derived using the Rayleigh Formula (28–5) is a more realistic value than that determined by UBC Formula (28–3), UBC Section 1628.8.3 relaxes the requirement that the value of T_B, for calculation of drift, may not exceed the value of T_A from UBC Formula (28–3) by 30% in seismic zone 4 and 40% in seismic zone 1,2 and 3.

Example 2–5 (Story drift)
Compare the actual value of the story drift ratio for the special moment-resisting steel frame detailed in Example 1–3 with the allowable value.

Solution
For a structure with a fundamental period exceeding 0.7 seconds, UBC Section 1628.8.2 limits the story drift to the minimum value given by

$$\Delta = 0.004h_s$$
$$= 0.0004 \times 15 \times 12$$
$$= 0.72 \text{ inches}$$

or

$$\Delta = 0.03h_s/R_w$$
$$= 0.03 \times 15 \times 12/12$$
$$= 0.45 \text{ inches ... governs}$$

Hence, the maximum allowable story drift ratio is

$$\Delta_R = \Delta/h_s$$
$$= 0.45/(15 \times 12)$$
$$= 0.0025$$

The actual drift ratio in the top story of the frame is

$$\Delta'_R = (4.0 - 1.9)/(15 \times 12)$$
$$= 0.012$$
$$> \Delta_R \text{ ... unsatisfactory}$$

Example 2–6 (Structural Engineering Examination 1988, Section A, weight 3.0 points)

GIVEN: A steel, special moment-resisting frame as shown in Figure 2–10.

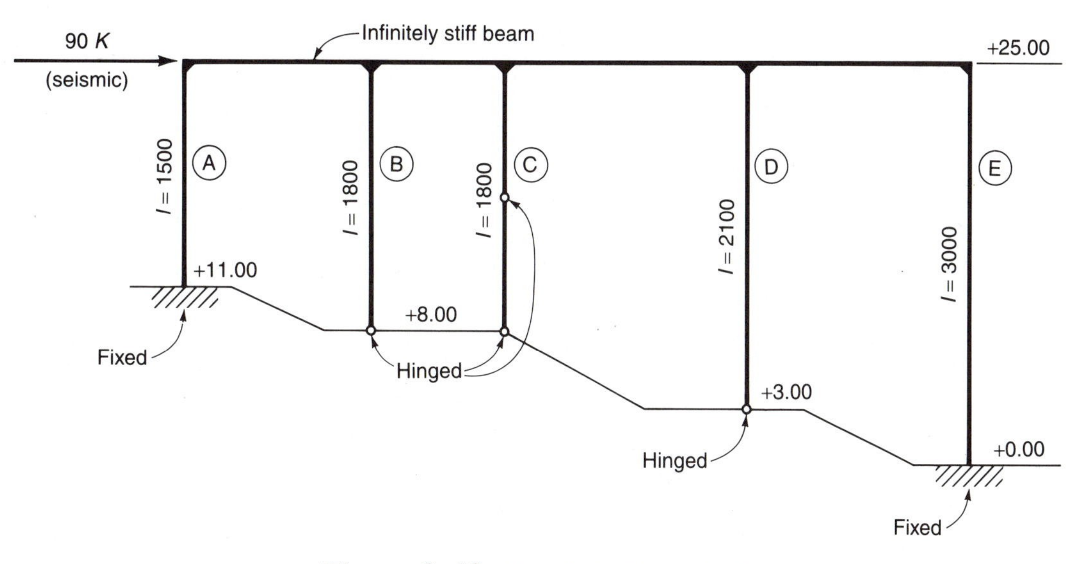

Figure 2–10 Details for Example 2–6

CRITERIA: Modulus of elasticity E = 29,000 kips per square inch
All columns are fixed at the top. For bottom condition, see Figure 2–10.
The continuous beam is infinitely stiff.

REQUIRED:
1. Calculate story drift due to the seismic load of 90 kips.
2. Determine whether your calculated drift is in compliance with the UBC requirements.

3. If non-structural masonry walls are required to fill in each bay, what are the minimum spacings you would recommend between wall and column, wall and beams, in order to comply with the UBC requirements.

SOLUTION

FRAME STIFFNESS

The frame stiffness may be obtained by summing the stiffnesses of the individual columns. The stiffness of column A, which is fixed at both ends, is given by

$$\begin{aligned} K_A &= 12\ EI/L^3 \\ &= 12E \times 1{,}500/(14^3 \times 1{,}728) \\ &= 6.56E/1{,}728 \end{aligned}$$

The stiffness of column B, which is hinged at one end is given by

$$\begin{aligned} K_B &= 3EI/L^3 \\ &= 3E \times 1{,}800/(17^3 \times 1{,}728) \\ &= 1.10E/1{,}728 \end{aligned}$$

Column C, which contains two hinges, constitutes a mechanism and its stiffness is

$$K_C = 0$$

The stiffness of column D, which is hinged at one end is given by

$$\begin{aligned} K_D &= 3EI/L^3 \\ &= 3E \times 2{,}100/(22^3 \times 1{,}728) \\ &= 0.59E/1{,}728 \end{aligned}$$

The stiffness of column E, which is fixed at both ends, is given by

$$\begin{aligned} K_E &= 12EI/L^3 \\ &= 12E \times 3{,}000/(25^3 \times 1{,}728) \\ &= 2.30E/1{,}728 \end{aligned}$$

The total frame stiffness, for lateral loading, is

$$\begin{aligned} K_T &= K_A + K_B + K_C + K_D + K_E \\ &= (6.56 + 1.10 + 0 + 0.59 + 2.30)E/1{,}728 \\ &= 10.55E/1{,}728 \end{aligned}$$

1. STORY DRIFT

The lateral displacement of the frame, or story drift, due to the seismic load V of ninety kips is given by

$$\begin{aligned} \Delta_H &= V/K_T \\ &= 90/K_T \\ &= 90 \times 1{,}728/(10.55 \times 29{,}000) \\ &= 0.51 \text{ inches} \end{aligned}$$

2. ALLOWABLE DRIFT

For a steel, special moment-resisting frame, the response factor R_W is obtained from UBC Table 16-N as 12. Assuming the building has a fundamental period of less than 0.7 seconds, UBC Section 1628.8.2 limits the allowable drift to the lesser of

$$\Delta_A = 0.005 \text{ (story height)}$$

or

$$\begin{aligned}\Delta_A &= 0.04 \text{ (story height)}/R_W \\ &= 0.04 \text{ (story height)}/12 \\ &= 0.0033 \text{ (story height), which governs.}\end{aligned}$$

Hence, the allowable drift is

$$\begin{aligned}\Delta_A &= 0.0033 \text{ (story height)} \\ &= 0.0033 \times 14 \times 12 \\ &= 0.56 \text{ inches} \\ &> 0.51 \text{ inches}\end{aligned}$$

Hence, the calculated drift is in compliance with UBC requirements.

3. BUILDING SEPARATION

It is required that all parts of a building be separated a sufficient distance to permit independent seismic motion without impact between adjacent parts. UBC Section 1631.2.4.2 specifies that the separation shall allow for $3(R_W/8)$ times the displacement due to the prescribed seismic forces.

Assuming that the displacements of the masonry walls are negligible, only the lateral and vertical displacements of the steel frame need be considered. The calculated lateral displacement of the frame is

$$\Delta_H = 0.51 \text{ inches}$$

and the required horizontal separation between columns and the masonry walls is

$$\begin{aligned}\Delta_{H(req)} &= \Delta_H \times 3R_W/8 \\ &= 0.51 \times 3 \times 12/8 \\ &= 2.3 \text{ inches}\end{aligned}$$

The calculated vertical displacement of the frame is

$$\begin{aligned}\Delta_V &= L_A - (L_A^2 - \Delta_H^2)^{1/2} \\ &= 14 \times 12 - [(14 \times 12)^2 - (0.51)^2]^{1/2} \\ &= 0.0007 \text{ inches}\end{aligned}$$

and the required vertical separation between beams and the masonry walls is

$$\begin{aligned}\Delta_{V(req)} &= \Delta_V \times 3R_W/8 \\ &= 0.00077 \times 3 \times 12/8 \\ &= 0.0035 \text{ inches}\end{aligned}$$

Use 0.5 inch minimum as specified in UBC Section 1631.2.4.2.

2.2.4 P–delta effects

The P–delta effects in a given story are caused by the eccentricity of the gravity load above that story, which produces secondary moments augmenting the sway moments in the story. If the ratio of secondary to primary moment exceeds 0.1 and, in seismic zones 3 and 4, the story drift ratio exceeds $0.02/R_W$, the effects of the secondary moments should be included in the analysis. As shown in Figure 2–11, the primary moment in the second story of the frame is

$$M_P = F_2 h_{S2}$$

and the secondary moment is

$$M_S = 2P_2\Delta_2$$

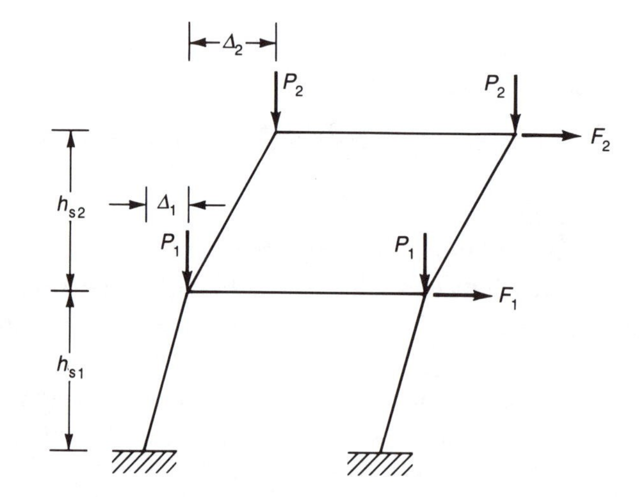

Figure 2–11 P–delta effects

Example 2–7 (P–delta effects)
For the steel frame detailed in Example 1–3, determine if it is necessary to consider P–delta effects.

Solution
For the bottom story, the story drift ratio is given by

$$\begin{aligned} \Delta_R &= \Delta_1/h_{S1} \\ &= 1.9/(15 \times 12) \\ &= 0.011 \end{aligned}$$

for the second story
$\Delta_R = 4.0/((15\times12)) = 0.022$

The limiting ratio for seismic zone 4 is

$$\begin{aligned} 0.02/R_W &= 0.02/12 \\ &= 0.00167 \\ &< \Delta_R \end{aligned}$$

The primary moment in the bottom story is

$$
\begin{aligned}
M_P &= (F_1 + F_2)h_{S1} \\
&= (68.41 + 53.99)15 \\
&= 1836 \text{ kip feet}
\end{aligned}
$$

The secondary moment in the bottom story is

$$
\begin{aligned}
M_S &= P_1\Delta_1 \\
&= 2 \times 0.4 \times 386.4 \times 1.9/12 \\
&= 49 \text{ kip feet}
\end{aligned}
$$

The ratio of the secondary to primary moment is

$$
\begin{aligned}
M_S/M_P &= 49/1836 \\
&= 0.027 \\
&< 0.1
\end{aligned}
$$

Hence, it is unnecessary to consider P-delta effects.

2.2.5 Diaphragm loads

The load acting on a horizontal diaphragm is given by UBC Formula (31-1) as

$$
\begin{aligned}
F_{px} &= (F_t + \Sigma F_i)w_{px}/\Sigma w_i \\
&\geq 0.35ZIw_{px} \\
&\leq 0.75ZIw_{px}
\end{aligned}
$$

where F_t = concentrated lateral force at roof level
F_i = lateral force at level i
w_i = total seismic dead load located at level i
w_{px} = seismic dead load tributary to the diaphragm at level x, not including walls parallel to the direction of the seismic load.

For a single story structure, this reduces to

$$
\begin{aligned}
F_p &= Vw_p/W \\
&= (ZIC/R_w)w_p
\end{aligned}
$$

Also, in a multistory structure, at the second floor level

$$
\begin{aligned}
(F_t + \Sigma F_i)/\Sigma w_i &= V/W \\
&= ZIC/R_W
\end{aligned}
$$

Example 2-8 (Diaphragm loads)
Determine the diaphragm loads for the two story structure detailed in Example 2-3.

Solution
Since the concentrated lateral force at roof level is zero, UBC Formula (31-1) reduces to

$$
F_{px} = w_{px}\Sigma F_i/\Sigma w_i
$$

The seismic dead load tributary to the diaphragm at level 2 is given by

$$\begin{aligned}
\text{Roof} &= 0.02 \times 100 \times 100 = 200 \text{ kips} \\
\text{Walls} &= 2 \times 0.10 \times 100 \times 15/2 = 150 \text{ kips} \\
w_{p2} &= 200 + 150 \\
&= 350 \text{ kips}
\end{aligned}$$

The seismic dead load tributary to the diaphragm at level 1 is given by

$$\begin{aligned}
\text{Second floor} &= 0.02 \times 100 \times 100 = 200 \text{ kips} \\
\text{Walls} &= 2 \times 0.10 \times 100 \times 15 = 300 \text{ kips} \\
w_{p1} &= 200 + 300 \\
&= 500 \text{ kips}
\end{aligned}$$

The relevant values are given in Table 2-3.

Level	Σw_i	ΣF_i	$\Sigma F_i/\Sigma w_i$	w_{px}	F_{px}
2	500	132	0.264	350	93
1	1300	238	0.183	500	92

Table 2–3 Diaphragm loads

The values of F_{px} are given in the Table and are shown on Figure 2–9. The values obtained for F_{px} lie within the stipulated minimum and maximum values which are

$$F_{px} \geq 0.35 \times 0.4 \times 1.0 \times w_{px} = 0.14\, w_{px}$$
$$F_{px} \leq 0.75 \times 0.4 \times 1.0 \times w_{px} = 0.30\, w_{px}$$

2.3 LATERAL FORCE RESISTANT SYSTEM

2.3.1 Basic Components

Figure 2–12 shows the lateral load resisting system in a basic, one–story structure. To transfer the seismic forces to the ground, vertical and horizontal resisting elements must be used to provide a continuous load path from the upper portion of the structure to the foundations[7]. The vertical components consist of shear walls, braced frames, and moment-resisting frames. The horizontal components consist of the roof and floor diaphragms, or the horizontal trusses, which distribute lateral forces to the vertical elements.

A diaphragm is considered flexible, in accordance with UBC Section 1628.5, when the midpoint displacement, under lateral load, exceeds twice the average displacement of the end supports. This is illustrated in Figure 2–13. The diaphragm may then be modeled as a simple beam between end supports, and the distribution of loading to the supports is independent of their relative stiffnesses and is proportional to the tributary areas supported. Flexible diaphragms consist of diagonally sheathed wood diaphragms, plywood sheathed diaphragms, and steel–deck diaphragms.

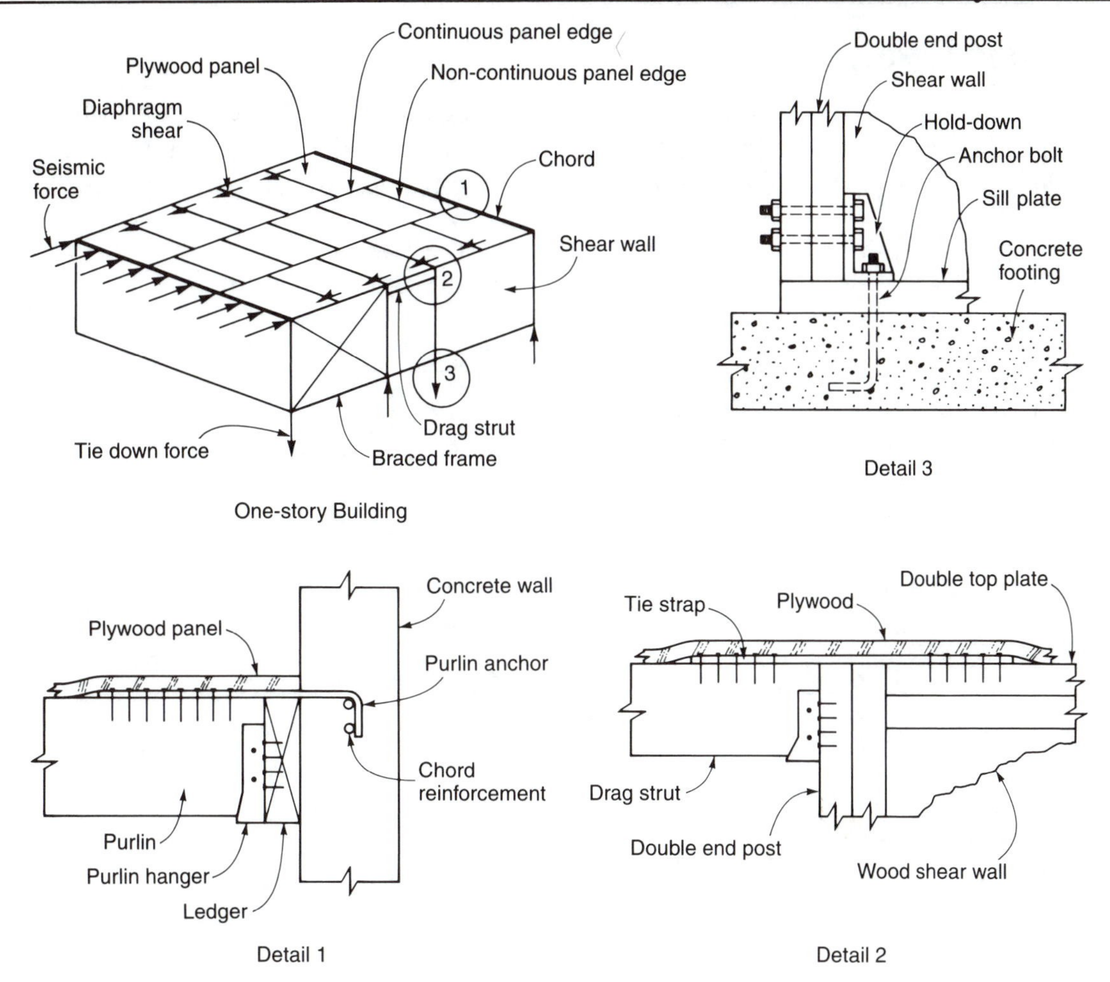

Figure 2–12 Load transfer elements

When the midpoint displacement of the diaphragm does not exceed twice the average story drift, the diaphragm is considered rigid. Allowance must then be made for the additional forces created by torsional effects with the diaphragm and supports assumed to undergo rigid body rotation. The distribution of loading to the supports is proportional to their relative stiffnesses and is independent of the tributary areas supported. Rigid diaphragms consist of reinforced concrete diaphragms, precast concrete diaphragms, and composite steel deck. Figure 2–14 provides a comparison between flexible and rigid diaphragms.

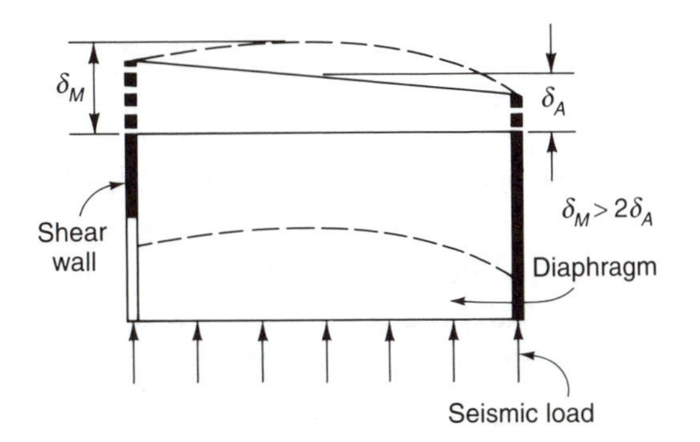

Figure 2–13 Flexible diaphragm

2.3.2 Plywood diaphragms

A plywood diaphragm is a flexible diaphragm which acts as a horizontal deep beam. The plywood sheathing forms the beam web to resist shear force, purlins act as web stiffeners, and the boundary members perpendicular to the load form the flanges to resist flexural effects[8,9,10,11]

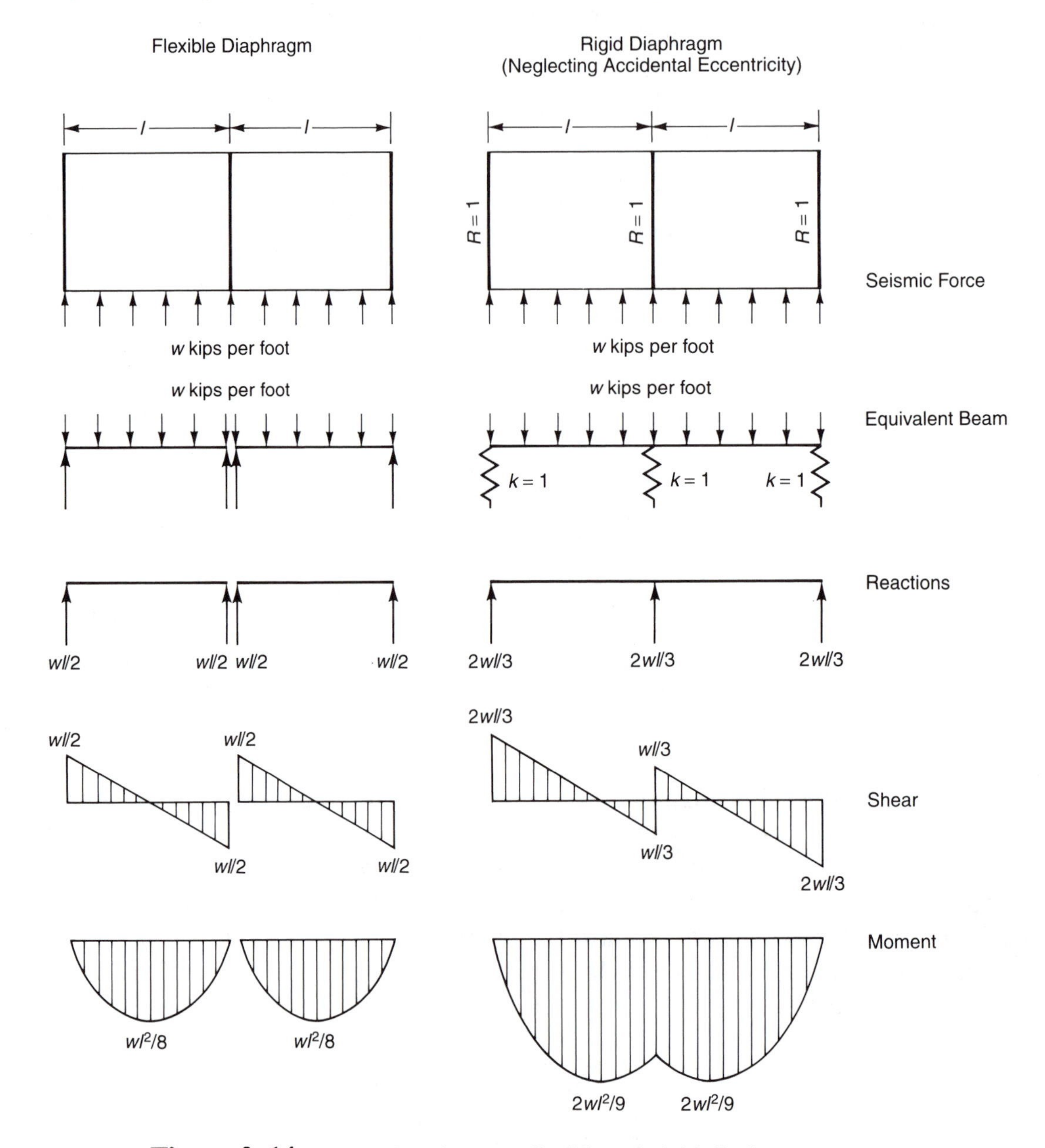

Figure 2–14 Comparison between flexible and rigid diaphragms

The boundary members acting as the flange or chord of the diaphragm may consist of the double top plate of a wood framed shear wall, a steel or wood ledger on the inside face of a concrete wall, or steel reinforcement in a masonry or concrete wall as shown in Figure 2–12. The contribution of the plywood sheathing to the flexural capacity of the deep beam is neglected and the chords are assumed to resist the total applied moment by developing axial forces which provide a couple equal and opposite to the moment. The axial force in a chord is given by

$$F_c = M_D/B_D$$

where M_D = bending moment in the diaphragm
B_D = distance between chord centers
≈ depth of diaphragm

The strength of the plywood sheathing is controlled by the shear strength of the plywood panel, by nail heads pulling through the panel face, by nails splitting panel edges, and by buckling of the panel. The design capacity of a plywood diaphragm depends on the sheathing thickness, grade, and orientation, the width of the framing members, the support of the panel edges, and the nail spacing, type, and penetration. This capacity has been determined experimentally[12,13] and design values are given in UBC Table 23-I-J-1. For cases not covered by this Table, additional design values may be calculated[11,12]. When all panel edges are supported by and are nailed to framing members the diaphragm is termed blocked. This increases the strength of the diaphragm and may be achieved by spacing purlins at eight feet centers and sub-purlins at two feet centers so as to support the edges of a four by eight feet panel. When this framing arrangement is not possible, blocking pieces may be nailed or clipped to the framing members as shown in Figure 2-15.

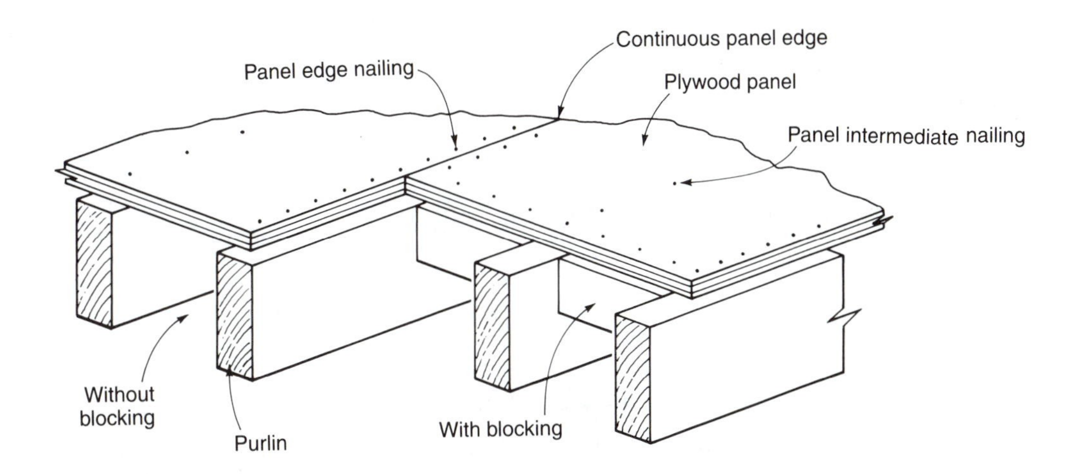

Figure 2-15 Plywood diaphragm

The unit shear stress in a diaphragm is given by

$$q = Q/B_D$$

where Q = shear force at the section considered
B_D = depth of diaphragm

For a given plywood panel grade, thickness, orientation, and edge support the required nail spacing may be obtained from UBC Table 23-I-J-1. Since the shear decreases in a uniformly loaded diaphragm from the end supports to midspan the nail spacing may be progressively increased. The strength of a diaphragm may be increased by increasing the grade and thickness of the plywood, reducing nail spacing, increasing the width of framing members, blocking all panel edges, and by staggering panel edges in the direction of the applied force.

A limit of four to one is imposed, in UBC Table 23–I–I, on the diaphragm aspect ratio in order to control deflection and ensure that the gravity load carrying capacity of bearing walls, perpendicular to the applied load, is not impaired. Diaphragm deflection depends on the flexural contribution, web shear contribution, chord splice slippage, and panel edge nail slippage. The value of the deflection is readily calculated[9,11] when required.

Example 2–9 (Nailing requirements)
Details of a single story industrial building, located in seismic zone 4, are shown in Figure 2–16. The weight of the wood roof is 15 pounds per square foot and the weight of the eight inch masonry walls is 75 pounds per square foot. The roof sheathing is ½-inch nominal Structural I grade plywood and the framing member dimensions are indicated on the roof plan. For north–south seismic loads, draw the required nailing diagram and determine the chord reinforcement required.

Solution
The relevant dead load tributary to the roof diaphragm in the north–south direction is due to the north and south wall and the roof dead load and is given by

Roof $= 15 \times 120 = 1800$ pounds per linear foot
North wall $= 75 \times 14^2/(2 \times 12.5) = 590$ pounds per linear foot
South wall $= 590$ pounds per linear foot

The total dead load tributary to the roof diaphragm is

$$w_p = (1800 + 590 + 590)\,256/1000 = 762 \text{ kips}$$

For a single story structure, the seismic load acting at the roof level is given by

$$V = (ZIC/R_W)w_p$$

where Z = 0.4 for zone 4 from UBC Table 16–I
I = 1.0 from UBC Table 16–K for a standard occupancy structure
C = 2.75 from UBC Section 1628.2.1
w_p = 762 kips, as calculated
R_W = 6 from UBC Table 16–N item 1.2.b
= maximum value allowed by UBC Section 1631.2.9 for a flexible diaphragm supporting a masonry wall

The diaphragm load is, then

$$V = (0.4 \times 1.0 \times 2.75/6)762 = 139.4 \text{ kips}$$

The shear force along the diaphragm boundaries at grid lines 1 and 9 is

$$Q = V/2 = 69.7 \text{ kips}$$

The unit shear along the diaphragm boundaries is

$$q_1 = Q/B_D$$
$$= 69.7 \times 1000/120$$
$$= 581 \text{ pounds per linear foot}$$

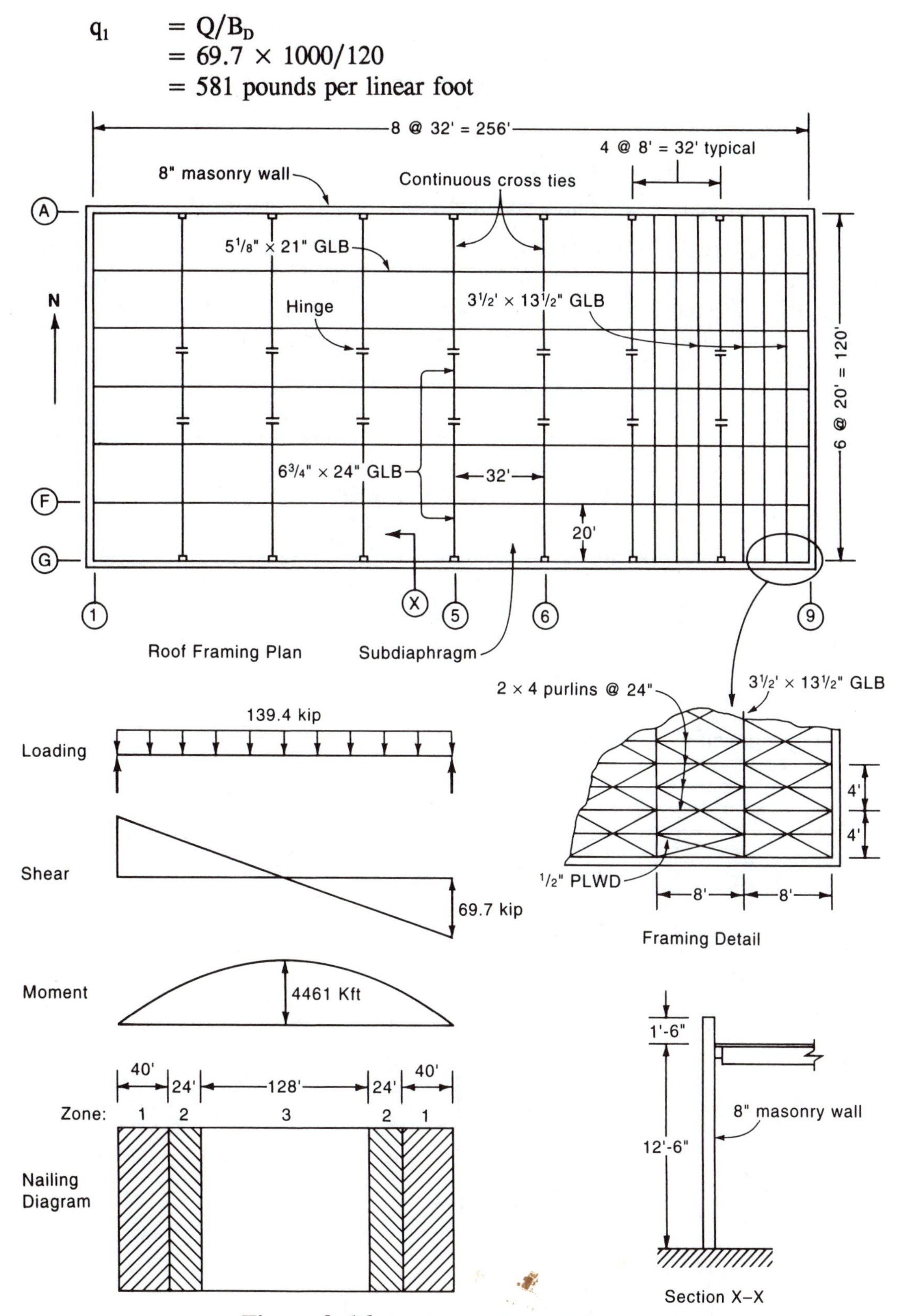

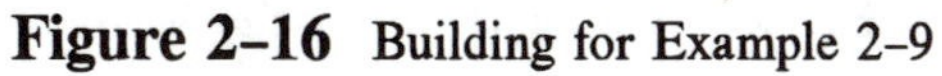

Figure 2–16 Building for Example 2–9

The nail spacing may be changed at the beam locations, shown in the Figure, and the unit shear a distance 40 feet from the boundary is given by

$$q_2 = q_1 \times 88/128$$
$$= 399 \text{ pounds per linear foot}$$

At 64 feet from the boundary, the unit shear is

$$q_3 = q_1 \times 64/128$$
$$= 291 \text{ pounds per linear foot}$$

The required nail spacing is obtained from UBC Table 23–I–J–1 with a case 4 plywood layout applicable, all edges blocked and with 3½-inch framing at continuous panel edges parallel to the load. Using ½-inch Structural I grade plywood and 10d nails with 1⅝-inch penetration, the nail spacing required in the three diaphragm zones are given in Table 2–4.

Zone	1	2	3
Diaphragm boundaries	2½"	4"	6"
Continuous panel edges	2½"	4"	6"
Other edges	4"	6"	6"
Intermediate members	12"	12"	12"
Capacity provided, plf	640	425	320
Capacity required, plf	581	399	291

Table 2–4 Nail spacing requirements

The bending moment at the midpoint of the north and south boundaries due to the north–south seismic force is

$$M_D = VL/8$$
$$= 139.4 \times 256/8$$
$$= 4461 \text{ kip feet}$$

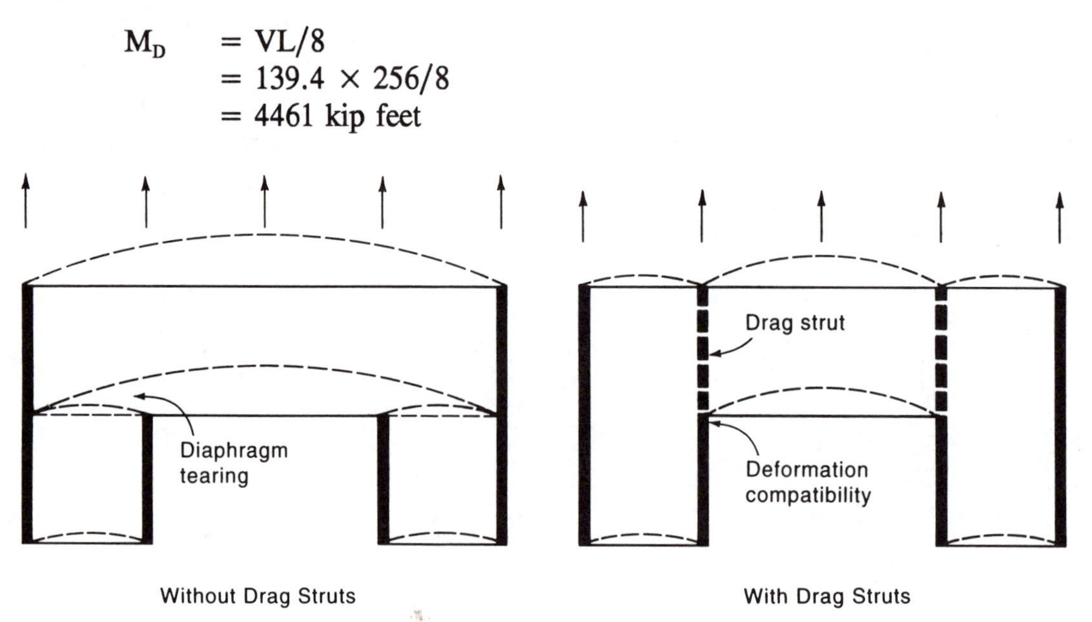

Figure 2–17 The function of collector elements

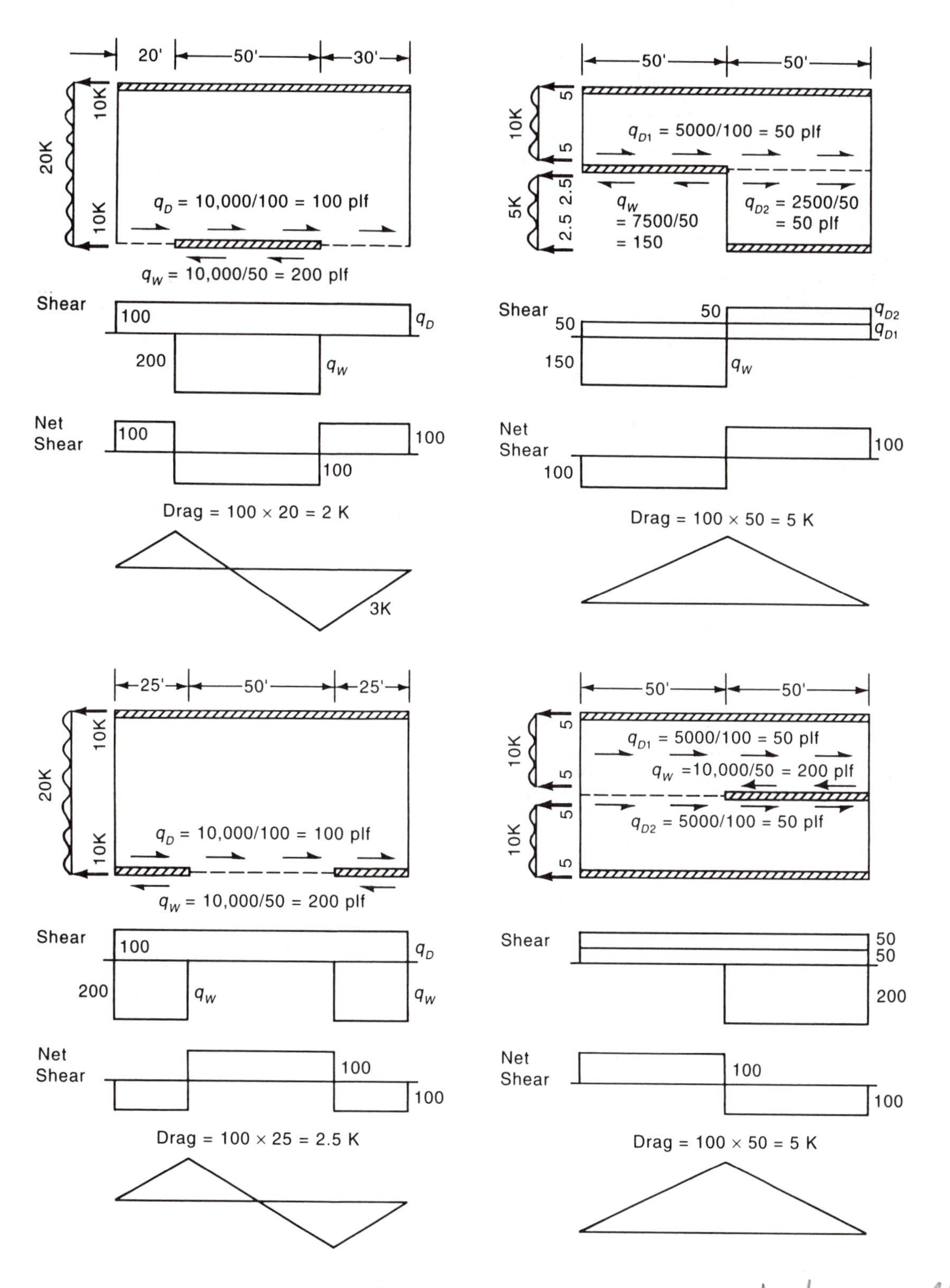

Figure 2–18 Typical drag force diagrams: wood shear walls

The corresponding chord force is

$$\begin{aligned} F_C &= M_D/B_D \\ &= 4461/120 \\ &= 37.2 \text{ kips} \end{aligned}$$

Using grade 60 deformed bar reinforcement, the allowable stress is given by UBC Section 2107.2.11 as

$$F_S = 24 \text{ kips per square inch}$$

A stress increase of one-third is allowed for seismic loading, in accordance with UBC Section 1603.5, and the area of reinforcement required is

$$\begin{aligned} A_S &= F/(F_S \times 1.33) \\ &= 37.2/(24 \times 1.33) \\ &= 1.17 \text{ square inches} \end{aligned}$$

Providing two number 8 bars gives an area of

$$\begin{aligned} A'_S &= 1.20 \text{ square inches} \\ &> A_S \ldots \text{ satisfactory} \end{aligned}$$

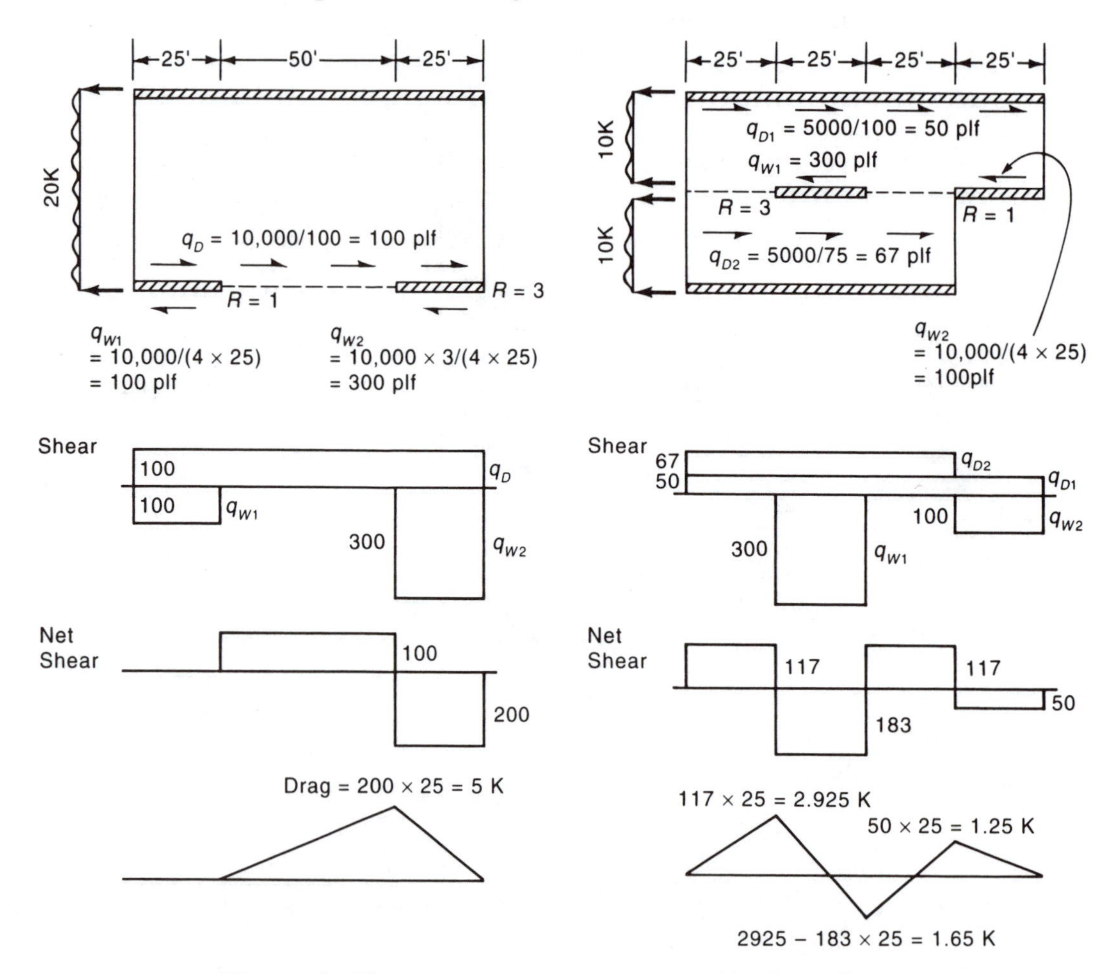

Figure 2-19 Masonry shear wall: typical drag force diagrams

2.3.3 Collector elements

Where shear walls are discontinuous, or reentrant corner irregularities are present, collector elements or drag struts are required to ensure deformation compatibility and prevent localized tearing of the diaphragm. This is illustrated in Figure 2–17. The drag strut transfers the shear originating in the unsupported portion of the diaphragm to the shear wall. Examples of typical drag force calculations involving wood shear walls, where the unit shear is constant in all walls, is shown in Figure 2–18. In the case of masonry or concrete shear walls, the shear resistance of the wall is proportional to its relative rigidity and this is shown in Figure 2–19. In the case of irregular structures, UBC Section 1631.2.9.6 allows no increase in stress in designing the connections of collector elements for seismic loads. Detail 2 of Figure 2–12 illustrates the use of a tie strap to connect a drag strut to a shear wall.

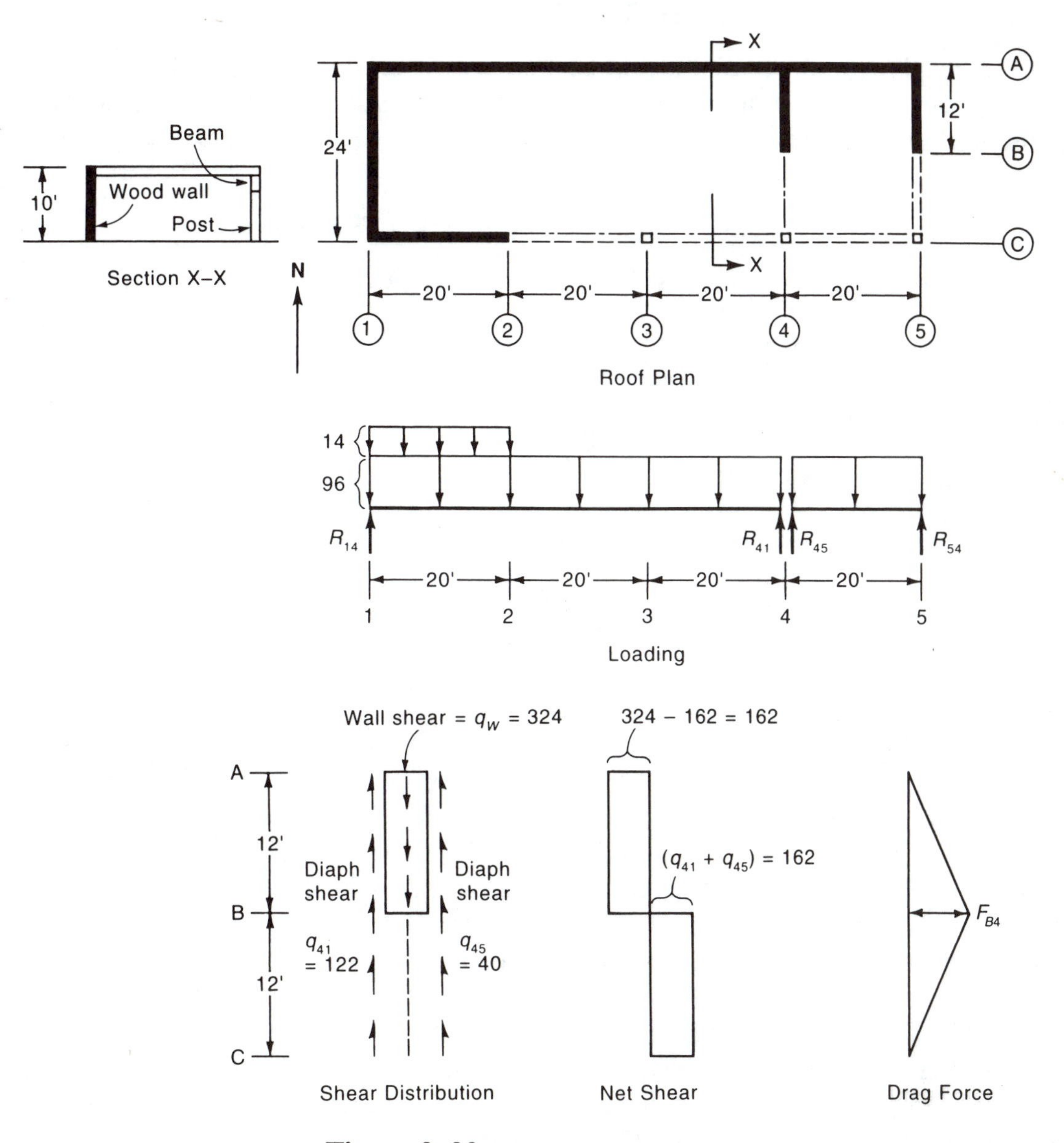

Figure 2–20 Building for Example 2–10

Example 2–10 (Drag force calculation)
Details of a single story wood structure are shown in Figure 2–20. The weight of the roof is 15 pounds per square foot and the weight of the shear wall is 12 pounds per square foot. The seismic load may be assumed to be given by $V = 0.229w_p$. Determine the drag force at the intersection of grids B and 4.

Solution
For a single story structure, the lateral force acting at the roof level for north–south seismic loading is given by

$$\text{Roof} = 0.229 \times 15 \times 24 = 82 \text{ pounds per linear foot}$$
$$\text{North wall} = 0.229 \times 12 \times 10/2 = 14 \text{ pounds per linear foot}$$
$$\text{South wall} = 14 \text{ pounds per linear foot}$$

For a flexible diaphragm, the shear wall at line 4 effectively subdivides the roof into two simply supported segments. These are spans 14 and 45. The seismic loads acting at the roof diaphragm level are shown in the Figure and the reactions at the shear walls are

$$R_{45} = 96 \times 20/2 = 960 \text{ pounds}$$
$$R_{41} = 96 \times 60/2 + 14 \times 20 \times 10/60 = 2927 \text{ pounds}$$

The diaphragm unit shears on either side of the shear wall at grid line 4 are

$$q_{45} = R_{45}/24 = 960/24 = 40 \text{ pounds per linear foot}$$
$$q_{41} = R_{41}/24 = 2927/24 = 122 \text{ pounds per linear foot}$$

The total unit shear in the shear wall at diaphragm level is

$$q_W = (R_{45} + R_{41})/12 = 324 \text{ pounds per linear foot}$$

The shear distribution and net shear at grid line 4 are shown in the Figure and the drag force at B4 is given by

$$F_{B4} = 12(q_{41} + q_{45}) = 12(122 + 40) = 1944 \text{ pounds}$$

2.3.4 Wood shear walls

A shear wall is a vertical diaphragm which acts as a vertical cantilever in transferring lateral forces from a horizontal diaphragm to the foundation. The construction details for a typical plywood sheathed shear wall are shown in Figure 2–21. The plywood sheathing forms the web of the cantilever to resist shear force, vertical studs acts as web stiffeners, and the end studs form the flanges to resist flexural effects. As in the case of a horizontal diaphragm, the design capacity of a shear wall depends on the thickness and grade of the plywood sheathing, the width of the framing members, support of the panel edges, and the spacing, penetration and type of nail used. This capacity has been determined experimentally[14]. Design values are provided in UBC Table 23–I–K–1 for walls with plywood sheathing on one side only, all panel edges blocked, and a minimum of two inch nominal framing members. To ensure satisfactory deflection of the shear wall, UBC Table 23–I–I imposes a maximum height–width ratio of 3½:1.

When plywood of equal shear capacity is applied to both faces of the shear wall, the allowable shear for the wall may be taken as twice the value permitted for one side, in accordance with UBC Section 2314.3. When the shear capacities are not equal, the allowable shear may be taken as the maximum value given by twice the permitted shear for the side with the lower capacity or equal to the permitted shear for the side with the higher capacity. When a combination of plywood and other sheathing materials is used, the allowable shear for the wall is taken as the value permitted for the plywood alone.

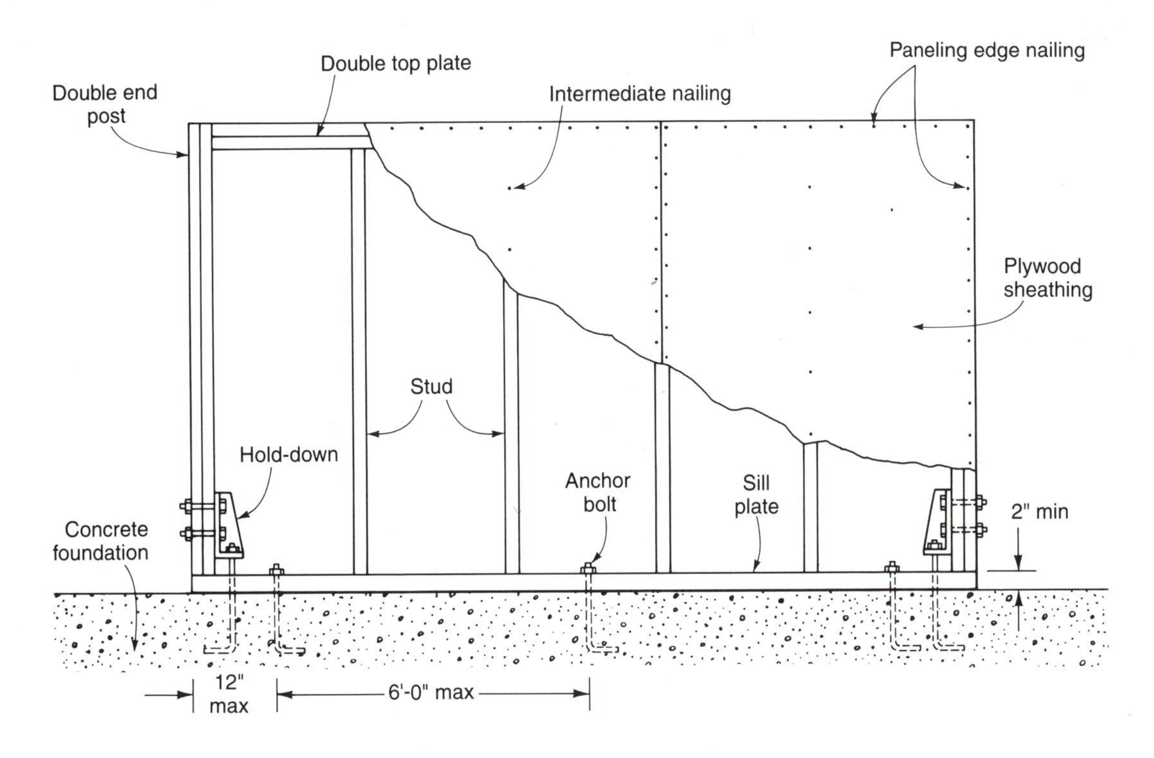

Figure 2–21 Shear wall details

For single story structures, UBC Section 2314.3 permits two inch nominal studs to be spaced at 24 inches on center when using sheathing not less than ⅜-inch thick . The minimum thickness of plywood permitted is 5/16-inch with two inch nominal studs spaced at 16 inches on center. For a two-story structure, UBC Table 23-I-R-3 requires 16 inch spacing for two inch nominal studs and 24 inch spacing for three inch nominal or larger studs. Because of excessive deflections, in accordance with UBC Section 2316.2, plywood sheathed shear walls may not be used to resist lateral forces contributed by concrete or masonry construction in structures exceeding two stories in height. In addition for two-story buildings of concrete or masonry construction the following limitations are imposed:

- Story-to-story wall heights shall not exceed twelve feet.
- Shear wall deflections shall not exceed 0.5 percent of the story height.
- In the lower story, the minimum thickness of plywood permitted is 15/32-inch.

In addition, SEAOC Section 5C.1 restricts the shear in the shear wall to fifty percent of that normally allowed.

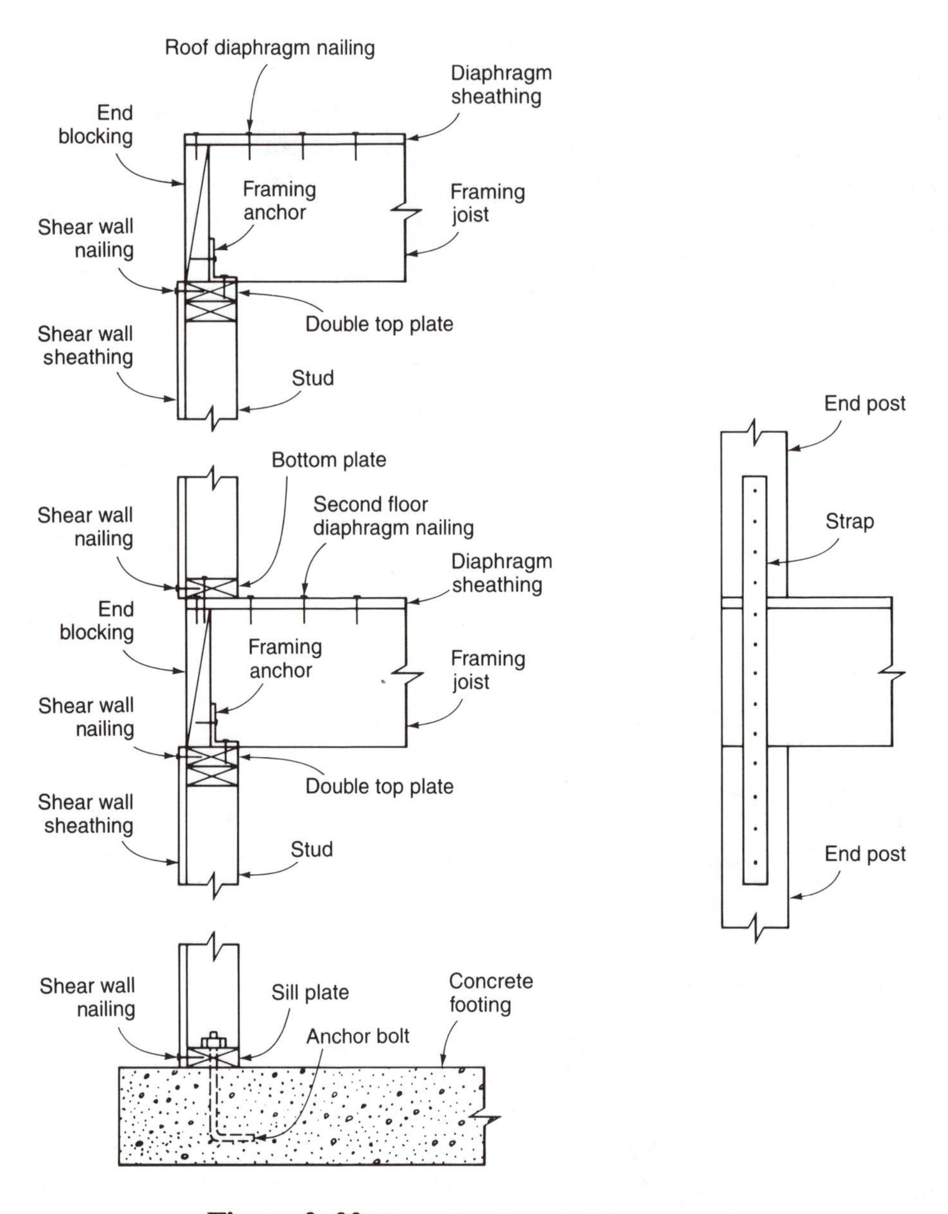

Figure 2–22 Lateral and vertical force transfer

Because of the light weight of timber framed construction, it is usually necessary to provide hold-downs at the ends of plywood sheathed shear walls to resist overturning. This is illustrated in detail 3 of Figure 2-12. An adequate bearing length is required for the bolts in the hold-down and is achieved with a double end post. In order to provide continuity for the top plate, a double plate is customarily used and, in accordance with UBC Section 2326.11.2, a minimum splice length of four feet must be provided between the two plates.

Transfer of the lateral force from the shear wall to the foundation is achieved by anchor bolts in the sill plate. UBC Section 1806.6 requires these to be not less than ½-inch diameter steel bolts embedded a minimum of seven inches into the concrete foundation and spaced a maximum of six feet apart. A bolt must be located not more than one foot from each end of the shear wall. UBC Section 2326.11.2 requires the sill plate to have a minimum thickness of two inches and a width not less than that of the floor studs in order to provide a nailing surface for the wall sheathing and to reduce the perpendicular-to-grain compressive stress in the plate. UBC Section 2317.4 requires the sill plate to be of treated wood, or wood naturally resistant to decay, when located on a concrete foundation which is in direct contact with earth. When an impervious membrane is placed between the foundation and the earth, untreated wood may be utilized.

The connection details required in a two-story, wood framed structure to ensure the transfer of lateral forces from the roof and second floor to the foundation are shown in Figure 2–22. The roof diaphragm shear is transferred to end blocking between framing joists and the boundary nail spacing is obtained from UBC Table 23–I–J–1. The shear is transferred from the end blocking to the top plate of the second story shear wall either by proprietary framing anchors, as shown, or by means of horizontal wood blocking. In the latter case, nailing requirements may be obtained from UBC Table 23–I–G. The nail spacing required at the shear wall edge is obtained from UBC Table 23–I–K–1. The total shear force at the bottom of the second story shear wall is due to the self-weight of the shear wall plus the roof diaphragm shear. This is transferred through the bottom plate to the second floor end blocking by nailing. The shear from the second floor diaphragm is also transferred to the end blocking by the diaphragm boundary nailing. Similarly, the accumulated forces are transferred to the sill plate of the first story shear wall. Finally, anchor bolts transfer the force in the sill plate to the concrete foundation. The allowable parallel-to-grain load on a steel bolt in the sill plate is, in accordance with UBC Section 2311.2, obtained as half the double shear value tabulated in UBC Table 23–I–F for a member twice the thickness of the plate.

To resist the uplift of the second story shear wall, the end posts of the second story and the first story shear walls are tied together with steel straps. The straps may be either bolted or nailed to the end posts as shown in Figure 2–22.

Example 2–11 (Shear wall nailing)
Determine the nailing requirements for the shear wall on grid line 4 in Example 2–10. The wall consists of ⅜-inch C–D sheathing grade plywood applied on one side of the studs which are at 16 inches on center. Nailing consists of 8d nails with 1½-inch penetration.
If the roof dead load on the shear wall is 60 pounds per square foot, determine the force developed in the hold-downs and the number of ½-inch diameter anchor bolts required in the 3 inch nominal Douglas Fir–Larch sill plate.

Solution
The dead load of the shear wall on grid line 4 is

$$W = 12 \times 10 \times 12 = 1440 \text{ pounds}$$

The seismic unit shear produced by the self weight of the wall is

$$q_s = 0.229W/L = 0.229 \times 1440/12 = 27 \text{ pounds per linear foot}$$

The total shear at the base of the wall is due to the self-weight of the shear wall plus the seismic force applied at the top of the wall by the diaphragm. The total unit shear in the wall is then

$$
\begin{aligned}
q &= q_W + q_S \\
&= 324 + 27 \\
&= 351 \text{ pounds per linear foot}
\end{aligned}
$$

The aspect ratio of the shear wall is

$$
\begin{aligned}
h/L &= 10/12 \\
&= 0.83 \\
&< 3.5
\end{aligned}
$$

Hence, this conforms to the requirements of UBC Table 23–I–I for plywood panels nailed on all edges. The allowable unit shear, in accordance with UBC Table 23–I–K–1, for ⅜-inch plywood applied to studs at 16 inches on center may be increased to the values shown for 15/32-inch plywood. The required spacing of 8d nails with 1½-inch penetration may be obtained from UBC Table 23–I–K–1. With 2-inch nominal Douglas Fir–Larch vertical studs and all panel edges backed with 2 inch nominal blocking, the required nail spacing is

- All panel edges — 4 inches
- Intermediate framing members — 12 inches
- Capacity provided — 380 pounds per linear foot
- Capacity required — 351 pounds per linear foot

The seismic load acting on the shear wall at roof diaphragm level is

$$
\begin{aligned}
F_W &= R_{41} + R_{45} \\
&= 2927 + 960 \\
&= 3887 \text{ pounds}
\end{aligned}
$$

The seismic load produced by the self-weight of the shear wall is

$$
\begin{aligned}
F_S &= 0.229W \\
&= 0.229 \times 1440 \\
&= 330 \text{ pounds}
\end{aligned}
$$

and this load acts at the midheight of the shear wall. The total seismic shear force at the base of the wall is

$$
\begin{aligned}
F &= F_W + F_S \\
&= 3887 + 330 \\
&= 4217 \text{ pounds}
\end{aligned}
$$

The vertical roof dead load acting on the shear wall is

$$
\begin{aligned}
W_R &= 60 \times 12 \\
&= 720 \text{ pounds}
\end{aligned}
$$

The loading on the shear wall is shown in Figure 2–23. The overturning moment acting on the wall is

$$
\begin{aligned}
M_O &= F_W \times h + F_S \times h/2 \\
&= 3887 \times 10 + 330 \times 5 \\
&= 40{,}520 \text{ pounds feet}
\end{aligned}
$$

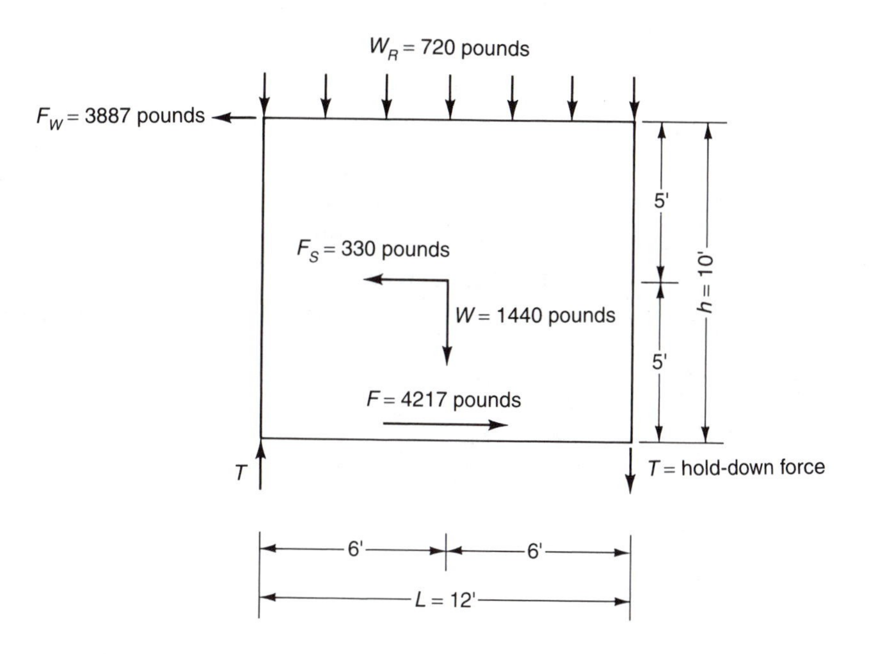

Figure 2-23 Forces acting on shear wall

In accordance with UBC Section 1631.1 the restoring moment is given by

$$\begin{aligned} M_R &= 0.85(W + W_R) \times L/2 \\ &= 0.85(1440 + 720) \times 6 \\ &= 11{,}016 \text{ pounds feet} \end{aligned}$$

Neglecting the short distance from the hold-down to the end of the shear wall, the hold-down force is given by

$$\begin{aligned} T &= (M_O - M_R)/L \\ &= (40{,}520 - 11{,}016)/12 \\ &= 2459 \text{ pounds} \end{aligned}$$

The allowable parallel-to-grain load on a ½-inch diameter anchor bolt in the 2½-inch thick Douglas Fir-Larch sill plate is given by UBC Section 2311.2 as half the double shear value tabulated in UBC Table 23-I-F for a five inch thick member. The allowable seismic load, in accordance with UBC Section 1603.5, is given by

$$\begin{aligned} P_a &= 1.33 \times 1270 \\ &= 1689 \text{ pounds} \end{aligned}$$

To conform to the spacing requirements of UBC Section 1806.6, a minimum of three anchor bolts are required. The seismic shear capacity of three bolts is

$$\begin{aligned} 3P_a &= 3 \times 1689 \\ &= 5067 \text{ pounds} \\ &> F \end{aligned}$$

Hence, three bolts are satisfactory.

2.3.5 Braced frames

Braced frames provide resistance to the lateral forces acting on a structure. The members of a braced frame act as a truss system and are subjected primarily to axial stress. Braced frames consist of two basic types, concentrically and eccentrically braced frames. Chevron bracing, cross bracing, K bracing, and diagonal bracing are classified as concentrically braced and are shown in Figure 2–24. Eccentrically braced frames are shown in Figure 2–25 and are restricted to all steel construction. In addition, concentrically braced frames are subdivided into two categories, ordinary braced frames and special concentrically braced frames.

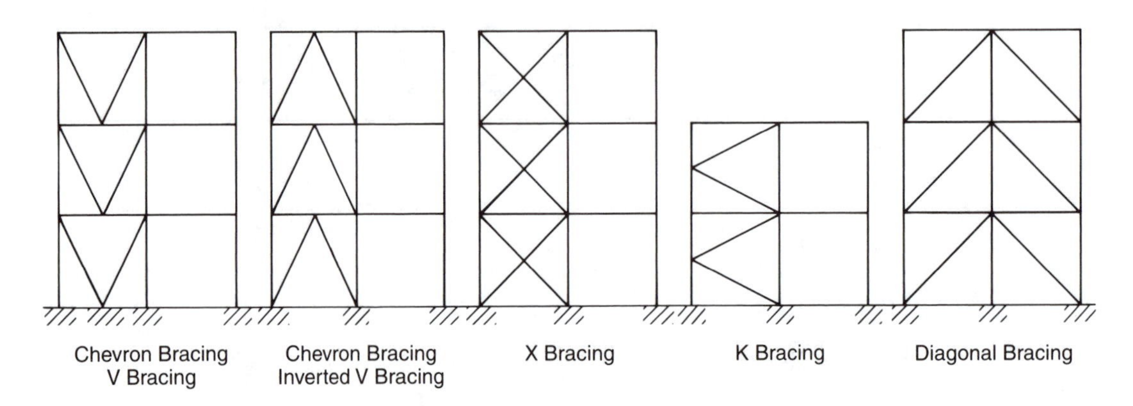

Figure 2–24 Concentrically braced frames

The inability to provide reversible inelastic deformation is the principal disadvantage of concentrically braced forms. After buckling, an axially loaded member loses strength and does not return to its original straight configuration. To reduce the possibility of this occurring during moderate level earthquakes, more stringent design requirements are imposed on bracing members in seismic zones 2, 3, and 4. UBC Section 2211.8.2 requires bracing members, of ordinary braced frames, to be designed with a slenderness ratio of

$$L/r \leq 720/(F_y)^{1/2}$$

where F_y = specified steel yield stress
L = length of the bracing member
r = governing radius of gyration

In addition, the compressive stress in a bracing member is restricted to a maximum value of

$$F_{as} = \beta F_a$$

where β = stress reduction factor
$= 1/(1 + kL/2rC_c)$
F_a = compressive stress normally allowed
$C_c = (2\pi^2 E/F_y)^{1/2}$

For special concentrically braced frames, the corresponding criteria are

$$L/r \leq 1000/(F_y)^{1/2}$$

$$\beta = 0$$

It is also required that neither the sum of the horizontal components of the compressive member forces nor the sum of the horizontal components of the tensile member forces, along a line of bracing, shall exceed seventy percent of the applied lateral force. This is to prevent an accumulation of inelastic deformation in one direction and precludes the use of tension-only diagonal bracing.

Chevron or V bracing which is loaded in the inelastic range causes large vertical deflections in the horizontal floor beam and the strength of the bracing members deteriorates rapidly with reversing load cycles. For this reason UBC Section 2211.8.4 requires that

- The braces shall be designed for 150 percent of the normally prescribed seismic force.
- The beam shall be continuous between columns.
- In the case of inverted V bracing, for structures exceeding one-story in height, the beam must be designed to carry all gravity loads without support from the bracing.

Inelastic deformation and buckling of K bracing or knee bracing members may produce lateral deflection of the connected columns causing collapse. Hence, in seismic zones 3 and 4, K bracing is prohibited. In seismic zone 2, K bracing is permitted with the same restrictions as apply to chevron bracing. For special concentrically braced frames, K bracing is not allowed.

Diagonal bracing and X bracing may be utilized provided that the requirement is observed to have the horizontal component of the compressive forces or the tensile forces not more than seventy percent of the applied lateral force. Hence, tension-only bracing may not be employed and the bracing must also be designed for compressive force.

For structures not exceeding two stories in height, the above requirements may be relaxed provided that the members are designed to resist forces of $3R_W/8$ times the normally prescribed seismic forces. Hence, for smaller structures tension-only diagonal bracing may be used.

The connections in a braced frame may be subjected to impact loading during an earthquake and, in order to avoid brittle fracture, must be designed for the lesser of

- Tensile strength of the bracing.
- $3R_W/8$ times the prescribed seismic force in the brace.
- The maximum force that can be transferred to the brace by the system.

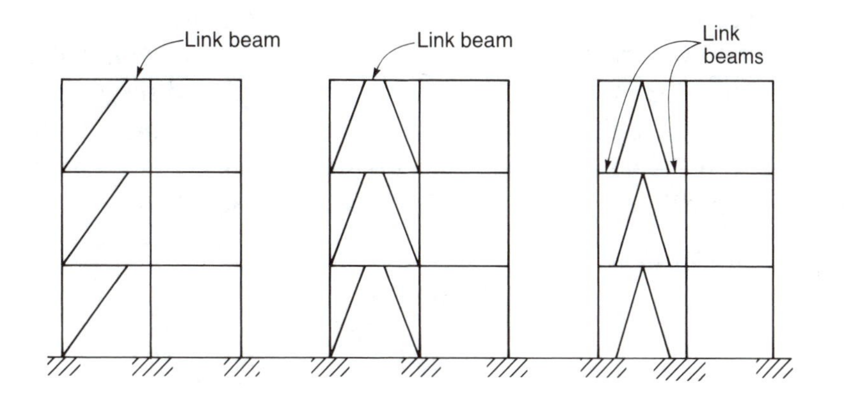

Figure 2–25 Eccentrically braced frames

The bracing member in an eccentrically braced frame[15,16], is connected to the beam so as to form a short link beam between the brace and the column or between two opposing braces. The link beam acts as a fuse to prevent other elements in the frame from being overstressed. After the elastic capacity of the system is exceeded, shear or flexural yielding of the link provides a ductile response comparable to that obtained in special moment-resisting frames. In addition, eccentrically braced frames may be designed to control frame deformations and minimize damage to architectural finishes during cyclical seismic loading. Eccentrically braced frames provide an economical system for the seismic design of structures.

2.3.6 Moment frames

Special moment-resisting frames provide a ductile structural system for resisting seismic forces after the elastic capacity of the system is exceeded. Inelastic energy absorption occurs by yielding of the members at plastic hinge positions. Because of the high degree of redundancy, risk of collapse caused by the premature failure of some elements is minimized.

Frame members are designed to avoid failure due to column buckling and special detailing requirements are specified to ensure the integrity of the joints during inelastic deformation. Compared with braced frame or shear wall structures, special moment-resisting frames have the advantage of providing unobstructed space. However, under high level seismic loading, the large deformations which occur may cause distress to architectural finishes.

2.4 STRUCTURAL ELEMENTS AND NONSTRUCTURAL COMPONENTS AND EQUIPMENT SUPPORTED BY STRUCTURES

2.4.1 Performance criteria

Structural elements include bearing walls and penthouses not framed as an extension to the main structure. Nonstructural components include nonbearing walls, cantilevered parapet walls, signs, ornamentation, and storage racks. Equipment includes chimneys, tanks, machinery, and equipment weighing 400 pounds or more. Design levels are specified for structural elements, components, equipment and their anchorage to ensure that life safety is not endangered and, in the case of safety related equipment, to ensure the continued function of essential facilities. The design force for elements is higher than that for buildings to take account of the dynamic response of the element to the motion of the structure, the amplified response of equipment, and the lack of redundancy and ductility in the component itself. This section of the Code is primarily concerned with the design of attachments which connect elements to the structure.

The design seismic force is given by UBC Formula (30–1) as

F_p = $ZI_pC_pW_p$ where,
Z = zone coefficient from UBC Table 16–I
I_p = importance factor from UBC Table 16–K
C_p = horizontal force factor from UBC Table 16–O
W_p = weight of element or component.

The seismic force is distributed over the height of an element in proportion to its mass distribution.

2.4.2 Anchorage of concrete and masonry walls

In order to prevent the separation of concrete and masonry walls from the floor or roof diaphragms, positive anchorage ties must be provided to resist the lateral force calculated from UBC Formula (30–1). In addition, UBC Table 16–O specifies that when flexible diaphragms are used, the values of C_p shall be increased by 50 percent over the central half of the diaphragm to allow for the increased acceleration response at midspan.

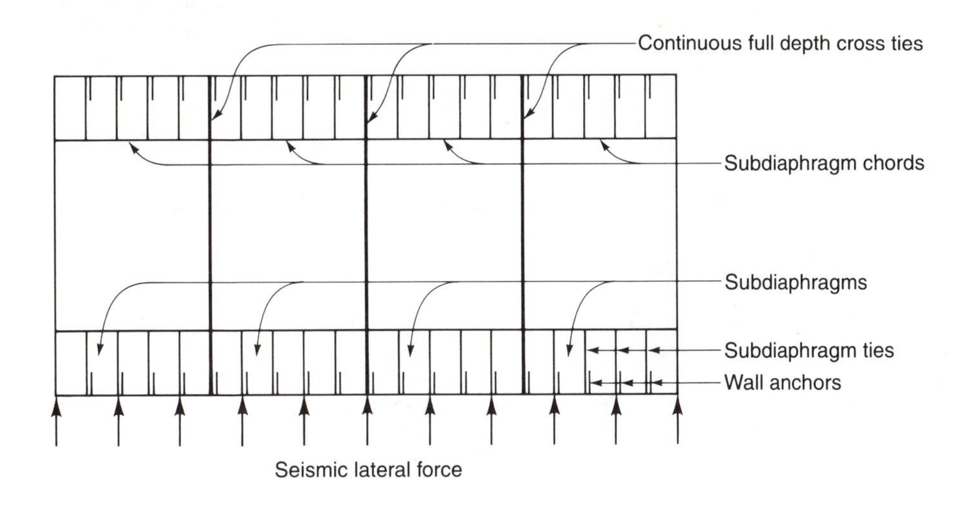

Figure 2–26 **Subdiaphragm details**

The minimum design force, in accordance with UBC Section 1611, may not be less than 200 pounds per linear foot. Also, for interior walls and partitions, UBC Section 1610.2 stipulates a minimum lateral load of 5 pounds per square foot. For the design of unbraced cantilever walls and parapets, a higher value of 2.0 is specified for C_p compared with a value of 0.75 for walls. This is to compensate for the poor seismic performance and lack of redundancy of parapets which may create a life hazard to the public. However, in determining the design anchorage force for a wall with a parapet, the value of $C_p = 0.75$ shall be taken for the entire wall plus parapet[1,17].

In order to distribute anchorage forces developed by concrete and masonry walls, continuous crossties shall be provided between diaphragm chords as stipulated in UBC Section 1631.2.9.4. To reduce the number of continuous, full depth ties required, subdiaphragms may be used to span between the continuous ties[1,9,17,18] as shown in Figure 2–26. The subdiaphragm must be designed for all criteria prescribed for the main diaphragm with the anchor ties running the full depth of the subdiaphragm to provide full development of the tie force. The subdiaphragm must act independently to transfer the wall anchorage force from the anchorage ties to the continuous full depth ties in the main diaphragm. Where the wall anchor spacing exceeds four feet, in accordance with UBC Section 1611, the wall must be designed to span between the anchors.

Example 2–12 (Subdiaphragm design)
Determine the force developed in wall anchorage ties located at eight feet on center along the wall on grid line G in Example 2–9. For the subdiaphragm bounded by grid lines F, G, 5, 6, calculate the chord forces and required nailing and determine the design force in the continuous crossties on grid lines 5 and 6.

Solution
The seismic loading on the concrete wall is given by UBC Formula (30–1) as

$F_p = ZI_pC_pW_p$

where
Z = 0.4 for zone 4 from UBC Table 16–I
I_p = 1.0 from UBC Table 16–K for a standard occupancy structure
$C_p = 1.5 \times 0.75$ from UBC Table 16–O item 1.1.b. and footnote 3.
W_p = 75 pounds per square foot, given wall weight

Then, the lateral force is given by

$F_p = 0.4 \times 1.0 \times 1.5 \times 0.75 \times 75$
$= 0.45 \times 75$
= 33.75 pounds square foot

The anchorage force at roof diaphragm level is given by

$p = F_p \times 14^2/(2 \times 12.5)$
= 265 pounds per linear foot
> 200

Hence, the minimum anchorage force of 200 pounds per linear foot of wall, stipulated in UBC Section 1611, does not govern and the force on each wall anchor for a spacing of eight feet is

$P = 8p$
$= 8 \times 265$
= 2120 Pounds

The function of the subdiaphragm is to transfer the wall anchorage force into the main diaphragm and the wall anchor force is transferred to the 3½–inch × 13½–inch glued–laminated subdiaphragm ties. The subdiaphragm aspect ratio is

$b/d = 32/20$
$= 1.6$
< 4

Hence the requirements of UBC Table 25–I-I are satisfied and the unit shear in the subdiaphragm is

$q = pb/2d$
$= 265 \times 32/(2 \times 20)$
= 212 pounds per linear foot
< 320

Hence, the capacity of the nailing in the main diaphragm is adequate. The design force in the continuous 6¾-inch × 24 inch glued-laminated crossties, at a spacing of 32 feet, is

$$
\begin{aligned}
P_t &= pb \\
&= 265 \times 32 \\
&= 8480 \text{ pounds}
\end{aligned}
$$

2.4.3 Components and equipment supported by structures

UBC Table 16-K stipulates that an importance factor of 1.5 shall be used for the design of equipment in essential facilities and for the design of containers enclosing toxic or explosive materials. This is to ensure the continued operation of equipment in essential facilities and prevent the escape of hazardous materials. In addition, a value of 1.5 shall be used for the anchorage of life-safety systems irrespective of the occupancy category.

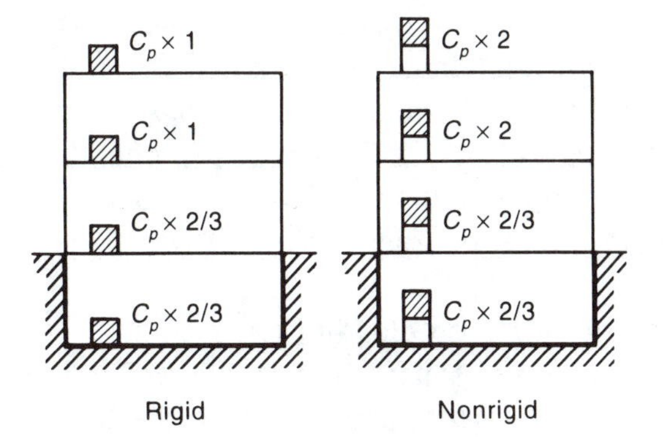

Figure 2-27 C_p modification factors

The value of C_p listed in UBC Table 16-O are applicable to components with a natural period not exceeding 0.06 seconds and which are considered to be rigid. For nonrigid components, which may be subjected to resonant behavior in the above grade stories of buildings, the listed values for C_p must be doubled. For components located at or below grade, which are not subject to the amplified response found higher in the structure, the listed values of C_p may be reduced by one-third. However, the design lateral force must not be less than that obtained for an independent structure by using UBC Formula (32-1). These requirements are illustrated in Figure 2-27.

Example 2-13 (Design seismic force on equipment)
An air conditioner located on the roof of a hospital in seismic zone 4 weighs two kips and has a natural period less than 0.06 seconds. Calculate the design seismic force.

Solution
The design seismic force is given by UBC Formula (30-1) as

F_p = $ZI_pC_pW_p$ where,
Z = 0.4 from UBC Table 16–I
I_p = 1.5 from UBC Table 16–K
C_p = 0.75 from UBC Table 16–O item 3.2
W_p = 2 kips, as given

The seismic force is then

F_p = $0.4 \times 1.5 \times 0.75 \times 1$
= 0.9 kips

2.4.4 Wall cladding

External cladding panels and their connections must be designed, in accordance with UBC Section 1631.2, to accommodate the actual anticipated inelastic deformation which is approximately ($3R_W/8$) times the elastic displacement determined from the prescribed design forces. To ensure that the connection remains ductile and capable of sustaining the applied loads, all fasteners in the connection must be designed for four times and the body of the connector must be designed for one and one third times the force calculated from UBC Formula (30–1). Allowable design stresses may be increased by one–third in accordance with UBC Section 1603.5. An additional stipulation, in UBC Table 16–K is that the importance factor I_p for the entire connection shall be 1.0 irrespective of the occupancy category.

Example 2–14 (Design seismic force on cladding)
A wall panel weighing 20 pounds per square foot is externally mounted on an office building in seismic zone 4. Calculate the design seismic force on the connectors and fasteners

Solution
The basic design seismic force normal to the panel is given by UBC Formula (30–1) as

F_p = $ZI_pC_pW_p$ where,
Z = 0.4 from UBC Table 16–I
I_p = 1.0 from UBC Table 16–K
C_p = 0.75 from UBC Table 16–O item 1.1.b.
W_p = 20 pounds per square foot, as given

The basic design lateral force is

F_p = $0.4 \times 1 \times 0.75 \times 20$
= 6 pounds per square foot

From UBC Section 1631.2.4.2 the design seismic load on fasteners is

F_F = $4 \times 6 = 24$ pounds per square foot

The design seismic load on the body of the connector is

F_C = $1.33 \times 6 = 8$ pounds per square foot

Example 2–15 (Structural Engineering Examination 1988, Section A, weight 4.0 points)

GIVEN: A 4 story multiuse building has a structural steel special moment-resisting frame and a nonbearing curtain wall. The building plan "A" and elevation "B" are shown in Figure 2–28.

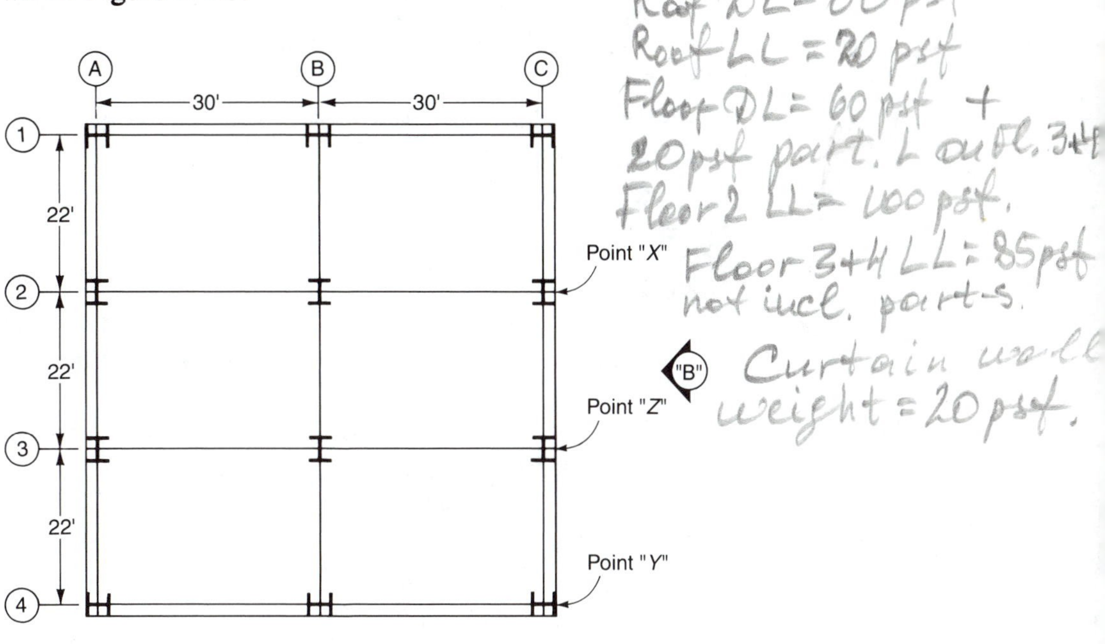

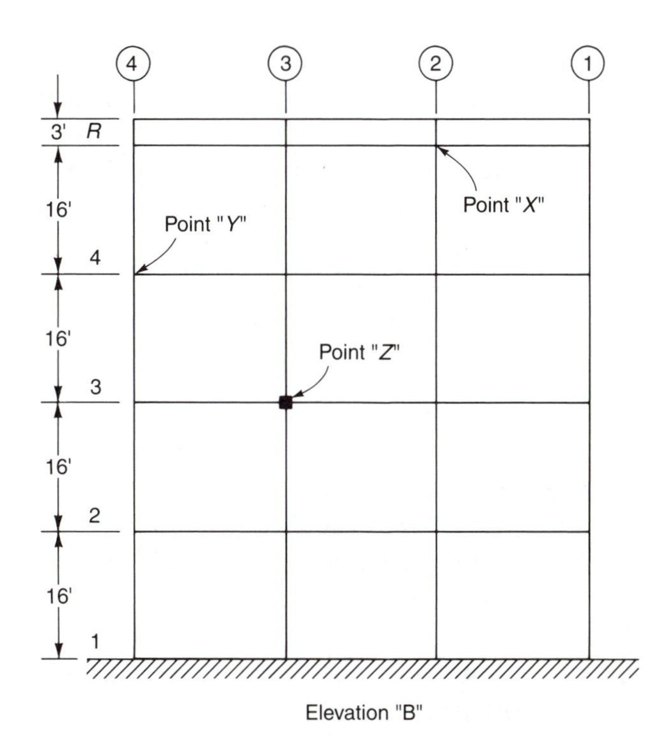

Figure 2–28 Details for Example 2–15

CRITERIA:
- Roof DL = 60 pounds per square foot
- Roof LL = 20 pounds per square foot (no snow load)
- Floor DL = 60 pounds per square foot + 20 pounds per square foot partition load on Floors 3 and 4
- Floor 2 live load = 100 pounds per square foot (public assembly use)
- Floors 3 and 4 live load = 85 pounds per square foot (office use, not including partitions)
- Average curtain wall weight = 20 pounds per square foot
- Basic wind speed area = 90 miles per hour
- Wind exposure B
- Seismic zone 4

Assumptions:
- The primary occupancy of the building is for office use
- Column weights are included in the floor dead load

REQUIRED:
1. Determine the following design loadings in pounds per square foot at point "X" on the parapet at column C-2 for
   a. Design wind loading
   b. Design seismic loading (normal to the cladding)
2. Determine the following design loadings in pounds per square foot at point "Y" on Floor 4 at column C-4 for
   a. Design wind loading (both inward and outward)
3. Determine the following design loadings in pounds per square foot at point "Z" on Floor 3 at column C-3 for
   a. Design wind loading (both inward and outward)
   b. Design seismic loading (normal to the cladding)
4. Determine the following design loadings in pounds per square foot at point "Z" on Floor 3 at column C-3 based on a 2 foot × 2 foot curtain wall panel for
   a. Design wind loading on a connection fastener (normal to the cladding)
   b. Design seismic loading on connection fastener (normal to the cladding)
5. Calculate the design dead load, live load, and total load at each floor on column C-2. Use live load reductions in accordance with the UBC.

SOLUTION

WIND FORCES[19,20,21,22]

Assume that the structure may be considered as enclosed and that pressures are required for cladding elements with a tributary area not exceeding ten square feet. Hence, the pressure coefficients may not be reduced and the design wind pressure is obtained from UBC Formula (18-1) as

$p = C_e C_q q_s I_w$

where
C_e = height, exposure and gust factor given in UBC Table 16-G
C_q = pressure coefficient given in UBC Table 16-H
q_s = wind stagnation pressure given in UBC Table 16-F
= 20.8 pounds per square foot for a wind speed of 90 miles per hour

	I_w	= importance factor given in UBC Table 16–K
		= 1.0
Then	p	= $20.8C_eC_q$

SEISMIC FORCES

The design seismic force on elements and components is obtained from UBC Formula (30–1) as

	F_p	= $ZI_pC_pW_p$
where	Z	= 0.4 for zone 4 from UBC Table 16–I
	I_p	= 1.0 from UBC Table 16–K
	C_p	= force factor given in Table 16–O
and	W_p	= weight of the curtain wall
		= 20 pounds per square foot
Then	F_p	= $8C_p$

1a. WIND LOAD AT POINT X

For exposure B and a roof of 64 feet, the value of C_e is 0.97. For a parapet, the value of C_q is 1.3 inward or outward and the design wind load at point X is

p	= 20.8 × 0.97 × 1.3
	= 26.2 pounds per square foot inward and outward

1b. SEISMIC LOAD AT POINT X

For a parapet, the value of C_p is 2.0 from UBC Table 16–O, item I.1a. and the design seismic load at point X is

F_p	= 8 × 2
	= 16 pounds per square foot normal to the parapet.

2. WIND LOAD AT POINT Y

For outward pressures, UBC Section 1620 indicates that the value of C_e shall be based on the mean roof height of 64 feet, giving a value for C_e of 0.97. For inward pressures, the value of C_e is based on the actual height of the element of 48 feet, giving a value for C_e of 0.88. The value of C_q for a wall corner element is 1.5 outward and 1.2 inward and is applied over a distance from the corner of the lesser of

	L	= 10 feet
or	L	= 0.1 × least width
		= 6 feet, which governs.

The design wind load at point Y is

	p	= 20.8 × 0.97 × 1.5
		= 30.3 pounds per square foot outward
and	p	= 20.8 × 0.88 × 1.2
		= 22.0 pounds per square foot inward

3a. WIND LOAD AT POINT Z

For outward pressures, the value of C_e based on a mean roof height of 64 feet, as indicated by UBC Section 1620, is 0.97. For inward pressures, the value of C_e based on the actual height of the element of 32 feet is 0.78. The value of C_q for a wall element is 1.2 inward and outward. The design wind load at point Z is

$$p = 20.8 \times 0.97 \times 1.2$$
$$= 24.2 \text{ pounds per square foot outward}$$

and

$$p = 20.8 \times 0.78 \times 1.2$$
$$= 19.5 \text{ pounds per square foot inward.}$$

3b. SEISMIC LOAD AT POINT Z

For a wall, the value of C_p is 0.75 from UBC Table 16–O, item 1.1.b. The design seismic load at point Z is

$$F_p = 8 \times 0.75$$
$$= 6 \text{ pounds per square foot normal to the cladding.}$$

4a. WIND LOAD ON CLADDING CONNECTION AT Z

The values of C_e and C_q for a cladding connection are the same as for the cladding and the design wind load of a connection at point Z is

$$p = 24.21 \text{ pounds per square foot outward}$$

and

$$p = 19.50 \text{ pounds per square foot inward}$$

4b. SEISMIC LOAD ON CONNECTION FASTENER AT Z

Connection fasteners, in accordance with UBC Section 1631.2.4.2, shall be designed for four time the seismic forces calculated by UBC Formula (30–1). The design seismic load on a fastener at point Z is

$$p = 6 \times 4$$
$$= 24 \text{ pounds per square foot normal to the cladding.}$$

5. DESIGN LOAD ON COLUMN C–2

LIVE LOAD REDUCTIONS

The column tributary area is given by

$$A = 22 \times 15$$
$$= 330 \text{ square feet}$$

Since the tributary area exceeds 150 square feet, both roof and floor live loads may be reduced as indicated in UBC Section 1606. The reduced roof live load is obtained from UBC Table 16–C, Method 1, as

$$L_R = 16 \text{ pounds per square foot}$$

On floors three and four, the total dead load including partitions is given by

$$D = 60 + 20$$
$$= 80 \text{ pounds per square foot}$$

and the live load is given by

$$L = 85 \text{ pounds per square foot}$$

The allowable reduction in live load at floor four is the lesser of

R = 60%, supporting load from two levels only

or (6–2) $R = 23.1(1 + D/L)$
$= 23.1[1 + (60 + 80)/(20 + 85)]$
$= 53.9\%$

or (6–1) $R = r(A - 150)$
$= 0.08(2 \times 330 - 150)$
$= 40.8\%$, which governs.

Hence the fourth floor live load is

$$L_4 = 330(20 + 85)(1 - 0.408) = 20{,}512 \text{ pounds}$$

The allowable reduction in live load at floor three is the lesser of

R = 60%, supporting load from three levels

or (6–2) $R = 23.1\,[1 + (60 + 2 \times 80)/(20 + 2 \times 85)]$
$= 49.8\%$, which governs

or $R = r(A - 150)$
$= 0.08(3 \times 330 - 150)$
$= 67.2\%$

Hence, the third floor live load is

$$L_3 = 330(20 + 2 \times 85)(1 - 0.498) = 31{,}475$$

At floor two, no reduction is allowed in the public assembly loading of 100 pounds per square foot.

COLUMN LOADING

Fourth floor to roof:

Roof dead load	= 60 × 330	=	19,800
Roof live load	= 16 × 330	=	5,280
Wall at fourth story	= 20 × 22 × 19	=	8,360
Total design load above fourth floor			33,440 pounds

Third floor to fourth floor:

Fourth floor dead load	= 80 × 330	=	26,400
Fourth floor live load addition	= 20,512 − 5,280	=	15,232
Wall at third story	= 20 × 22 × 16	=	7,040
Total design load above third floor			82,112 pounds

Second floor to third floor;			
Third floor dead load	$= 80 \times 330$	=	26,400
Third floor live load addition	$= 31{,}475 - 20{,}512$	=	10,963
Wall at second story	$= 20 \times 22 \times 16$	=	7,040
Total design load above second floor			126,515 pounds
First floor to second floor:			
Second floor dead load	$= 60 \times 330$	=	19,800
Second floor live load addition	$= 100 \times 330$	=	33,000
Wall at first story	$= 20 \times 22 \times 16$	=	7,040
Total design load above first floor			286,355 pounds

2.5 SELF-SUPPORTING NONBUILDING STRUCTURES

2.5.1 General requirements

The UBC Code presents seismic design procedures for the following nonbuilding systems:

- Structures which are similar to building structures.
- Systems which are rigid.
- Ground supported tanks.
- Other structures such as hoppers, silos, and tanks supported on legs.

Compared with building structures, nonbuilding systems have fewer nonstructural elements and reduced damping capacity and redundancy. Because of these differences, the fundamental period may not be determined by Method A using UBC Formula (28-3) but must be obtained by Method B or other rational method[23]. The value obtained need not be limited to a specific percentage of the Method A value as is required for building structures by UBC Section 1628.2.2.

The drift limitation imposed on building structures in UBC Section 1628.8 is required to control damage to finishes. This is not necessary in nonbuilding systems. However, P-delta effects should be considered, in accordance with UBC Section 1628.9, as this influences the stability of the system. In determining the value of the seismic dead load, all dead load plus the normal operating contents of containers should be included.

2.5.2 Systems similar to building structures

Nonbuilding structures consisting of one or more levels at which the mass is concentrated and which may be classified in accordance with the building structures in UBC Table 16-N are designed by the provisions of UBC Section 1627. The design base shear is given by UBC Formula (28-1) as

$$V = (ZIC/R_W)W$$

and the seismic force coefficient is given by UBC Formula (28-2) as

$$C = 1.25S/T^{2/3}$$

The response modification factor R_W is selected from UBC Table 16-N.

2.5.3 Rigid systems

For a rigid structure, which is defined as having a natural period of less than 0.06 seconds, the design lateral force is given by UBC Formula (32-1) as

$$V \quad = 0.5ZIW$$

The lateral force must be distributed over the height of the structure in accordance with the distribution of the mass. This method is applicable to equipment mounted directly on a concrete pad foundation at grade level[24].

Example 2-16 (Transformer on pad foundation)
The rigid transformer shown in Figure 2-29 weighs 15 kips and is attached to a concrete foundation pad with four ¾-inch diameter A325 anchor bolts. Determine the adequacy of the bolts if the transformer is located in seismic zone 4 and the strength of the concrete in the pad is f'_c = 2000 pounds per square inch.

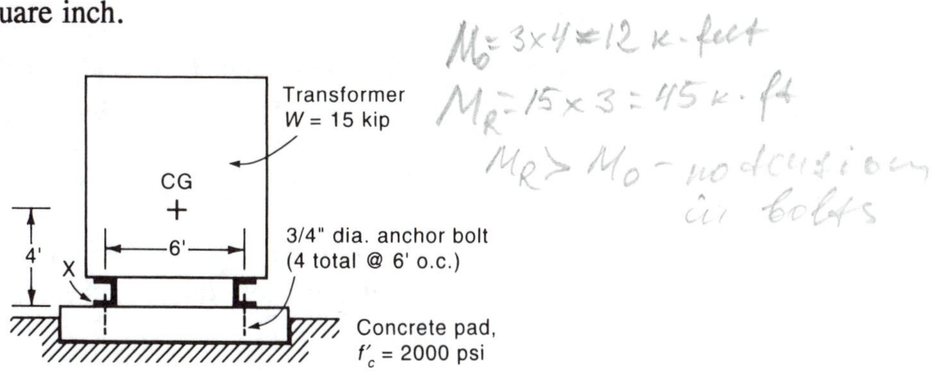

Figure 2-29 Details for Example 2-16

Solution
The seismic lateral force for a rigid component mounted at grade is given by UBC Formula (32-1) as

$$V \quad = 0.5ZIW$$

where Z = 0.4 from UBC Table 16-I for zone 4
I = 1.0 from UBC Table 16-K for a non-essential component
W = 15 kips, as given

Then, the lateral force is

$$V = 0.5 \times 0.4 \times 1 \times 15$$
$$= 3 \text{ kips}$$

The shear force on each bolt is

$$P_v = 3/4$$
$$= 0.75 \text{ kips}$$

From UBC Table 19-E, the allowable shear for a ¾-inch diameter bolt in 2000 pound per square inch concrete, after allowing for the one-third increase in stress in accordance with UBC Section 1603.5, is given by

$$P'_v = 1.33 \times 2.94$$
$$= 3.91 \text{ kip}$$
$$> P_v \text{ ... satisfactory}$$

Taking moments about axis X, the overturning moment is given by

$$M_O = 4V = 4 \times 3 = 12 \text{ kip feet}$$

The restoring moment is

$$M_R = 3W = 3 \times 15 = 45 \text{ kip feet}$$
$$> M_O$$

Hence, no tension is developed in the bolts.

2.5.4 Ground supported tanks

Tanks founded at or below grade may be designed as rigid structures considering the entire weight of the tank and its contents. The design lateral force may be determined from UBC Formula (32-1) as

$$V = 0.5ZIW$$

Alternatively, an approved national design standard may be used or a response spectrum modal analysis may be employed to include the inertial effects of the contained fluid.

2.5.5 Other Structures

For other nonbuilding structures, the design base shear is given by UBC Formula (28-1) as

$$V = (ZIC/R_W)W$$

where the value of R_W is obtained from UBC Table 16-P. The value of the ratio C/R_W shall be not less than 0.4 as stipulated in UBC Section 1632.5. The R_W values given in UBC Table 16-P are lower than the values given in UBC Table 16-N to reflect the reduced redundancy of nonbuilding structures.

The vertical distribution of force may either be determined by UBC Formula (28-8) or from a dynamic analysis.

Example 5-17 (Grain bin supported on legs)
A grain bin is supported on braced frames founded on grade in seismic zone 3. Determine the seismic lateral force in the north-south direction if the stiffness of the braced frame in this direction is 500 kips per inch and the fully loaded bin weighs 50 kips.

Solution
The natural period is given by

$$T = 2\pi(m/k)^{1/2}$$
$$= 0.32(W/k)^{1/2}$$

where $W = 50$ kips, as given
$k = 500$ kips per inch, as given

The natural period is, then

$$T = 0.32(50/500)^{1/2}$$
$$= 0.10 \text{ seconds}$$
$$> 0.06$$

Hence, the structure is nonrigid and, from UBC Section 1628.2.1 the value of the force coefficient is given by

$$C = 2.75$$

For a nonrigid structure, the lateral force is given by UBC Formula (28–1) as

$$V = (ZIC/R_W)W$$

where $Z = 0.3$ from UBC Table 16–I for zone 3
$I = 1.0$ from UBC Table 16–K for a non-essential structure
$C = 2.75$ as calculated
$R_W = 4$ from UBC Table 16–P item 7
$C/R_W = 2.75/4$
$= 0.69$
> 0.4 as required by UBC Section 1632.5

Then, the total lateral force is

$$V = (0.3 \times 1 \times 2.75/4)50$$
$$= 10.3 \text{ kips}$$

2.6 TORSION AND RIGIDITY

2.6.1 Rigid diaphragms

As stipulated by UBC Section 1628.5, a diaphragm is considered rigid when its midpoint displacement, under lateral load, is less than twice the average displacement at its ends. It is then assumed that the diaphragm and shear walls undergo rigid body rotation and this produces additional torsional forces in the shear walls.

2.6.2 Torsional moment

The center of rigidity is that point about which a structure tends to rotate when subjected to an eccentric force. In the case of a seismic force, this acts at the center of mass of the structure and torsional moment is the product of seismic force and the eccentricity of the center of mass with respect to the center of rigidity. The calculated location of the center of mass may not be exact due to the distribution of structure weight being imprecisely known. Similarly inaccuracies in calculated rigidity of the shear walls, and the neglect of nonstructural components such as partitions and stairs, lead to the inexact location of the center of rigidity[25].

To account for these uncertainties, UBC Section 1628.5 specifies that the center of mass be assumed to be displaced from its calculated position, in each direction, a distance equal to five percent of the building dimension perpendicular to the direction of the seismic force. This accidental eccentricity is amplified where torsional irregularity exists, as defined in UBC Table 16–M. The amplification factor is given by UBC (28–9) as

$$A_x = (\delta_{max}/1.2\delta_{avg})^2 \leq 3.0$$

where δ_{max} = maximum displacement at level x

δ_{avg} = average of displacements at extreme points of the structure at level x

In addition, UBC Section 1603.3.3 stipulates that torsionally induced forces shall be neglected when of opposite sense to the in-plane forces. Figure 2–30 illustrates the analysis required for torsional effects.

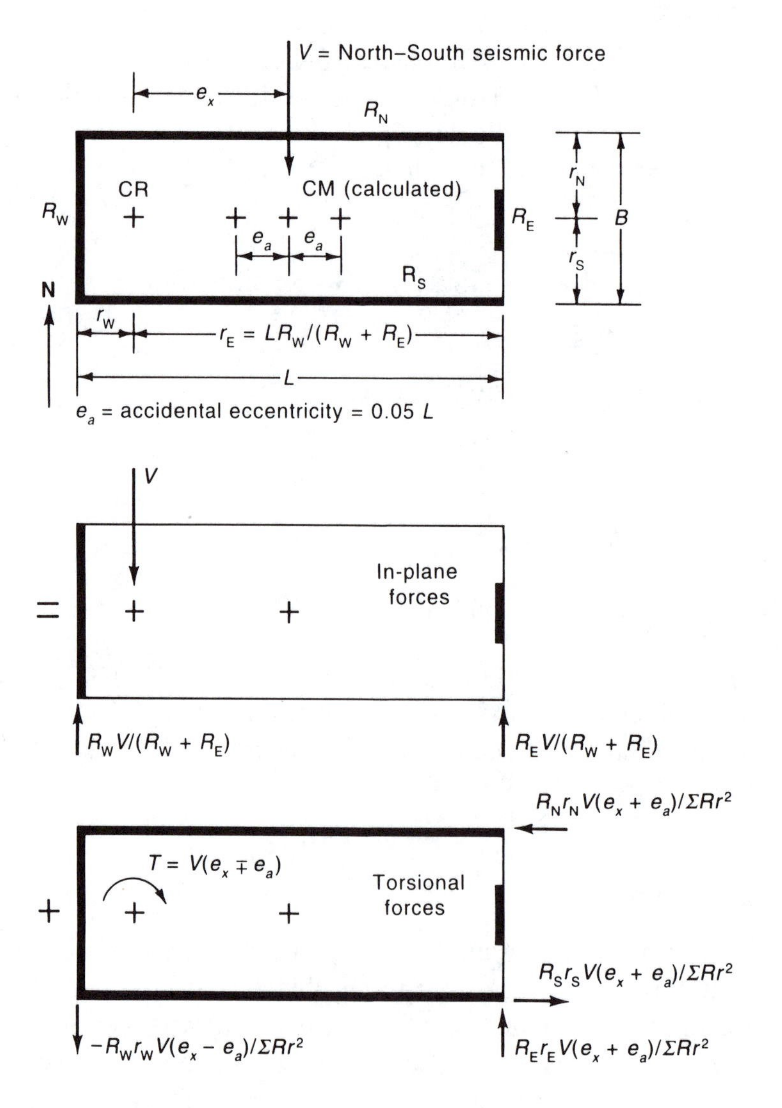

Figure 2–30 Torsional effects

2.6.3 Center of mass and center of rigidity

The location of the center of rigidity is obtained by taking statical moments of the wall rigidities about a convenient origin. For seismic loads in the north–south direction, the north–south walls, which have no stiffness in this direction, are omitted and only east and west walls are considered. From Figure 2–30, the center of rigidity is located a distance from the east wall given by

$$r_E = \Sigma R_y x / \Sigma R_y$$
$$= (R_W \times L + R_E \times 0)/(R_W + R_E)$$
$$= R_W L/(R_W + R_E)$$

The center of rigidity is located a distance from the south wall given by

$$r_S = \Sigma R_x y / \Sigma R_x$$
$$= (R_N \times B + R_S \times 0)/(R_N + R_S)$$
$$= R_N B/(R_N + R_S)$$

The polar moment of inertia of the walls is given by

$$J = \Sigma r^2 R$$
$$= r_N^2 R_N + r_S^2 R_S + r_E^2 R_E + r_W^2 R_W$$

The center of mass is obtained by taking statical moments of the wall weights about a convenient origin. When the lateral seismic force is determined at roof diaphragm level in a single story building, the force in the north–south direction does not include an allowance for the mass of the east and west shear walls. Hence, in locating the center of mass for north–south forces, the east and west walls are omitted. From Figure 2–30, for this situation, the distance of the center of mass from the east wall is given by

$$\bar{x} = \Sigma W_{NS} x / \Sigma W_{NS}$$
$$= (W_R \times L/2 + W_N \times L/2 + W_S \times L/2)/(W_R + W_N + W_S)$$

where W_R = weight of roof
W_N = weight of north wall
W_S = weight of south wall

In this instance for equal lengths and distribution of mass in the north and south walls and in the roof, the center of mass lies midway between the east and west walls and

$$\bar{x} = L/2$$

The total shear force at the base of the east and west walls is then given by the sum of the shear due to the in–plane forces, torsional forces and the force due to the self weight of the wall.

In a single story building when the lateral force is determined at base level (i.e., the base shear), and for multistory structures where the lateral force includes an allowance for the mass of all walls, it is appropriate to include all walls in the calculation of the center of mass. From Figure 2–30, for this situation, the center of mass is located a distance from the east wall given by

$$\bar{x} = \Sigma Wx/\Sigma W$$
$$= (W_R \times L/2 + W_N \times L/2 + W_S \times L/2 + W_E \times 0 + W_W \times L)/(W_R+W_N+W_S+W_E+W_W)$$

The total shear force at the base of the east and west walls is then given by the sum of the shear due to the in-plane forces and the torsional forces.

2.6.4 Torsional effects

The eccentricity between the center of mass and the center of rigidity is shown in Figure 2-30 as

$$e_x = r_E - \bar{x}$$

The accidental eccentricity is given by

$$e_a = 0.05L$$

The total eccentricity is

$$e_T = e_x \pm e_a$$

The torsional moment for north-south seismic load is given by

$$T_{NS} = Ve_T$$
$$= V(e_x \pm e_a)$$

The total force in the east wall, when V represents the seismic force at diaphragm level, is given by

$$F = F_S + F_T + F_W$$

where the in-plane shear force is

$$F_S = VR_E/(R_E + R_W)$$

the torsional shear force is

$$F_T = T_{NS}r_E R_E/J$$

the shear force due to the self weight of the wall is

$$F_W = (ZIC/R_W)W_W$$

For the west wall, since torsional forces are of opposite sense to the in-plane forces, the torsional forces are neglected and the total design force is

$$F = F_S + F_W$$

When V represents the total base shear, or the lateral force at story level in a multistory structure, the total force in the east wall, for north-south seismic load is

$$F = F_S + F_T$$

Since the base shear includes an allowance for the mass of all walls, the additional term, F_W, is not required. For the west wall, neglecting torsional forces, the total design force is given by

$$F = F_S$$

Example 2–18 (Structure with rigid diaphragm)
Determine the shear in the east and west shear walls of the building in Example 2–2 for the calculated base shear in the north–south direction. The roof consists of a concrete slab and the relative rigidities of the walls are

$$R_1 = R_2 = R_3 = 1.0$$
$$R_4 = 0.6$$

Solution
The base shear, calculated in Example 2–2, allows for the mass of the roof and all walls. Hence, in locating the center of mass, it is appropriate to include the weight of the roof and all walls. From the symmetry of the structure, for a north–south seismic load, the center of mass is located midway between wall 2 and wall 4 and its distance from wall 4 is

$$\bar{x} = 40/2 = 20 \text{ feet}$$

The position of the center of mass in the orthogonal direction is not relevant to the question. In locating the center of rigidity for a north–south seismic load, wall 1 and wall 3, which have no stiffness in the north–south direction, are omitted. Taking moments about wall 4, the distance of the center of rigidity from wall 4 is given by

$$\begin{aligned} r_4 &= \Sigma R_y x/\Sigma R_y \\ &= (1.0 \times 40 + 0.2 \times 0)/(1.0 + 0.2) \\ &= 33.33 \text{ feet} \end{aligned}$$

The distance of the center of rigidity from wall 2 is

$$\begin{aligned} r_2 &= 40 - 33.33 \\ &= 6.67 \text{ feet} \end{aligned}$$

In locating the center of rigidity for an east-west seismic load, wall 2 and wall 4, which have no stiffness in the east-west direction are omitted. Due to the symmetry of the structure, the center of rigidity is located midway between wall 1 and wall 3 and

$$r_1 = r_3 = 10 \text{ feet}$$

The polar moment of inertia of the walls is

$$\begin{aligned} J &= \Sigma r^2 R \\ &= r_1^2 \times R_1 + r_2^2 \times R_2 + r_3^2 \times R_3 + r_4^2 \times R_4 \\ &= 10^2 \times 1.0 + 6.67^2 \times 1.0 + 10^2 \times 1.0 + 33.33^2 \times 0.2 \\ &= 467 \text{ square feet} \end{aligned}$$

The sum of the wall rigidities for a seismic load in the north–south direction is

$$\Sigma R_y = R_2 + R_4 = 1.0 + 0.2 = 1.2$$

For a seismic load in the north–south direction, the eccentricity is

$$e_x = r_4 - \bar{x} = 33.33 - 20 = 13.33 \text{ feet}$$

Accidental eccentricity, in accordance with UBC Section 1628.5 is

$$e_a = \pm 0.05 \times L = \pm 0.05 \times 40 = \pm 2 \text{ feet}$$

An accidental displacement of the center of mass to the west is the most critical and gives a total eccentricity of

$$e_T = e_x + e_a = 13.33 + 2.0 = 15.33 \text{ feet}$$

The counterclockwise torsional moment acting about the center of rigidity is

$$T = Ve_T = 11.43 \times 15.33 = 175 \text{ kip feet}$$

The force produced in a wall by the base shear acting in the north–south direction is the sum of the in–plane shear force and the torsional shear force. The shear force due to the wall self weight is not required as this is already included in the value of the base shear. The in–plane shear force is

$$F_S = VR_y/\Sigma R_y = 11.43R_y/1.2 = 9.525R_y$$

The torsional shear force is

$$F_T = TrR/J = 175rR/467 = 0.375rR$$

with a negative value indicating that the torsional force is opposite in sense to the in–plane force. The total force in a wall is

$$F = F_S + F_T$$

The forces produced in wall 2 and wall 4 are given in Table 2–5

Wall	R_y	F_S	rR	F_T	F'_T	F
2	1.0	9.53	6.67	-2.50	-2.72	9.53
4	0.2	1.91	6.67	2.50	2.72	4.63

Table 2–5 Forces in walls 2 and 4

Taking the algebraic sum of F_S and F_T, including negative values of F_T, gives net wall forces of:

$$F_2 = 9.53 - 2.50 = 7.03 \text{ kips}$$
$$F_4 = 1.91 + 2.50 = 4.41 \text{ kips}$$

The relative displacement of a wall is given by

$$\delta = F/R$$

The relative displacements of wall 2 and wall 4 are:

$$\delta_2 = 7.03/1.0 = 7.03$$
$$\delta_4 = 4.41/0.2 = 22.05$$

The ratio of the maximum displacement of wall 4 to the average displacement of wall 2 and wall 4 is

$$\begin{aligned}\mu &= 2\delta_4/(\delta_2 + \delta_4)\\ &= 2 \times 22.05/29.08\\ &= 1.52\\ &> 1.20\end{aligned}$$

This constitutes a torsional irregularity as defined in UBC Table 16–M and the accidental eccentricity must be amplified, as specified in UBC Section 1628.6, by the factor

$$\begin{aligned}A_x &= (\mu/1.2)^2\\ &= (1.52/1.2)^2\\ &= 1.60\\ &< 3.00\end{aligned}$$

The revised accidental eccentricity is

$$\begin{aligned}e'_a &= A_x e_a\\ &= 1.60 \times 2.00\\ &= 3.20 \text{ feet}\end{aligned}$$

The revised total eccentricity for a displacement of the center of mass to the west is

$$\begin{aligned}e'_T &= e_x + e_a\\ &= 13.33 + 3.20\\ &= 16.53 \text{ feet}\end{aligned}$$

The revised torsional moment is

$$T' = V_T'$$
$$= 11.43 \times 16.53$$
$$= 191 \text{ kip feet}$$

The revised torsional shear force is

$$F_T' = T'rR/J$$
$$= 191\ rR/467$$
$$= 0.408rR$$

The revised total force in a wall is

$$F = F_S + F_T'$$

The torsional shear force is neglected, in accordance with UBC Section 1603.3.3, when of opposite sense to the in-plane force and the final design forces in wall 2 and wall 4 are shown in Table 2-5.

2.6.5 Shear wall rigidity

The rigidity of a concrete or masonry shear wall is the force required to produce unit displacement at the top edge of the wall. This is most readily obtained as the reciprocal of the deflection of the wall due to unit load applied at the top edge. The deflection of a wall or pier due to a unit applied load is given by

$$\delta = \delta_F + \delta_S$$

where δ_F = deflection due to flexure
$= 4(H/L)^3/Et$ for a cantilever pier
$= (H/L)^3/Et$ for a pier fixed at top and bottom
H = height of pier
L = length of pier
E = modulus of elasticity of pier
t = thickness of pier
δ_S = deflection due to shear
$= 1.2H/GA$
$= 3(H/L)/Et$
G = rigidity modulus of pier
$= 0.4E$
A = cross sectional area of pier
$= tL$

The rigidity, or stiffness, of a pier is given by

$$R = 1/\delta$$

The rigidity of a wall with openings is most accurately determined by the following technique[24].

The deflection of the wall is first obtained as though it is a solid wall. From this is subtracted the deflection of that portion of the wall which contains the opening. The deflection of each pier, formed by the openings, is now added back. The relevant calculations are readily performed by hand or may be obtained by using the calculator program in Appendix III.

Example 2–19 (Wall rigidity)
Determine the rigidity of the masonry wall shown in Figure 2–31. The wall is 8 inches thick, with a modulus of elasticity of E_m = 1,500,000 pounds per square inch, and may be considered fixed at the top and bottom.

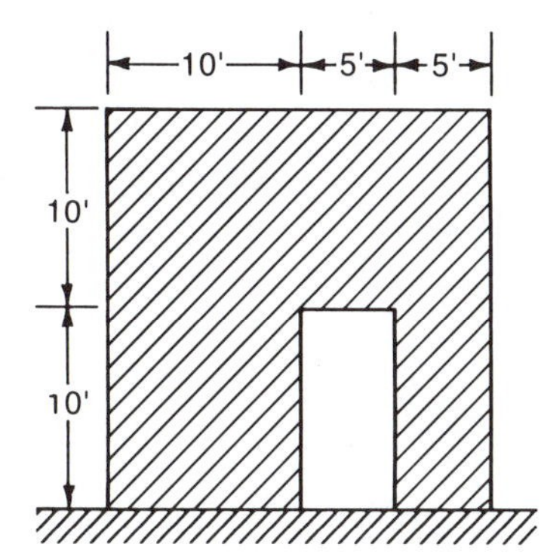

Figure 2–31 Masonry wall rigidity

Solution
The relevant details are given in Table 2–6.
The actual rigidity of the wall is

$$\begin{aligned} R &= 0.182Et \\ &= 0.182 \times 1500 \times 8 \\ &= 2185 \text{ kips per inch} \end{aligned}$$

Pier	H	L	Type	$(H/L)^3 = Et\delta_F$	$3H/L = Et\delta_s$	$Et(\delta_F+\delta_S) = Et\delta$	R/Et
Wall	20	20	Fixed	1	3	4.000	–
1+2+4	10	20	Fixed	-0.125	-15.000	-1.625	–
1	10	10	Fixed	1	3	–	0.250
2	10	5	Fixed	8	6	–	0.071
1+2	–	–	–	–	–	3.115	←0.321
Total	–	–	–	–	–	5.490→	0.182

Table 2–6 Rigidity of masonry wall

Example 2–20 (Structural Engineering Examination 1993, Section A, weight 8.0 points)

GIVEN: A roof plan of a single story building as shown in Figure 2–32

- The diaphragm design force is 225 kips, evenly distributed over the roof area. The center of mass is located as shown on Figure 2–32.

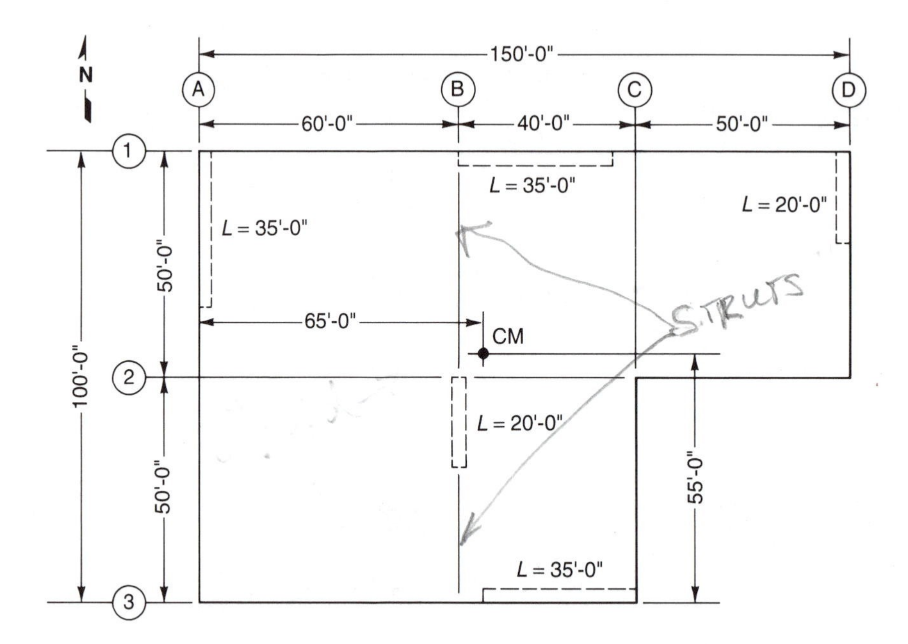

Figure 2–32 Details for Example 2–20

CRITERIA: Materials:

- All wall are 6 inches thick.
- Modulus of elasticity E = 1,000,000 pounds per square inch.

Assumptions:

- The lateral force resisting system consists of shear walls in both directions.
- The building has no torsional irregularity as defined in UBC Table 16–M.
- The story height is 14 feet.
- Consider the lateral forces in the North–South direction only.

REQUIRED: 1. Assume the diaphragm is rigid and the walls are cantilevered.

a. Given the relatively rigidity of the walls as follows:

Wall on line A	R = 0.81
Wall on line B	R = 0.42
Wall on line D	R = 0.42
Wall on line 1	R = 1.00
Wall on line 3	R = 1.00

Determine the center of rigidity of the wall group.

b. Determine the maximum design loads for the shear walls given in 1a.

2. Assume the diaphragm is flexible. Determine the lateral force distributed to each wall.
3. a. Determine the rigidity of the wall on line B.
   b. Determine the deflection of the wall on line A for a diaphragm load of 2.5 kips per foot, assuming its rigidity to be 2500 kips per inch.
   c. Given the midpoint deflection of the diaphragm between Line A and Line B is 0.2 inch due to the uniform lateral load of 2.5 kips per foot, determine whether the diaphragm is considered by code to be flexible or rigid, given the deflections of walls A and B under the given load are 0.04 inches and 0.08 inches, respectively.
4. Considering the diaphragm for forces in both directions, on a sketch of the building plan, indicate schematically the locations of the chords and collectors required to complete the design.

SOLUTION

1a. CENTER OF RIGIDITY

In locating the center of rigidity for seismic loads in the east–west direction, walls A, B, and D, which have no stiffness in this direction, are omitted. By inspection, the center of rigidity lies midway between walls 1 and 3.

In locating the center of rigidity for seismic loads in the north–south direction, walls 1 and 2 are omitted. Taking moments about the bottom left hand corner of the diaphragm, the distance of the center of rigidity from line A is

$$\begin{aligned} x_r &= (0.25R_A + 60R_B + 149.75R_D)/(R_A + R_B + R_D) \\ &= (0.25 \times 0.81 + 60 \times 0.42 + 149.75 \times 0.42)/1.65 \\ &= 53.51 \text{ feet} \end{aligned}$$

1b. APPLIED TORSION

The seismic load acts through the center of mass and for seismic loads in the north–south direction, the eccentricity about the center of rigidity is

$$\begin{aligned} e_x &= \bar{x} - x_r \\ &= 65 - 53.51 \\ &= 11.49 \text{ feet} \end{aligned}$$

To comply with UBC Section 1628.5, accidental torsion must be considered and this amounts to five percent of the diaphragm dimension perpendicular to the direction of the seismic force. The accidental eccentricity for seismic loads in the north–south direction is given by

$$\begin{aligned} e_a &= \pm 0.05 \times 150 \\ &= \pm 7.5 \text{ feet} \end{aligned}$$

As the problem statement indicates that the building has no torsional irregularity as defined in UBC Table 16–M, amplification of this eccentricity, as specified in UBC Section 1628.6, is unnecessary. The net eccentricity is

$$
\begin{aligned}
e &= e_x + e_a \\
&= 11.49 \pm 7.5 \\
&= 18.99 \text{ or } 3.99 \text{ feet}
\end{aligned}
$$

The torsional moment acting about the center of rigidity is

$$
\begin{aligned}
T &= Ve \\
&= 225 \times 18.99 \text{ or } 225 \times 3.99 \\
&= 4273 \text{ or } 898 \text{ kip feet}
\end{aligned}
$$

STRUCTURE PROPERTIES

The distance r of each wall from the center of rigidity is

$$
\begin{aligned}
r_A &= 53.51 - 0.25 \\
&= 53.26 \text{ feet} \\
r_B &= 60 - 53.51 \\
&= 6.49 \text{ feet} \\
r_D &= 149.75 - 53.51 \\
&= 96.24 \text{ feet} \\
r_1 &= 50 - 0.25 \\
&= 49.75 \text{ feet} \\
r_3 &= 50 - 0.25 \\
&= 49.75 \text{ feet}
\end{aligned}
$$

The polar moment of inertia of the walls is

$$
\begin{aligned}
J &= \Sigma r^2 R \\
&= 53.26^2 \times 0.81 + 6.49^2 \times 0.42 + 96.24^2 \times 0.42 + 49.75^2 \times 2 \\
&= 11{,}156 \text{ square feet}
\end{aligned}
$$

The sum of the shear wall rigidities for seismic loads in the north–south direction is

$$
\begin{aligned}
\Sigma R_y &= R_A + R_B + R_D \\
&= 1.65
\end{aligned}
$$

WALL FORCES

The force produced in a wall by the diaphragm design force acting in the north–south direction is the sum of the in-plane shear force and the torsional shear force. The in-plane shear force is

$$
\begin{aligned}
F_S &= VR_y/\Sigma R_y \\
&= 225R_y/1.65 \\
&= 136.36R_y \text{ kips}
\end{aligned}
$$

The torsional shear force is

$$
\begin{aligned}
F_T &= TrR/J \\
F_{T1} &= 4273rR/11{,}156 \\
&= 0.383rR \text{ kips} \\
F_{T2} &= 898rR/11{,}156 \\
&= 0.080rR \text{ kips}
\end{aligned}
$$

The total force in a wall is

$$F_F = F_S + F_T$$

and the torsional shear force F_T, when of opposite sense to the in-plane shear force F_S, is neglected in accordance with UBC Section 1603.5. The wall forces are shown in Table 2-7.

Wall	R_y	F_S	rR	F_{T1}	F_{T2}	F_F
A	0.81	110.45	43.14	-16.52	-16.52	110.45
B	0.42	57.27	2.73	1.05	0.22	58.32
D	0.42	57.27	40.42	15.48	3.23	72.75
1	0.00	0.00	49.75	19.05	3.98	19.05
3	0.00	0.00	49.75	19.05	3.98	19.05

Table 2-7 Wall forces

2. FLEXIBLE DIAPHRAGM

For a flexible diaphragm no torsion occurs and only the in-plane shear force is relevant. Shear walls A, B, and D form supports for the simple spans AB and BD. The problem statement indicates that the seismic force is evenly distributed over the roof area which is given by

$$A = 150 \times 100 - 50 \times 50$$
$$= 12{,}500$$

The force per unit area is

$$v = 225 \times 100/12{,}500$$
$$= 18 \text{ pounds per square foot}$$

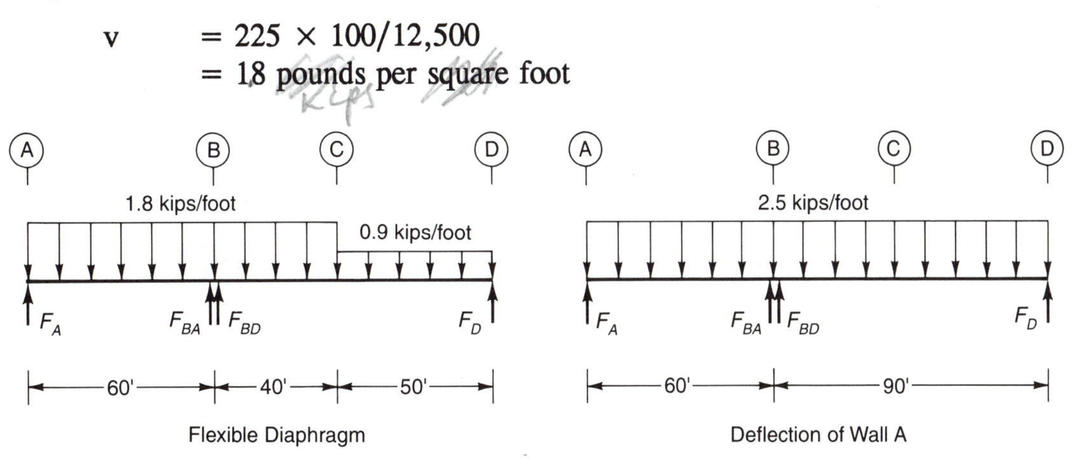

Figure 2-33 Diaphragm analysis for Example 2-20

The applied loading on the shear walls is shown in Figure 2-33 and the forces on the shear walls are

$$F_A = wL/2$$
$$= 1.8 \times 60/2$$
$$= 54 \text{ kips}$$

F_D = 0.9 × 90/2 + 0.9 × 40 × 20/90
= 48.5 kips

F_B = F_{BA} + F_{BD}
= 54 + 0.9 × 90/2 + 0.9 × 40 × 70/90
= 122.5 kips

3a. RIGIDITY OF WALL B

The rigidity of a cantilever wall is derived as the reciprocal of the deflection of the wall due to a unit load applied at the top edge. This deflection is given by

δ = $\delta_F + \delta_S$

where δ_F = deflection due to flexure
= $4(H/L)^3/Et$

δ_S = deflection due to shear
= $3(H/L)/Et$

The rigidity of wall B may be obtained by using the calculator program in Appendix III as indicated in Table 2-8.

H	L	$4(H/L)^3 = Et\delta_F$	$3(H/L) = Et\delta_S$	$Et(\delta_F + \delta_S) = Et\delta$	R/Et
14	20	1.37	2.10	3.47	0.288

Table 2–8 Rigidity of wall B

The actual rigidity of wall B is

R_B = 0.288Et
= 0.288 × 1000 × 6
= 1728 kips per inch

3.6 DEFLECTION OF WALL A

For a flexible diaphragm with a uniform lateral load of 2.5 kips per foot, as shown in Figure 2-33, the force on wall A is

F_A = wL/2
= 2.5 × 60/2
= 75 kips

The deflection of wall A due to this force is

δ_A = F_A/R_A
= 75/2500
= 0.030 inches

3c. DETERMINATION OF DIAPHRAGM FLEXIBILITY

For diaphragm AB the given displacements of the end supports are

$$\delta_A = 0.04 \text{ inches}$$
$$\delta_B = 0.08 \text{ inches}$$

The average displacement of the end supports is

$$\delta_a = (\delta_A + \delta_B)/2 = (0.04 + 0.08)/2 = 0.06 \text{ inches}$$

The midpoint displacement of diaphragm AB is given as

$$\delta_m = 0.2 \text{ inches} > 2\delta_a$$

Hence, in accordance with UBC Section 1628.5 the diaphragm may be considered flexible.

4. CHORDS AND COLLECTORS

The required chords and collectors are shown in Figure 2-34.

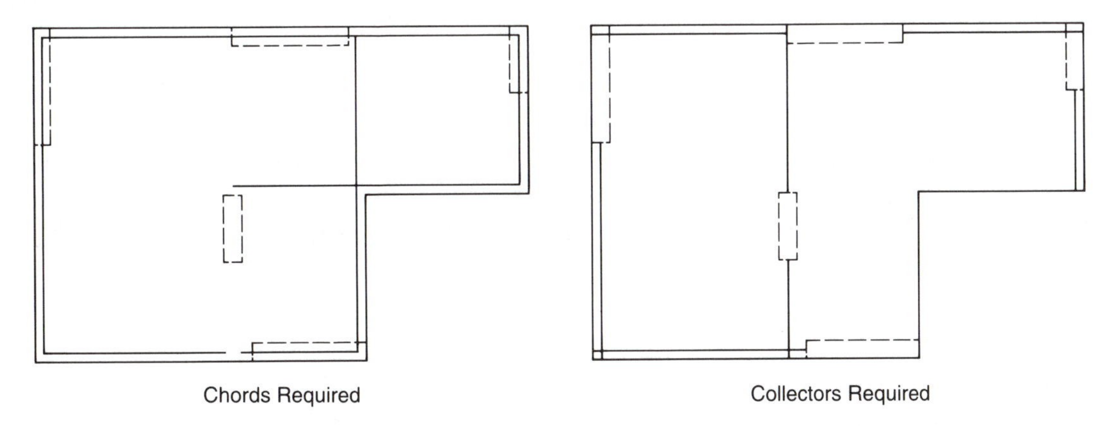

Figure 2-34 Chords and collectors for Example 2-20

2.6.6 Braced frame stiffness

The stiffness of a braced frame is the force required to produce unit displacement at the top of the frame. This is obtained as the reciprocal of the deflection of the frame due to a unit, horizontal, virtual force applied at the top. The horizontal displacement at the point of application of the load is determined by the virtual work method[26,27] by evaluating the expression

$$\delta = \Sigma u^2L/AE$$

where u is the force in a member due to the unit virtual load, L is the length of the member, A is the sectional area of the member, E is the modulus of elasticity of the member and the summation extends over all the members of the frame. The stiffness is, then,

$$k = 1/\delta = 1/(\Sigma u^2L/AE)$$

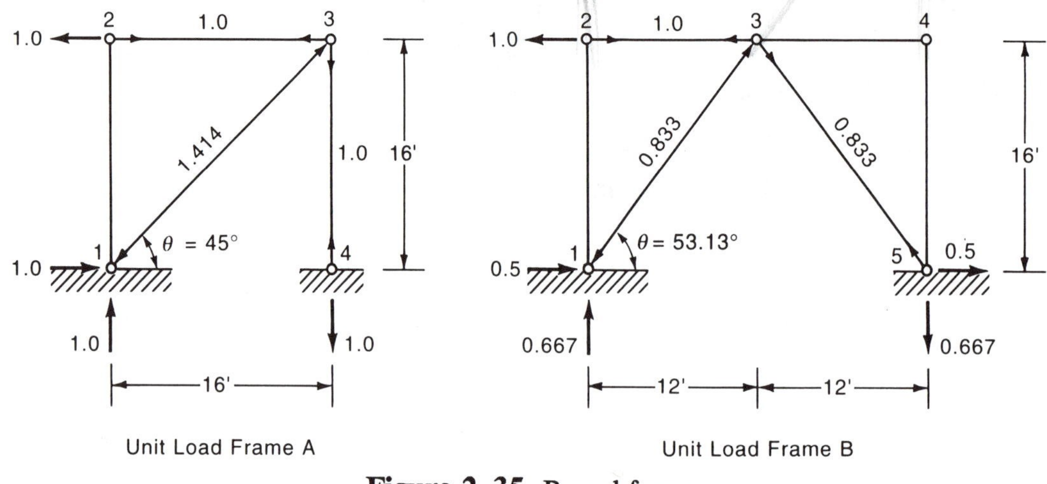

Figure 2–35 Braced frames

For frame A shown in Figure 2–35, if the elastic shortening of the beam and columns may be neglected, only the diagonal brace contributes to the virtual work summation and the expression for frame stiffness simplifies to

$$k = \cos^2\theta(AE/L)$$

where L, A and E are the length, area and modulus of elasticity of the brace and θ is its angle of inclination to the horizontal.

For frame B, shown in Figure 2–35, there are no forces in the columns due to unit horizontal load and when the elastic shortening of the beam may be neglected and, when both diagonal braces have identical areas, lengths, and moduli of elasticity, the expression for the frame stiffness simplifies to

$$k = \Sigma\cos^2\theta(AE/L)$$

and the summation extends over the diagonal braces only.

Example 2–21 (Braced frame)
Calculate the stiffness of frame A and frame B shown in Figure 2–35. The beam in both frames may be considered infinite in area and the columns are W8 × 31. The diagonal brace in Frame A is TS 7 × 7 × ¼ and in frame B the braces are TS 6 × 6 × 3/16

Solution
The stiffness of frame A is obtained by applying a unit virtual load to the frame, as shown in Figure 2–35. The horizontal displacement at the point of application of the load is determined by the virtual work method by evaluating the expression

$$\delta = \Sigma u^2L/AE$$

where u is the force in a member due to the unit virtual load, L is the length of the member, A is the sectional area of the member and E is the modulus of elasticity. The summation extends over all the members with the exception of the beam which is considered to have negligible elastic shortening. The relevant values are given in Table 2–9.

Member	L	A	u	u^2L/A
Brace 13	22.62	6.590	-1.414	6.860
Column 34	16.00	9.130	1.000	1.750
Total	–	–	–	8.610

Table 2-9 Displacement of frame A

The horizontal displacement of point 2 is given by

$$\begin{aligned}\delta &= \Sigma u^2L/AE \\ &= 8.61 \times 12/29{,}000 \\ &= 0.00356 \text{ inches}\end{aligned}$$

The lateral stiffness of frame A, which is defined as the force required to produce unit horizontal displacement, is given by

$$\begin{aligned}k &= 1/\delta \\ &= 1/0.00356 \\ &= 280.7 \text{ kips per inch}\end{aligned}$$

The stiffness of frame B is obtained by applying a unit load at joint 2 which produces the forces shown in Figure 2-35. Since the elastic shortening of the beam may be neglected, only the diagonal members of the frame contribute to the stiffness which is given by

$$k = \Sigma \cos^2\theta\,(AE/L)$$

where θ is the angle of inclination of the diagonal member to the horizontal and the summation extends over the diagonal members only. The stiffness frame of B is

$$\begin{aligned}k &= 2 \times 4.27 \times 29{,}000 \times \cos^2(53.13^\circ)/(20 \times 12) \\ &= 371.5 \text{ kips per inch}\end{aligned}$$

Example 2-22 (Structural Engineering Examination 1993, Section A, weight 6.0 points)

GIVEN: A water tank is supported on a braced steel frame as shown in Figure 2-36
Case 1: On the ground as shown in Elevation 1.
Case 2: On the top of the roof as shown in Elevation 2.
Operating weight tributary to the frame is 10.0 kips.

CRITERIA: Seismic zone 4

- $I = 1.0$
- $R_w = 3.0$
- Soil type 3
- Height to the top of the roof is 56 feet

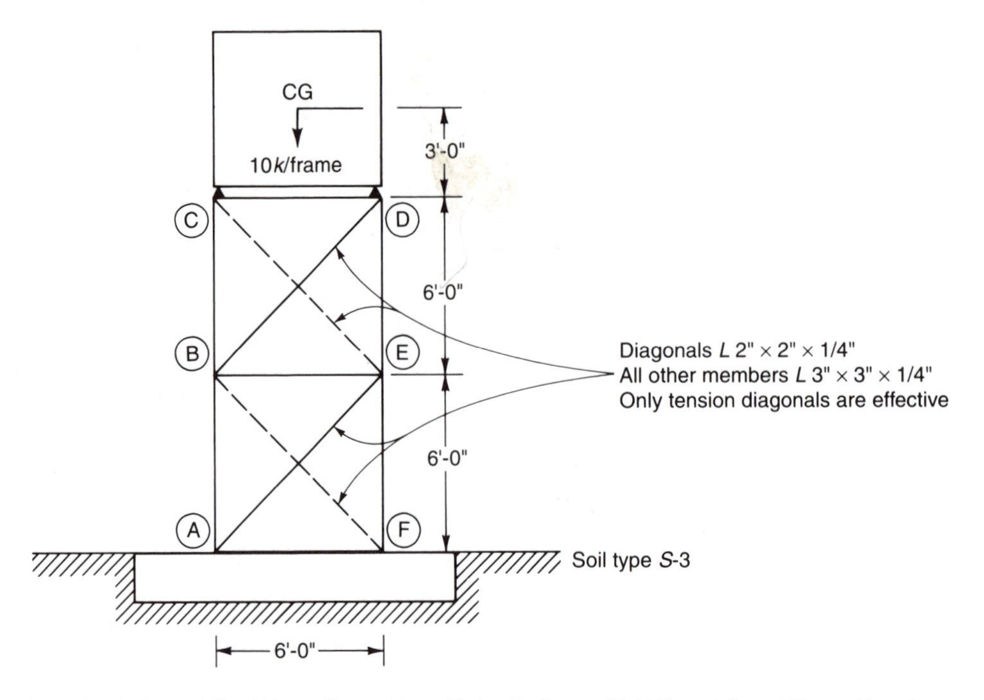

Elevation 1: Braced Steel Frame Supporting a Water Tank on a Rigid Foundation at Ground Level.

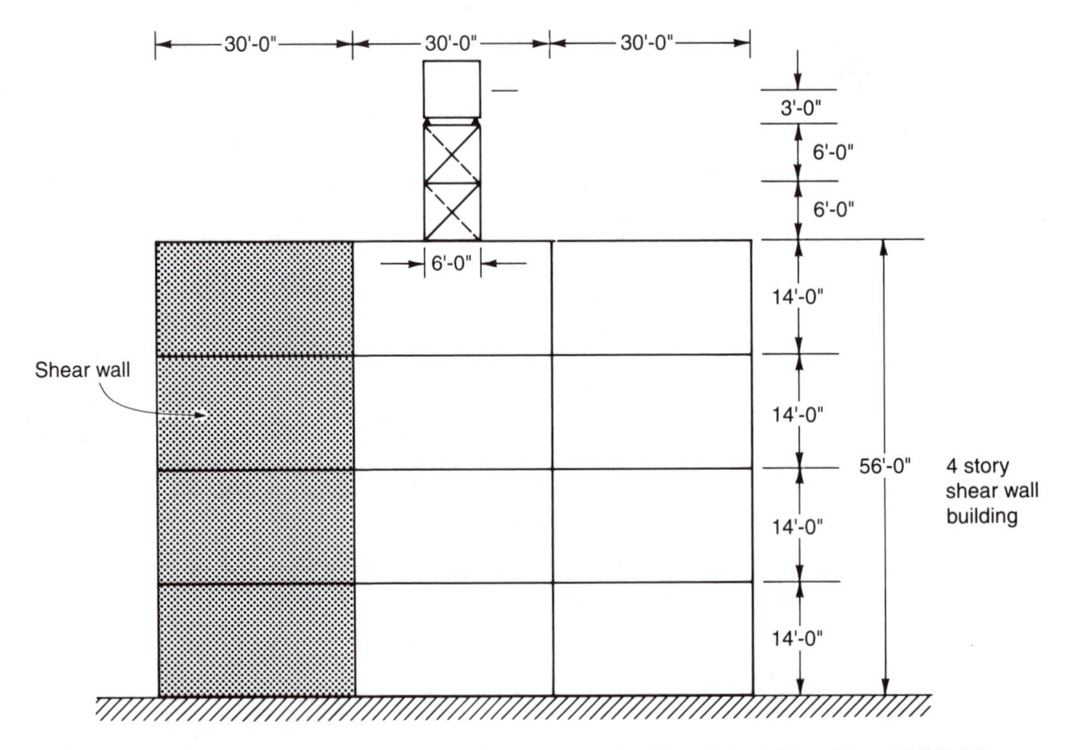

Elevation 2: Braced Steel Frame Supporting a Water Tank at Roof Level of a 4 Story Shear Wall Building. For Braced Frame Detail See Elevation 1.

Figure 2–36 Details for Example 2–22

Materials:
- Structural steel ASTM A 36
- Modulus of elasticity E= 29,000 kips per square inch

Assumptions:
- Only tension diagonals are effective
- Neglect Hydrodynamic effects
- All joints in the truss frame are pin jointed
- Neglect Weight of the truss members

REQUIRED:
1. By a rational method determine the horizontal deflection at the top of the frame at point D, shown in Elevation 1, for a unit horizontal load of 1 kip at the center of gravity of the tank.
2. Determine the period of the frame shown in Elevation 1 assuming the deflection in item 1 above is 0.05 inches.
3. Determine the seismic force on the frame when the frame is supported on a rigid foundation as in Elevation 1 assuming the period of the frame is 0.2 seconds.
4. Determine the period of the four story shear wall building in Elevation 2, assuming ΣA_e = 80.0 square feet and the length of each shear wall in the first story in the direction parallel to the applied force is 30 feet.
5. What is the design seismic shear force on the braced steel frame for the water tank in Elevation 2.

SOLUTION

1. HORIZONTAL DEFLECTION

The member forces P due to a one kip horizontal force acting at the center of gravity of the tank are shown in Figure 2-37 and tabulated in Table 2-10. The member forces u due to a one kip horizontal force acting at point D are shown in Figure 2-37 and are tabulated in Table 2-10. The horizontal displacement of point D due to a one kip horizontal force acting at the center of gravity of the tank is determined by the virtual work method by evaluating the expression:

Member	P	u	L	A	PuL/A
AB	1.50	1.00	6.00	1.440	6.250
DE	-1.50	-1.00	6.00	1.440	6.250
EF	-2.50	-2.00	6.00	1.440	20.830
BE	-1.00	-1.00	6.00	1.440	4.170
AE	1.41	1.41	8.49	0.938	18.100
BD	1.41	1.41	8.49	0.938	18.100
Total	–	–	–	–	73.700

Table 2-10 Displacement of point D

$$\delta = \Sigma PuL/AE$$

where L is the length, A is the sectional area, and E is the modulus of elasticity of the members. The relevant values are given in Table 2-10. The horizontal displacement of point D is given by

$$\begin{aligned}\delta &= \Sigma PuL/AE \\ &= 73.7 \times 12/29{,}000 \\ &= 0.030 \text{ inches}\end{aligned}$$

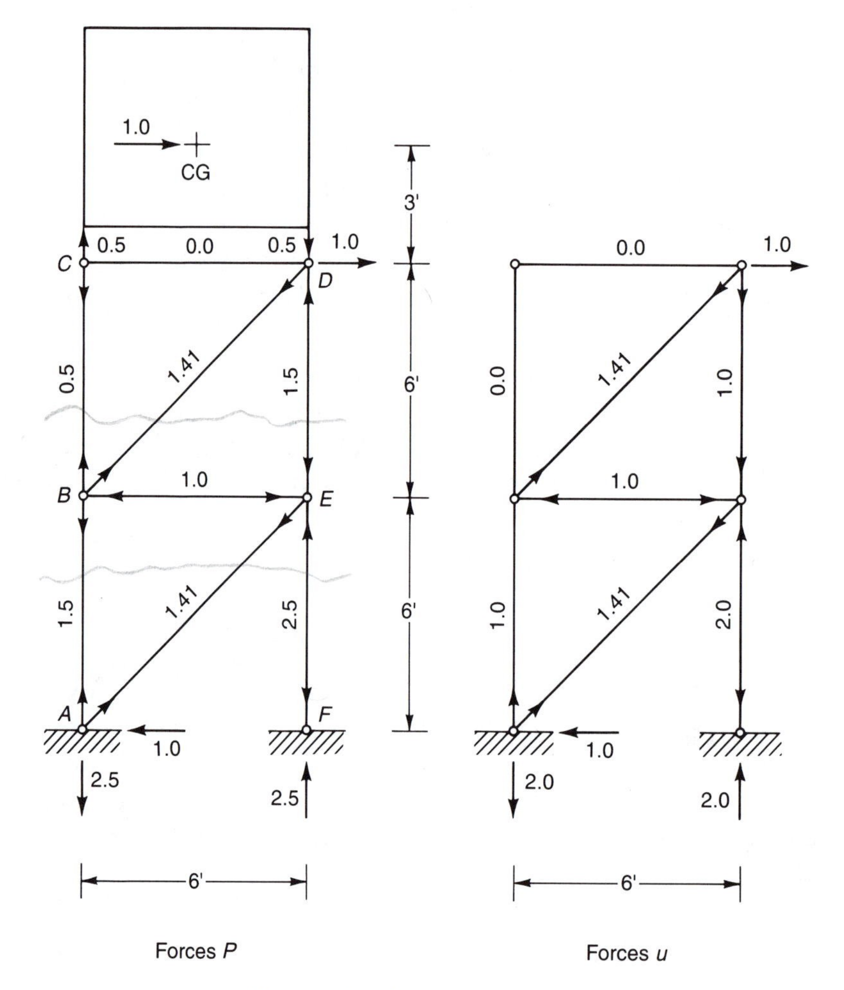

Figure 2-37 Frame forces

2. FRAME PERIOD

The stiffness of the frame is the magnitude of the horizontal force which, when applied to the center of gravity of the tank, produces a unit horizontal displacement of point D. Assuming a displacement of point D of 0.05 inches, due to a one kip force acting at the center of gravity of the tank, the frame stiffness is given by

$$\begin{aligned}k &= 1/0.05 \\ &= 20 \text{ kips per inch}\end{aligned}$$

The natural period of the frame is given by

$$T = 0.32(W/k)^{1/2}$$
$$= 0.32(10/20)^{1/2}$$
$$= 0.226 \text{ seconds}$$

3. SEISMIC FORCE FOR GROUND SUPPORTED FRAME

The force coefficient is given by UBC Formula (28–2) as

$$C = 1.25S/T^{2/3}$$

where $T = 0.2$ seconds as given

$S = 1.5$ from UBC Table 16–J for soil type 3

Then

$$C = 1.25 \times 1.5/0.2^{2/3}$$
$$= 5.48$$
$$> 2.75$$

Hence, use the upper limit specified in UBC Section 1628.2.1 of

$$C = 2.75$$

Also

$$C/R_W = 2.75/3.0$$
$$= 0.917$$
$$> 0.4$$

as required by UBC Section 1632.5.

Hence $C = 2.75$

is an acceptable value for the force coefficient.

Since the natural period of the frame exceeds 0.2 seconds, the frame is flexible as defined by UBC Section 1630.2, and the seismic force in accordance with UBC Section 1632.5 is obtained from UBC Formula (28–1) as

$$V = (ZIC/R_w)W$$

when $Z = 0.4$ for zone 4 from UBC Table 16–I

$I = 1.0$ as given

$C = 2.75$ as calculated

$R_W = 3.0$ as given

$W = 10$ kips as given

Then

$$V = (0.4 \times 1.0 \times 2.75/3)10$$
$$= 3.67 \text{ kips}$$

4. BUILDING PERIOD

The combined effective area of the shear walls in the first story is given by UBC Formula (28–4) as

$$A_c = \Sigma A_e[0.2 + (D_e/h_n)^2]$$

where $A_e = 80$ square feet as given

$D_e = 30$ feet as given

$h_n = 56$ feet as given

and D_e/h_n = 0.54
< 0.9 as required by UBC Section 1628.2.2

Then A_c = $80[0.2 + (30/56)^2]$
= 38.96 square feet

The value of the numerical coefficient C_t is obtained from UBC Section 1628.2.2 as

$C_t = 0.1(A_c)^{1/2}$
= 0.016

Alternatively, the value of C_t for a shear wall building may be obtained directly from UBC Section 1628.2.2 as

$C_t = 0.020$

The natural period of the building is given by UBC Formula (28–3) as

$T = C_t(h_n)^{3/4}$
$= 0.020(56)^{3/4}$
= 0.409 seconds

5. SEISMIC FORCE FOR FRAME ON ROOF OF BUILDING

For flexibly supported equipment mounted above grade, the seismic shear force is given by UBC Formula (30–1) as

$V = ZIC_pW$

where Z = 0.4 for zone 4 from UBC Table 16–I
I = 1.0 as given
W = 10 kips as given
C_p = (2 × 0.75) from UBC Table 16–O item 3.1 and Section 1630.2

Then V = 0.4 × 1.0 × (2 × 0.75) × 10
= 6 kips

2.7 SEISMICALLY ISOLATED BUILDINGS

2.7.1 Advantages of seismic isolation

Base isolation substantially reduces the transmission of earthquake ground motions to the structure thus producing a seismic load which is smaller than in a fixed base structure[1,28]. This allows the structure to be designed to remain essentially elastic during the design basis earthquake of a ten percent probability of exceedence in fifty years. Thus, almost complete protection of the building frame and non–structural components is provided and minimum disruption to the business activity results. In addition, the isolation system is designed to withstand the maximum capable earthquake with a ten percent probability of exceedence in 100 years, which provides a recurrence interval of 950 years.

2.7.2 General design principles

By reducing the stiffness of the system, the natural period is increased resulting in a reduced response to the earthquake motion. In addition, the provision of increased damping provides increased energy dissipation thus further reducing the response. This is shown, qualitatively, in Figure 2–38. In order to ensure essentially elastic response of the structure, the values adopted for the response modification factor, R_{WI}, are up to one quarter the values of R_W used for fixed base structures.

The system may be designed using either a static lateral analysis procedure or a dynamic procedure. The static procedure may be used when the structure is located in seismic zone 3 or 4, not less than 15 kilometers from any active fault, and on soil type S_1, or S_2. The structure must be of a regular configuration, not exceeding four stories or 65 feet in height and with an isolated period greater than three times the fixed base period but not greater than three seconds.

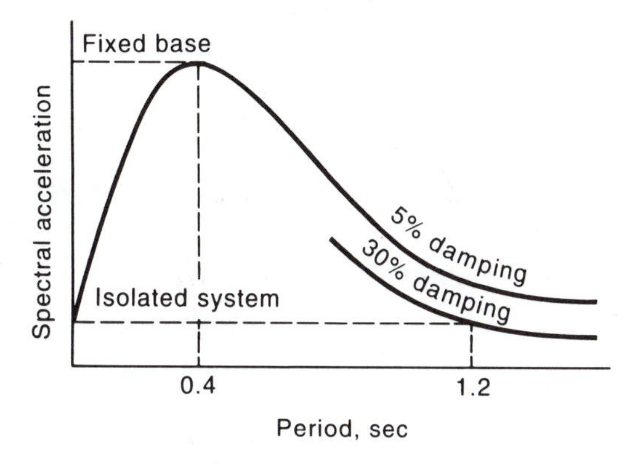

Figure 2–38 Response spectra

2.7.3 Static lateral force procedure

The UBC code presents formulae for determining the effective period and displacement of the isolated structure. Hence, the total design lateral shear on the structure above the isolated system may be calculated. This may not be taken as less than the base shear for a fixed-base structure with the same period, the design wind load, or the force required to activate the isolation system. The total lateral shear is then distributed over the height of the structure, in accordance with UBC Formula (54–7), to provide an inverted triangle load pattern.

Example 2–23 (Lateral force on seismically isolated structure)
A four story eccentrically braced steel frame of regular configuration is located in zone 4, 15 kilometers from an active fault, with a maximum capable earthquake magnitude of 8.0 on the Richter scale, on soil type S_2. The structure is 100 feet square in plan, 60 feet high with 15 feet story height, and the weight at each floor and roof level is 2000 kips. Assume that the effective period of the isolated structure is two seconds and that the isolation system has an effective damping ratio of 30 percent, equal minimum and maximum stiffness values and requires a force of 0.05W to fully activate the system. Determine the vertical distribution of lateral force on the structure.

Solution

The relationship between the effective period of the isolated structure, T_I, and the minimum stiffness of the isolation system is given by UBC Formula (54–2) as

$$T_I = 2\pi(W/gk_{min})^{1/2}$$

where W = total structure weight = 8000 kips
g = 386.4 inches per second per second
T_I = 2 seconds as given

Hence, the value of the minimum stiffness is

$$k_{min} = (2\pi/2)^2 \times 8000/386.4$$
$$= 204 \text{ kips per inch}$$
$$= k_{max} \text{ as given}$$

The design displacement is given by UBC Formula (54–1) as

$$D = 10ZNS_IT_I/B$$

where Z = 0.4 for zone 4 from UBC Table 16–I
N = 1.0 from UBC Table A–16–D for a location of 15 kilometers from an active fault with a maximum capable earthquake magnitude of 8.0 on Richter scale
S_I = 1.4 from UBC Table A–16–C for soil type S_2
B = 1.7 from UBC Table A–16–E for a damping ratio of 30 percent

Hence, the design displacement is

$$D = 10 \times 0.4 \times 1.0 \times 1.4 \times 2/1.7$$
$$= 6.59 \text{ inches}$$

The total design lateral force on the structure above the isolation system is given by UBC Formula (54–6) as

$$V_S = k_{max} D/R_{WI}$$

where R_{WI} = 3.0 from UBC Table A–16–F for an eccentrically braced frame

Then, the design lateral force is

$$V_S = 204 \times 6.59/3$$
$$= 448 \text{ kips}$$

The seismic force required to activate the isolation system is given as

$$V_I = 0.05W$$
$$= 0.05 \times 8000$$
$$= 400 \text{ kips}$$
$$< V_S$$

The design wind pressure is given by UBC Formula (18–1) as

$$P = C_eC_qq_sI_w$$

where C_e = 0.95 from UBC Table 16–G for exposure B and a 60 foot high building
C_q = 1.4 from Table 16–H, method 2, for an enclosed 60 foot high building
q_s = 12.6 from UBC Table 16–F for wind speed of 70 miles per hour
I_w = 1.0 from UBC Table 16–K for a standard occupancy structure

Hence, the design wind pressure is

$$q_s = 0.95 \times 1.4 \times 12.6 \times 1.0$$
$$= 16.76 \text{ pounds per square foot}$$

Then, the total lateral force due to wind is

$$V_W = 16.76 \times 100 \times 60/1000$$
$$= 100 \text{ kips}$$
$$< V_S$$

For a fixed-base structure with a fundamental period of two seconds, the force coefficient is given by the UBC Formula (28–2) as

$$C = 1.25S/T^{2/3}$$

where S = 1.2 from UBC Table 16–J for soil type S_2

Hence, the force coefficient is

$$C = 1.25 \times 1.2/2^{2/3}$$
$$= 0.94$$
$$< 2.75 \text{ as required by UBC Section 1628.2}$$

The base shear is given by UBC Formula (28–1) as

$$V_F = (ZIC/R_W)W$$

where Z = 0.4 for zone 4 from UBC Table 16–I
I = 1.0 from UBC Table 16–K for a standard occupancy structure
R_W = 10 from UBC Table 16–N for an eccentrically braced frame
W = 8000 kips, as given

Then, the seismic base shear is

$$V_F = (0.4 \times 1.0 \times 0.94/10)W$$
$$= 0.038W$$
$$= 304 \text{ kips}$$
$$< V_S$$

Hence, the governing value for the seismic lateral force, in accordance with UBC Section 1654.4.3 is

$$V_S = 448 \text{ kips.}$$

The design lateral force at level x in the structure is given by UBC Formula (54–7) as

$$F_S = V_S w_x h_x / \Sigma\, w_i h_i$$
$$= 448 w_x h_x/300{,}000$$
$$= 0.00149 w_x h_x$$

The relevant values are shown in Table 2–11

Level	w_x	h_x	$w_x h_x$	F_x
Roof	2000	60	120,000	179
Floor 4	2000	45	90,000	134
Floor 3	2000	30	60,000	90
Floor 2	2000	15	30,000	45
Total	–	–	300,000	448

Table 2–11 Vertical force distribution

References

1. Structural Engineers Association of California. *Recommended lateral force requirements and commentary.* Sacramento, CA, 1990.

2. Nester, M.R. and Porusch, A.R. A rational system for earthquake risk management. *Proceedings of the 1991 convention of the Structural Engineers Association of California.* Sacramento, CA, 1992.

3. Shah, H.C. Damage assessment and retrofit. *New England seismic design.* Boston Society of Civil Engineers, Boston, 1991.

4. Chia–Ming, U. and Maarouf, A. Evaluation of displacement amplification factor for seismic design codes. *Proceedings of the SMIP92 seminar on seismological and engineering implications of recent strong–motion data.* California Department of Conservation, Division of Mines and Geology, Sacramento, CA, 1992.

5. Paz, M. *Structural dynamics.* Van Nostrand Reinhold, New York, 1991.

6. Chopra, A.K. *Dynamics of structures.* Earthquake Engineering Research Institute, Berkeley, CA, 1980.

7. Building Seismic Safety Council. *Improving the seismic safety of new buildings.* Washington, D.C., 1986.

8. American Plywood Association. *Diaphragms.* APA Design/Construction Guide, Tacoma, WA, 1989.

9. Applied Technology Council. *Guidelines for the design of horizontal wood diaphragms.* Berkeley, CA, 1981.

10. Brandow, G.E. *UBC diaphragm requirements.* Structural Engineering Association of Southern California Design Seminar. Los Angeles, CA, 1992.

11. Porter, M.L. et al. *Assembly of existing diaphragm data.* TCCMAR Report 5.2-1. Iowa State University, Ames, IA, 1990.

12. Tissell, J.R. *Horizontal plywood diaphragm tests.* Laboratory Report 106. American Plywood Association, Tacoma, WA, 1967.

13. Tissell, J.R. and Elliott, J.R. *Plywood diaphragms.* Research Report 138. American Plywood Association, Tacoma, WA, 1986.

14. Adams, N.R. *Plywood shear walls.* Laboratory Report 105. American Plywood Association, Tacoma, WA, 1976.

15. Teal, E.J. Design of eccentric braced frames. *Proceedings of the 1988 seminar of the Structural Engineers Association of Southern California.* Los Angeles, CA, November 1988.

16. Ishler, M. *Seismic design practice for eccentrically braced frames.* Structural Steel Educational Council, Moraga, CA, 1992.

17. Sheedy, P. Anchorage of concrete and masonry walls. *Building Standards,* October 1983 and April 1984. International Conference of Building Officials, Whittier, CA.

18. Coil, J. *Subdiaphragms.* Structural Engineering Association of Southern California Design Seminar. Los Angeles, CA, 1991.

19. Bush, V.R. *Handbook to the Uniform Building Code.* International Conference of Building Officials, Whittier, CA, 1988, pp. 156–163.

20. American Society of Civil Engineers. *Minimum design loads for buildings and other structures.* New York, 1990.

21. Mehta, K.C. et al. *Guide to the use of the wind load provisions of ASCE 7-88.* American Society of Civil Engineers, New York, 1991.

22. Structural Engineers Association of Washington. *Wind commentary to the Uniform Building Code.* Seattle, WA, 1993.

23. Bachman, R.E. C_p and nonbuilding structures: 1988 UBC and Blue Book overview and perspective. *Proceedings of the 1988 seminar of the Structural Engineers Association of Southern California.* Los Angeles, CA, November 1988.

24. Corps of Engineers. *Seismic design for buildings.* NAVFAC, Technical manual P-355. Washington, D.C., 1982.

25. De la Llera, J.C. and Chopra, A.K. Evaluation of code-accidental torsion provisions using earthquake records from three nominally symmetric-plan buildings. *Proceedings of the SMIP92 seminar on seismological and engineering implications of recent strong-motion data.* California Department of Conservation, Division of Mines and Geology, Sacramento, CA, 1992.

26. Tuma, J.J. *Structural analysis.* McGraw-Hill, New York, 1969.

27. Williams, A. *The analysis of indeterminate structures.* Hart Publishing Company, New York, 1968.

28. Kircher, C.A. and Bachman, R.E. Guidelines for design criteria for base isolation retrofit of existing buildings. *Proceedings of the 1991 convention of the Structural Engineers Association of California.* Sacramento, CA, 1992.

3

Dynamic lateral force procedure for buildings

3.1 BUILDING CONFIGURATION REQUIREMENTS

3.1.1 Structural framing systems

The static lateral force procedure is applicable to structures which are of essentially regular construction. That is, a structure possessing a uniform distribution of mass and stiffness and without irregular features which will produce a concentration of torsional stresses. When these conditions are satisfied, the static force procedure provides a reasonable envelope of the forces and deformations due to the actual dynamic response.

UBC Tables 16–L and 16–M define possible vertical and plan structural irregularities[1] and detail additional requirements which must be satisfied if the irregularities are present. These are illustrated in Figures 3–1 and 3–2.

3.1.2 Selection of lateral force procedure

Vertical structural irregularities produce loads at various levels which differ significantly from the distribution of base shear which is assumed in the static lateral force procedure. As specified in UBC Section 1627.8, a dynamic analysis is necessary under the following conditions:

- All structures 240 feet or more in height except for structures in seismic zone 1 and for standard occupancy structures, as defined in UBC Table 16–K, in seismic zone 2.
- Structures exceeding 5 stories or 65 feet in height having vertical irregularities Type 1 (stiffness), Type 2 (mass), or Type 3 (geometric) except for structures in seismic zone 1 and standard occupancy structures, as defined in UBC Table 16–K, in seismic zone 2.

- Structures exceeding 5 stories or 65 feet in height in seismic zones 3 and 4 not having the same structural system throughout their height, except as permitted by UBC Section 1628.3.2.
- All structures located on soil profile type S_4 which have a period exceeding 0.7 seconds except for structures in seismic zone 1 and for standard occupancy structures, as defined in UBC Table 16–K, in seismic zone 2.

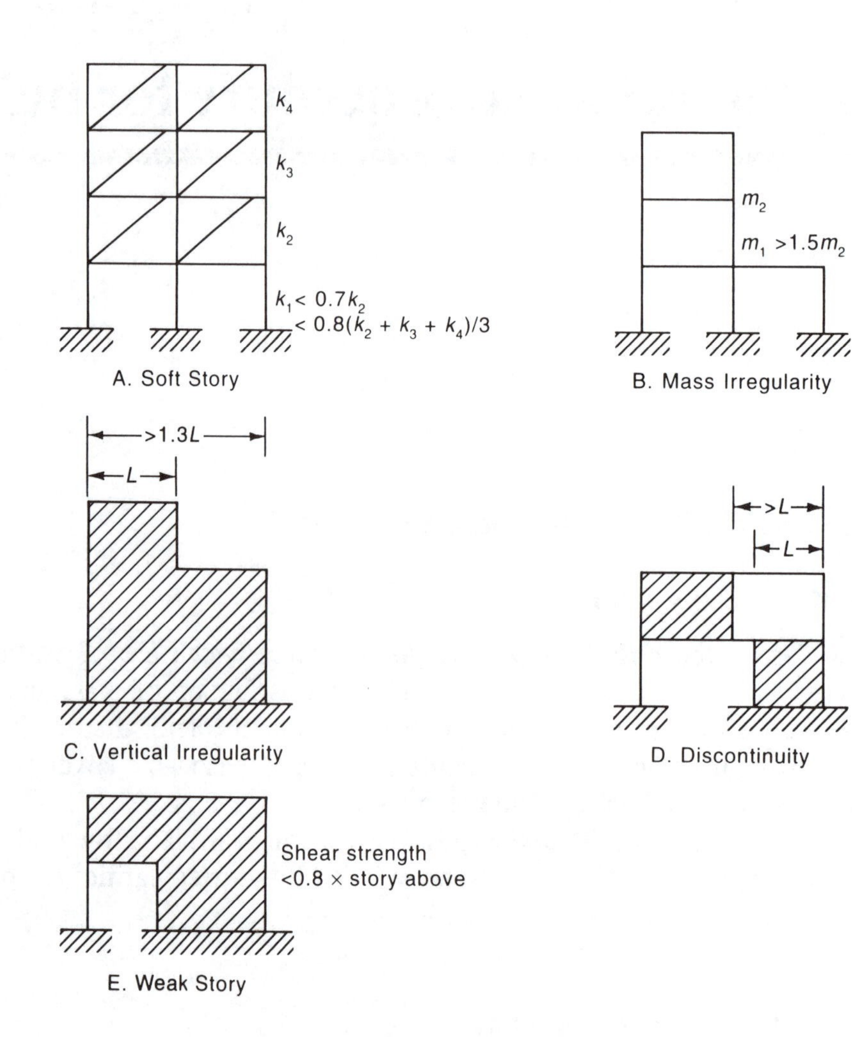

Figure 3–1 Vertical structural irregularities

Example 3–1 (Structural irregularities)
A three story office building with special moment-resisting frames in the north–south direction and eccentrically braced frames in the east–west direction is located in seismic zone 4 on soil type S3. The dead load W at each level and the stiffness k and shear strength v in the north–south direction, are indicated in Figure 3–3. Determine the vertical and plan irregularities for the building and indicate additional code requirements and procedures required for each.

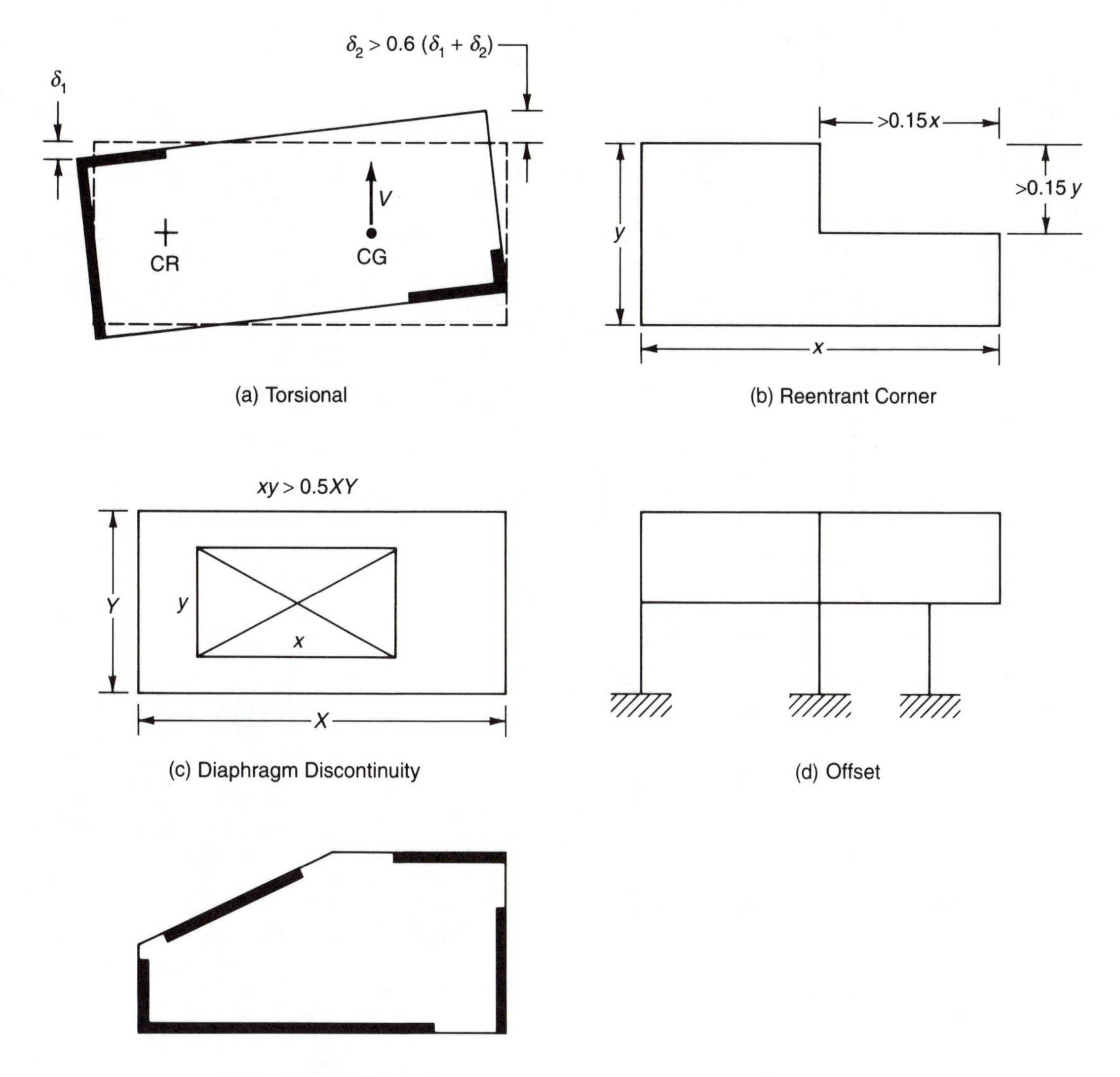

Figure 3–2 Plan structural irregularities

Solution

- The total stiffness of the second story in the north–south direction is

$$k_2 = 2 \times 200$$
$$= 400 \text{ kips per inch}$$

The total stiffness of the first story in the north–south direction is

$$k_1 = 2 \times 80 + 2 \times 50$$
$$= 260 \text{ kips per inch}$$
$$< 70\% \times k_2$$

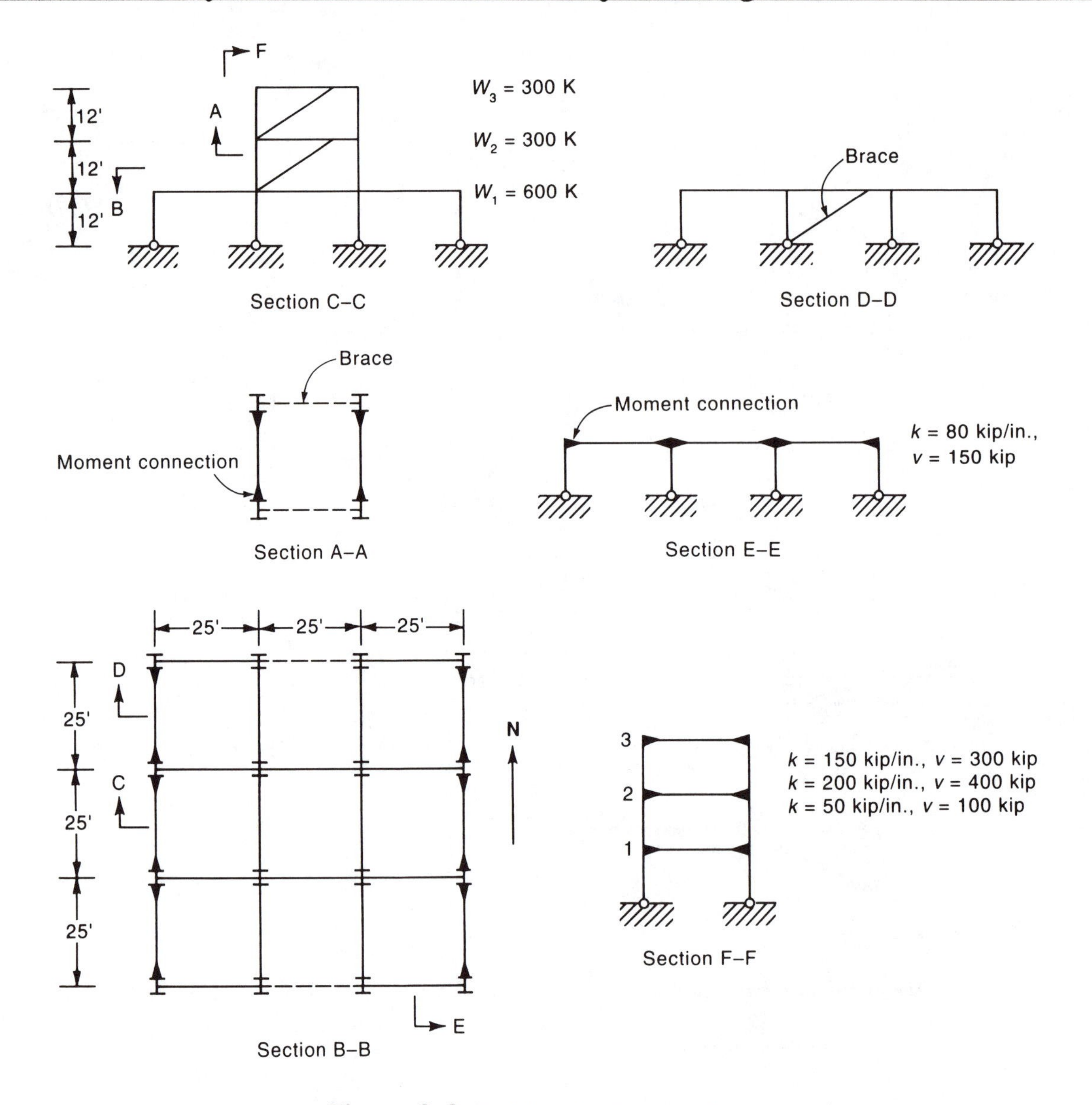

Figure 3–3 Building for Example 3–1

Hence, the first story constitutes a soft story and is considered a vertical irregularity Type 1 in UBC Table 16–L. Since the structure does not exceed five stories and is less than 65 feet in height, UBC Section 1627.8.2.3. permits the structure to be designed using the static lateral force procedure. However, if the dynamic force procedure is used, UBC Section 1629.5.3 specifies that the design base shear used shall not be less than that calculated by the static force procedure. The ten percent reduction allowed for regular structure is not permitted for irregular structures. In addition, UBC Section 1809.4 specifies that, for regular structures, the force, F_t, at the top of the structure determined from UBC Formula (28–7) may be omitted when determining the overturning moment at the base. For an irregular structure, the force, F_t, shall be included when determining the overturning moment.

- The effective weight of the second story is

$$W_2 = 300 \text{ kips}$$

The effective weight of the first story is

$$\begin{aligned} W_1 &= 600 \text{ kips} \\ &> 150\% \times W_1 \end{aligned}$$

Hence, this constitutes a weight irregularity Type 2 in UBC Table 16-L. The additional code requirements are identical with those given for the vertical irregularity Type 1.

- The horizontal dimension of the moment-resisting frame in the second story is

$$L_2 = 25 \text{ feet}$$

The horizontal dimension of the moment-resisting frame in the first story is

$$\begin{aligned} L_1 &= 75 \text{ feet} \\ &> 130\% \times L_2 \end{aligned}$$

Hence, this constitutes a vertical geometric irregularity Type 3 in UBC Table 16-L. The additional code requirements are identical with those given for the vertical irregularity Type 1.

- The total shear strength of the second story in the north-south direction is

$$\begin{aligned} v_2 &= 2 \times 400 \\ &= 800 \text{ kips} \end{aligned}$$

The total shear strength of the first story in the north-south direction is

$$\begin{aligned} v_1 &= 2 \times 150 + 2 \times 100 \\ &= 500 \text{ kips} \\ &< 80\% \times v_2 \end{aligned}$$

Hence, the first story constitutes a weak story and is considered a vertical irregularity Type 5 in UBC Table 16-L. Since the structure exceeds two stories and is more than 30 feet in height, the first story is required by UBC Section 1627.9.1 to be designed for a lateral force of $3(R_W/8)$ times the normal design force in the north-south direction.

- The lateral force resisting system in the east-west direction above the first story consists of braced frames located on the tower section walls. In the first story, the brace lines are on the outer walls of the base. This out-of-plane offset of the vertical elements constitutes a plan irregularity Type 4 in UBC Table 16-M. The UBC Section 1631.2.9.6 specifies that, for this irregularity, connection of diaphragm and collectors to the vertical elements shall be designed without considering the usual one-third increase in allowable stress. In addition, UBC Section 1628.7.2 requires that the first story columns, under the tower section, which support the discontinuous braced frames in story two and above, shall be especially designed and detailed for the load combinations

$$1.0DL + 0.8LL + (3R_W/8)E$$

and $$0.85DL + (3R_W/8)E$$

where

DL = force due to dead load

LL = force due to live load

E = force due to seismic load

3.2 APPLICATION OF MODAL ANALYSIS TO BUILDINGS

3.2.1 Modal analysis advantages

The modal analysis procedure is suitable for calculating the response of complex multiple-degree-of-freedom structures to earthquake motion. The structural response is modeled as the maximum response of a number of single-degree-of-freedom oscillators each representing a specific mode of vibration of the actual structure. Combining the responses of the individual modes produces the equivalent external forces and base and story shears which may then be used in the same manner as in the static lateral force procedure. The modal analysis procedure has the advantage[2] of determining the actual distribution of lateral forces, from the actual mass and stiffness distribution over the height of an irregular structure, which may differ appreciably from the simplified linear distribution assumed in the static lateral force method. In addition, it accounts for the effects of the higher modes of response of a structure some of which may contribute significantly to the overall response of the structure.

Example 3–2 (Structural Engineering Examination 1988, Section A, weight 4.0 points)

GIVEN: A portion of a canopy structure is to be constructed over the ambulance entrance to a hospital. Its mathematical model is shown in Figure 3–4. The beam can be considered infinitely rigid and the columns axially inextensible implying a deformation pattern as shown by the dashed lines.

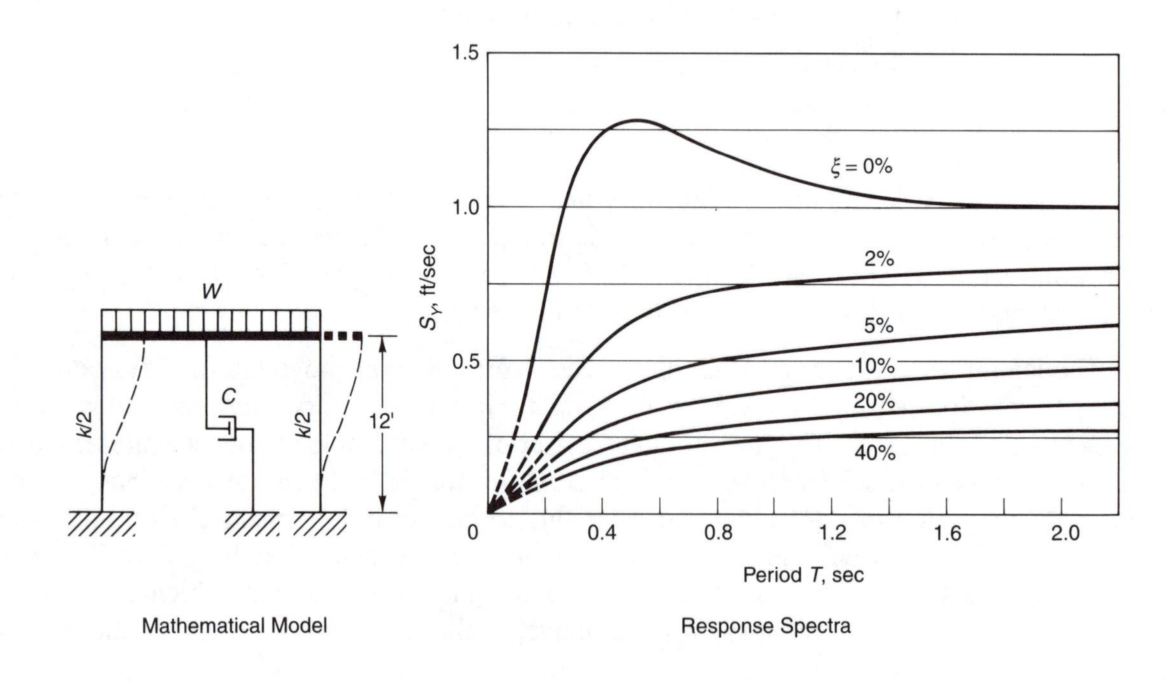

Figure 3–4 Details for Example 3–2

CRITERIA: Response Spectrum : Design pseudo-velocity (S_v) response spectrum as shown in Figure 3-4.

Material Properties:
- A36 steel

Frame Properties:
- Weight = W = 926 kips
- Lateral stiffness = k = 148 kips per inch
- Viscous damping coefficient = c = 1.88 kips-seconds per inch

Seismic Parameters:
- Seismic zone 3
- Response factor = R_W = 4
- Site Coefficient = S = 1.5
- Gravitational acceleration = g = 386 inches per sec^2

Assumptions:
- Frame behaves as a single-degree-of-freedom system.

REQUIRED:
1. Calculate the moment of inertia I for the columns that corresponds to the given lateral stiffness.
2. Calculate the frame's natural period of vibration using UBC Formula (28-5).
3. Calculate the UBC seismic base shear.
4. Calculate the exact (theoretical) undamped natural circular frequency and period of vibration.
5. Calculate the damping ratio.
6. From the response spectrum, determine the pseudo-velocity.
7. Calculate the corresponding pseudo-acceleration and the maximum relative displacement.
8. Calculate the resulting maximum base shear.

SOLUTION

1. COLUMN INERTIA

Since the frame stiffness is 148 kips per inch, the stiffness of each column is

$$\begin{aligned} k &= 148/2 \\ &= 74 \text{ kips per inch.} \end{aligned}$$

The moment of inertia of each column, which is considered fixed at both ends, is given by

$$\begin{aligned} I &= kL^3/12E \\ &= 74 \times 12^3 \times 1{,}728/12 \times 29{,}000 \\ &= 635 \text{ in}^4 \end{aligned}$$

2. PERIOD OF VIBRATION

The canopy may be considered a nonbuilding structure and, in accordance with UBC Section 1632.1.4, the fundamental period may be obtained by means of the Rayleigh method. From UBC Formula (28–5), the fundamental period for a multiple-degree-of-freedom system is

$$T = 2\pi(\Sigma w_i\delta_i^2/g\Sigma f_i\delta_i)^{1/2}$$

For a single-degree-of-freedom system of weight W with a lateral force F producing a lateral displacement δ this formula reduces to

$$\begin{aligned} T &= 2\pi(W\delta^2/gF\delta)^{1/2} \\ &= 2\pi(W\delta^2/gk\delta^2)^{1/2} \\ &= 0.32(W/k)^{1/2} \\ &= 0.32 \times (926/148)^{1/2} \\ &= 0.80 \text{ seconds} \end{aligned}$$

3. BASE SHEAR

For a nonbuilding structure, as defined in UBC Section 1632.5, the force coefficient is given by UBC Formula (28–2) as

$$C = 1.25S/T^{2/3}$$

where the site coefficient S is 1.5. Thus, the force coefficient is

$$\begin{aligned} C &= 1.25 \times 1.5/(0.8)^{2/3} \\ &= 2.18 \end{aligned}$$

which is less than 2.75 as required by UBC Section 1628.2.1 and is greater than $0.4R_W$ as required by UBC Section 1632.5.1. From UBC Formula (28–1) the base shear is

$$V = (ZIC/R_W)W$$

where Z = 0.3 for zone 3 from UBC Table 16–I

I = 1.25 for an essential facility as defined in UBC Table 16–K

C = 2.18 as calculated

R_W = 4 as given

and W = 926 kips as given

Then

$$\begin{aligned} V &= (0.3 \times 1.25 \times 2.18/4)926 \\ &= 0.204 \times 926 \\ &= 189.25 \text{ kips} \end{aligned}$$

4. UNDAMPED CIRCULAR FREQUENCY

The undamped circular frequency is given by[3]

$$\omega = (k/m)^{1/2}$$

where m = 926/386 is the mass of the structure.

Then

$$\begin{aligned} \omega &= (148 \times 386/926)^{1/2} \\ &= 7.85 \text{ radians per second} \end{aligned}$$

The period of vibration is

$$T = 2\pi/\omega$$
$$= 2\times 3.142/7.85$$
$$= 0.80 \text{ seconds}$$

5. DAMPING RATIO
The damping ratio is given by

$$\xi = c/2m\omega$$

where c = damping coefficient = 1.88 kips-seconds per inch, as given.
Then

$$\xi = 1.88 \times 386/(2 \times 926 \times 7.85)$$
$$= 0.05$$
$$= 5\%$$

6. PSEUDO-VELOCITY
From the response spectra, for a period of 0.8 seconds and a damping ratio of 5%, the pseudo-velocity is obtained as

$$S_v = 0.5 \text{ feet per second}$$

7. PSEUDO-ACCELERATION
The pseudo-acceleration is given by

$$S_a = \omega S_v$$
$$= 7.85 \times 0.5$$
$$= 3.93 \text{ feet per second per second}$$

Assuming elastic behavior, the maximum relative displacement is given by

$$S_d = S_v/\omega$$
$$= 0.5/7.85$$
$$= 0.0637 \text{ feet}$$
$$= 0.764$$

8. BASE SHEAR
Assuming elastic behavior, the maximum base shear is given by

$$V = kS_d$$
$$= 148 \times 0.764$$
$$= 113 \text{ kips}$$

3.2.2 Modal analysis procedure

The stages necessary in the modal analysis procedure consist of selecting the appropriate ground motion response spectrum, applying a dynamic analysis technique to a mathematical model of the structure, combining the response of a sufficient number of modes to ensure a ninety percent participation of the mass of the structure, and scaling the results to ensure consistency with the static lateral force procedure.

The normalized design spectrum presented in the UBC and shown in Figure 1-20 may be used after scaling by the appropriate zone factor Z and the response modification factor R_W to provide the requisite response spectrum envelopes. Alternatively, site specific design spectra, as illustrated in Figure 1-6, may be utilized after scaling by the response modification factor R_W to obtain the input spectrum. The design spectrum should be smoothed to eliminate reduced response at specific periods and correspond to a ten percent probability of exceedence in fifty years and a damping ratio of five percent.

Two methods of dynamic analysis are given in UBC Section 1629.4. The response spectrum technique uses an appropriate response spectrum to calculate the peak modal response of all necessary modes. The time-history technique determines the structural response through numerical integration over short time increments for a site specific, time-dependent, seismic input motion which is representative of actual earthquake motions.

Since the modal maximums do not all occur simultaneously or act in the same direction, a statistical combination of these values is necessary. The square-root-of-the-sum-of-the-squares method[2,3] is acceptable for a two-dimensional analysis when the ratio of the period of any higher mode to any lower mode is 0.75 or less. The higher modes do not contribute significantly to the total response of the structure and an acceptable procedure, as specified in UBC Section 1629.5.1, is to utilize a sufficient number of modes to ensure that 90 percent of the participating mass of the structure is included. The total structure weight is given by

$$W = \Sigma W_i$$

where W_i = weight at floor level i

The effective weight for a given mode is defined by

$$\begin{aligned} W^E &= (\Sigma W_i\phi_i)^2/\Sigma W_i\phi_i^2 \\ &= P\Sigma W_i\phi_i \\ &= P^2\Sigma W_i\phi_i^2 \\ &= gV/S_a \end{aligned}$$

where ϕ_i = mode shape component for node point i for a given mode
S_a = spectral acceleration for the given mode
V = base shear for the given mode
P = participation factor for the given mode
$= \Sigma W_i\phi_i/\Sigma W_i\phi_i^2$

For normalized mode shapes, these expressions reduce to

$$\begin{aligned} P &= \Sigma W_i\phi_i/g \\ W^E &= gP^2 \\ &= (\Sigma W_i\phi_i)^2/g \end{aligned}$$

The relationship between effective weight and total structure weight is given by[2,3,4,5]

$$\Sigma W^E = W$$

where ΣW^E = sum of the effective weights for all modes.

This provides a method of satisfying the requirement that sufficient modes are included in the analysis to ensure that 90 percent of the structural mass participates in the derivation of the response parameters. Thus, sufficient modes may be defined to ensure that the sum of their effective weights is

$$\Sigma W^E \geq 0.9W$$

By this means, a minimum of 90 percent of the structural mass participates in the determination of the response parameters.

To ensure consistency with the basic design principles adopted in the static lateral force procedure, a minimum value is stipulated in UBC Section 1629.5.3 for the base shear derived by a dynamic analysis and all corresponding response parameters must be scaled accordingly. For a regular structure, the base shear determined by a dynamic analysis shall be increased to 90 percent of that obtained by the static lateral force procedure when the period is determined by method B of UBC Section 1628.2.2. However, the value adopted for the base shear shall not be less than 80 percent of that obtained when the period is determined by method A. For an irregular structure, the base shear adopted shall not be less than that obtained by the static lateral force procedure. The base shear derived by a dynamic analysis need not exceed the above values and, if required, the corresponding response parameters may be decreased accordingly. The displacements determined by using the scaled values of the base shear shall be increased, in accordance with UBC Sections 1631.2.4 and 1631.2.11, by the factor $(3R_W/8)$ to account for inelastic deformation.

Example 3-3 (Response spectrum analysis)
For the two story building of Example 1-3, determine the number of modes which must be combined to ensure that all significant modes have been included in the analysis and determine the design base shear and required scaling factors.

Solution
For design purposes, in accordance with UBC Section 1629.2.4, the site specific response curves of Figure 1-6, which were used in Example 1-3, must be scaled down by the response modification factor, R_W. For a special moment-resisting steel frame, this is obtained from UBC Table 16-N as

$$R_W = 12$$

Hence, the diagonal matrix of the design values of the spectral accelerations for the two periods, T_1 = 1.16 seconds and T_2 = 0.48 seconds, previously calculated is

$$[S_a] = \begin{bmatrix} 0.25 & 0.00 \\ 0.00 & 0.84 \end{bmatrix} g/12$$

The row vector of base shears for both modes is

$$[V] = [65.93 \quad 38.44]\ 1/12$$

The total structure weight is

$$W = (m_{11} + m_{22})g$$
$$= (0.4 + 0.4)386.4$$
$$= 309 \text{ kips}$$

The effective weight for each mode is given by

$$W^E = gV/S_a$$

In matrix notation, the row vector of effective weights is

$$[W^E] = [263 \quad 46]$$

and
$$\Sigma W^E = 263 + 46$$
$$= 309 \text{ kips}$$
$$= \text{total structure weight}$$

As a percentage of the total structural weight, the effective modal weights are

$$100[W^E]/W = [85\% \quad 15\%]$$

Hence, in order to obtain the effects from 90 percent of the total structure weight as required by UBC Section 1629.5.1, both modes must be included in the analysis. The resultant dynamic design value of the base shear is obtained from the value calculated in Example 1-3 as

$$V_D = 76.32/12$$
$$= 6.36 \text{ kips}$$

Applying the principles of the static lateral force procedure method A, to the structure, the fundamental period is given by UBC Formula 28-3 as

$$T_A = C_t(h_n)^{3/4}$$

where C_t = 0.035 for a steel moment-resisting frame from UBC Section 1628.2.1
h_n = roof height of 30 feet

Then, the fundamental period is

$$T_A = 0.035(30)^{3/4}$$
$$= 0.45 \text{ seconds}$$

The seismic force coefficient is given by UBC Formula (28-2) as

$$C_A = 1.25\ S/T_A^{2/3}$$

where S = 1.0 from UBC Table 16-J for a rock site

Then the force coefficient is

$$C_A = 1.25 \times 1.0/(0.45)^{2/3}$$
$$= 2.13$$

The value of the base shear is given by UBC Formula (28-1) as

$$V_A = (ZIC/R_W)W$$

where $Z = 0.4$ for zone 4 from UBC Table 16–I
$I = 1.0$ from UBC Table 16–K for a standard occupancy structure
$R_w = 12$ from UBC Table 16–N for a special moment-resisting steel frame

Then the seismic base shear is

$$V_A = (0.4 \times 1 \times 2.13/12)309$$
$$= 0.071 \times 309$$
$$= 21.94 \text{ kips}$$

Applying method B to the structure, the fundamental period is given by Example 1–3 as

$$T_1 = 1.16 \text{ seconds}$$

In accordance with UBC Section 1628.2.2.2, for a structure in zone 4, this is limited to a value of

$$T_B = 1.3T_A$$
$$= 1.3 \times 0.45$$
$$= 0.59 \text{ seconds}$$

The corresponding value of the seismic force coefficient is given by UBC Formula 28–2 as

$$C_B = 1.25 \times 1.0/(0.59)^{2/3}$$
$$= 1.79$$

The corresponding value of the base shear is given by UBC Formula (28–1) as

$$V_B = (0.4 \times 1 \times 1.79/12)309$$
$$= 0.060 \times 309$$
$$= 18.41 \text{ kips}$$

The minimum allowable base shear for a regular structure, in accordance with UBC Section 1629.5.3, is given by

$$V = 0.9V_B$$
$$= 16.57 \text{ kips}$$

or

$$V = 0.8V_A$$
$$= 17.55 \text{ kips ... controls}$$

Hence, the scaling factor for the dynamic analysis results is

$$r = 0.8V_A/V_D$$
$$= 17.55/6.36$$
$$= 2.8$$

Example 3–4 (Structural Engineering Examination 1991, Section A, weight 7.0 points)

GIVEN: A five-story special moment-resisting steel frame building. The building mass distribution and story heights are shown in Figure 3–5. The building fundamental mode shape has been determined by computer analysis and is shown. The fundamental period has been calculated by rational analysis to be $T = 0.90$ seconds. Site specific response spectra data are given in Table 3–1.

Period	S_a	S_v	S_d	Freq.
Seconds	Inch/Sec/Sec	Inches/Sec	Inches	Rad/Sec
0.10	308.80	4.91	0.08	62.83
0.20	386.00	12.29	0.39	31.42
0.30	386.00	18.43	0.88	20.94
0.40	386.00	24.57	1.56	15.71
0.50	386.00	30.72	2.44	12.57
0.60	347.40	33.18	3.17	10.47
0.70	301.10	33.55	3.74	8.98
0.80	258.60	32.93	4.19	7.85
0.90	231.60	33.18	4.75	6.98
1.00	208.40	33.17	5.28	6.28
1.50	154.40	36.86	8.80	4.19
2.00	115.80	36.86	11.73	3.14
2.50	96.50	38.40	15.28	2.51

Table 3–1 Site specific response spectra data

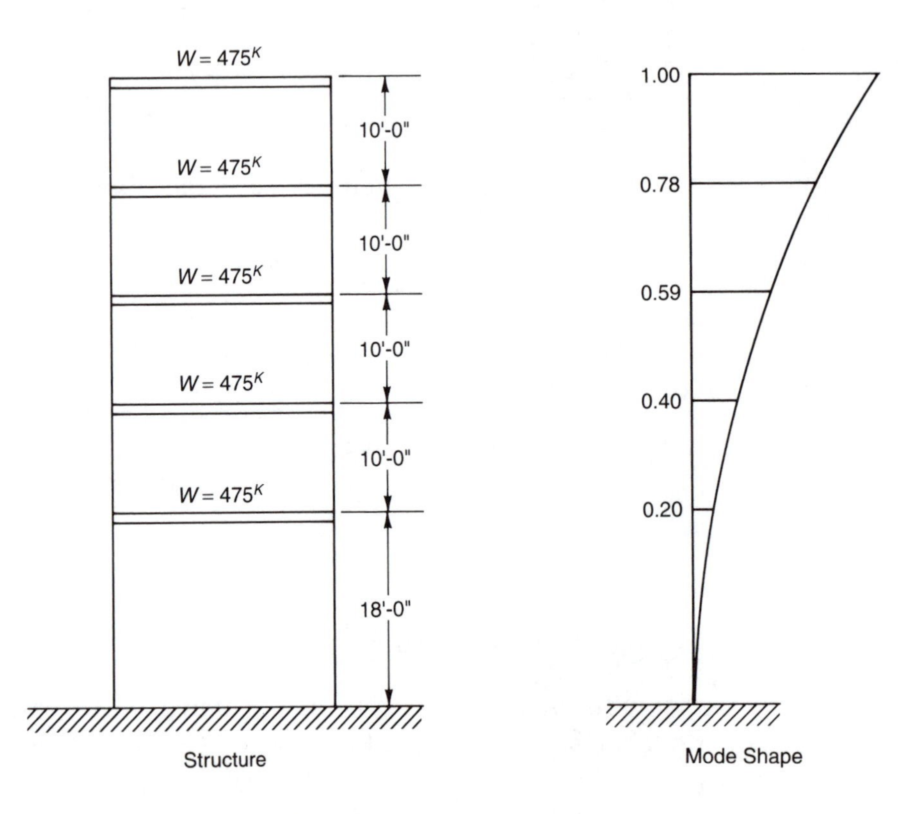

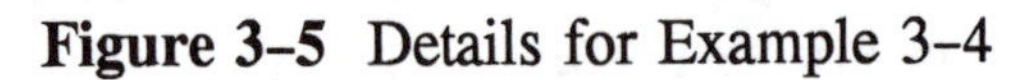

Figure 3–5 Details for Example 3–4

CRITERIA: Seismic zone 4
Soil type 2
Importance factor 1.0

Assumptions:
- Only the fundamental (first) mode response need be considered for dynamic analysis.

REQUIRED:
1. Using the given mass distribution, mode shape, building period, and site specific response spectrum data; determine the total shear at the base of the building, using dynamic lateral force procedures.
2. Using the given data, determine the design lateral force (shear at the base) using the static force procedure.
3. Using the given site specific response spectrum, determine the minimum design story shear distribution in accordance with UBC Section 1629.4 of the code.
4. Using the response spectrum data, determine the expected displacement at the top of the building and the story drift ratios.

SOLUTION

1. DYNAMIC PROCEDURES

For a moment-resisting steel frame, the response modification factor is obtained from UBC Table 16-N as

$$R_W = 12$$

For the given fundamental period of 0.90 seconds, the site specific spectral acceleration is obtained from the given data as

$$S_a = 231.6 \text{ inches per sec}^2$$

The effective modal weight is given by

$$W^E = (\Sigma W_i\phi_i)^2/\Sigma W_i\phi_i^2$$

where W_i = weight at floor level i
ϕ_i = mode shape component at floor level i

From Table 3-2 the effective modal weight is obtained as

$$W^E = 1411^2/1{,}024$$
$$= 1944$$

Assuming elastic behavior the total base shear is

$$V_D = W^E S_a$$
$$= 1944 \times 231.6/386.4$$
$$= 1165 \text{ kips}$$

The design base shear is given by UBC Section 1629.2.4 as

$$V_u = V_D/R_W = 1165/12 = 97.11 \text{ kips}$$

Level	W_i	ϕ_i	$W_i\phi_i$	$W_i\phi_i^2$	F_i	V_i
Roof	475	1.00	475	475	39.25	39.25
Floor 5	475	0.78	371	289	30.66	69.91
Floor 4	475	0.59	280	165	23.14	93.05
Floor 3	475	0.40	190	76	15.70	108.75
Floor 2	475	0.20	95	19	7.85	116.60
Total	2375	-	1411	1024	116.60	-

Table 3–2 Effective modal weight

2. STATIC PROCEDURE

In the static force procedure, UBC Section 1628.2 allows two procedures for calculating the value of the fundamental period and each of the values may be used to determine the base shear.

METHOD A

The fundamental period is given by UBC Formula (16–3) as

$T_A = C_t(h_n)^{3/4}$

where C_t = 0.035 for a steel moment-resisting frame

and h_n = the roof height of fifty eight feet

Then $T_A = 0.035(58)^{3/4}$

$= 0.7356$ seconds

The force coefficient is given by UBC Formula (16–2) as

$C_A = 1.25\ S/T_A^{2/3}$

where S = 1.2 from UBC Table 16–J, for soil type 2

Hence $C = 1.25 \times 1.2/(0.7356)^{2/3}$

$= 1.84$

< 2.75

and $C/R_W = 1.84/12$

$= 0.153$

> 0.075

Hence C = 1.84 is an acceptable value for the force coefficient.

The base shear is given by UBC Formula (16–1) as

$V_A = (ZIC/R_W)W$

where Z = 0.4 for zone 4 from UBC Table 16–I

I = 1.0 as given

W = total dead load of 2375 kips

Then $V_A = (0.4 \times 1 \times 1.84/12)\ 2375$
$= 0.0613 \times 2375$
$= 145.7$ kips

METHOD B

In accordance with UBC Section 1628.2.2, for a structure in zone 4, the fundamental period T_B obtained by method B is limited to a value of

$T_B = 1.3T_A$
$= 1.3 \times 0.7356$
$= 0.956$ seconds
> 0.9 seconds

Hence $T_B = 0.9$ seconds is acceptable

Using the given value 0.9 seconds, the lateral force coefficient is given by UBC Formula (16–2) as

$C = 1.25S/T_B^{2/3}$
$= 1.25 \times 1.2/(0.9)^{2/3}$
$= 1.61$

The base shear is given by UBC Formula (16–1) as

$V_B = (ZIC/R_W)W$
$= (0.4 \times 1 \times 1.61/12)\ 2{,}375$
$= 0.0537 \times 2{,}375$
$= 127.5$ kips

3. STORY SHEAR DISTRIBUTION

For a regular structure, UBC Section 1629.5.3 requires that the design base shear, based on a dynamic analysis procedure, be scaled up to the greater value given by

$V = 0.90V_B$
$= 0.90 \times 127.5$
$= 114.7$ kips

or $V = 0.80V_A$
$= 0.80 \times 145.7$
$= 116.6$ kips, which governs.

Then, the design lateral force at each level is given by

$F_i = VW_i\phi_i/\Sigma W_i\phi_i$
$= 116.6W_i\phi_i/1{,}411$
$= 0.0826W_i\phi_i$

The design force at each level and the shear V_i at each story is shown in Table 3–2.

4. EXPECTED DISPLACEMENTS

For the given fundamental period of 0.90 seconds, the site specific spectral displacement is

S_d = 4.75 inches

Assuming elastic behavior, the displacement at each level is given by

$x_i = P\phi_i S_d$

where P = participation factor

$= W^E/\Sigma W_i\phi_i$

$= 1944/1411$

$= 1.38$

Then $x_i = 1.38 \times 4.75\phi_i$

$= 6.56\phi_i$

The drift at a given story is defined as the relative displacement of the upper and lower floor levels at that story and is given by

$\Delta_i = x_i - x_{(i-1)}$

The drift ratio at a given story is defined as the ratio of the story drift to the height of that story and is given by

$R_i = \Delta_i/h_i$

These values are shown in Table 3–3.

Level	Story	h_i	ϕ_i	x_i	Δ_i	R_i%
Roof	–	–	1.00	6.56	–	–
Floor 5	5	10	0.78	5.12	1.44	1.20
Floor 4	4	10	0.59	3.87	1.25	1.04
Floor 3	3	10	0.40	2.62	1.25	1.04
Floor 2	2	10	0.20	1.31	1.31	1.09
Floor 1	1	18	–	–	1.31	0.61

Table 3–3 Story displacements

The design displacements are given by UBC Section 1631.2.4 as

$x_u = x_i \times (3R_W/8) \times 1/R_W$

$= 0.375x_i$

Similarly, the design drifts are given by

$\Delta_u = 0.375\Delta_i$

Example 3–5 (Structural Engineering Examination 1989, Section A, weight 7.0 points)

GIVEN: A three-story office building, using ductile moment-resisting frames, symmetrical in both directions as shown in Figure 3–6.

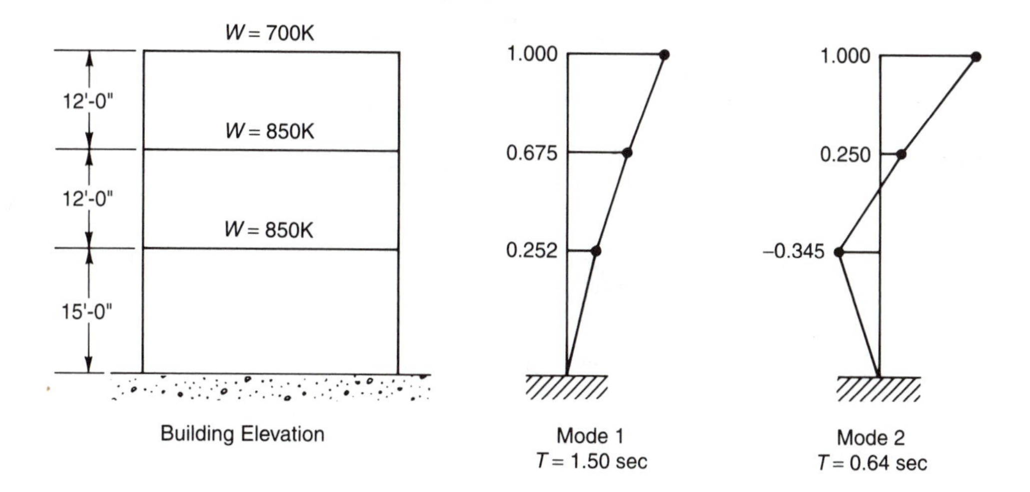

Figure 3–6 Details for Example 3–5

CRITERIA: Seismic zone 4.

REQUIRED: 1. (STATIC METHOD) Calculate the following using the relevant sections of the UBC:

a. Find the seismic base shear using S = 1.5.
b. Distribute the seismic base shear over the height of the building.
c. Calculate the diaphragm force for each level.

2. (DYNAMIC METHOD) Calculate the following using the two mode shapes given in Figure 3–6:

a. The effective weight in each mode.
b. The story force and shear in each mode.
c. The combined story force using the square-root-of-the-sum-of-the-squares method.

Useful equations:

For $T \leq 0.65$ seconds : S_a = Spectral Acceleration = 0.3g
For $T > 0.65$ seconds : S_v = Spectral velocity = 12 inches per second
W^E = Effective Weight = $(\Sigma W_i \phi_i)^2 / \Sigma W_i \phi_i^2$
V = Total shear = $S_a W^E$
V_i = Story Shears = $(W_i \phi_i / \Sigma W_i \phi_i) V$
S_a = $S_v / 61.5T$

SOLUTION

1a. SEISMIC BASE SHEAR

From UBC Section 1628.1, the structure period is given by UBC Formula (28–3) as

	T	$= C_t(h_n)^{3/4}$
where	C_t	= 0.035 for a steel moment-resisting frame
and	h_n	= the roof height of thirty nine feet
Then	T	$= 0.035(39)^{3/4}$
		= 0.546 seconds

The force coefficient is given by UBC Formula (28–2) as

	C	$= 1.25S/T^{2/3}$
where	S	= 1.5 as given
Then	C	$= 1.25 \times 1.5/(0.546)^{2/3}$
		= 2.81
		> 2.75 hence use the upper limit of
	C	= 2.75

The base shear is given by UBC Formula (28–1) as

	V	$= (ZIC/R_w)W$
where	Z	= 0.4 for zone 4 from UBC Table 16–I
	I	= 1.0 from UBC Table 16–K for a standard occupancy structure
	C	= 2.75 as calculated
	R_W	= 12 from UBC Table 16–N for a steel moment-resisting space frame
	W	= total dead load
		= 700 + 850 + 850
		= 2400 kips
Then	V	$= (0.4 \times 1 \times 2.75/12)2400$
		$= 0.092 \times 2400$
		= 220 kips

1b. BASE SHEAR DISTRIBUTION

The base shear is distributed over the height of the structure in accordance with UBC Formula (28–8)

	F_x	$= (V - F_t)w_xh_x/\Sigma w_ih_i$
where	F_x	= the lateral force at level x
	V	= the base shear of 220 kips, as calculated
	F_t	= the additional concentrated force at roof level
		= 0, as T is less than 0.7 seconds
	w_x	= the structure dead load at level x
	h_x	= the height in feet above the base to level x
	Σw_ih_i	$=$ the sum of the product $w \times h$ at all levels
		= 63,000 kips feet, from Table 3–4
Then	F_x	$= Vw_xh_x/\Sigma w_ih_i$
		$= 220w_xh_x/63{,}000$
		$= 0.00349w_xh_x$

and the force at each level, and the shear at each story V_x is shown in Table 3–4.

Level	w_x	h_x	$w_x h_x$	F_x	V_x
Roof	700	39	27,300	95.33	-
3rd Floor	850	27	22,950	80.14	95.33
2nd Floor	850	15	12,750	44.53	175.47
1st Floor	-	-	-	-	220.00
Total	2400	-	63,000	220.00	-

Table 3–4 Vertical force distribution

1c. DIAPHRAGM FORCE

The diaphragm force[6] at level x is given by UBC Formula (31–1) as

$$F_{px} = w_{px} \Sigma F_i / \Sigma w_i$$

where F_{px} = the diaphragm force at level x
w_{px} = weight of the diaphragm and tributary elements at level x
ΣF_i = the sum of the lateral forces at level x and above
Σw_i = the sum of the structure dead load w_x at level x and above

and

$$F_{px} \leq 0.75 Z I w_{px} = 0.30 w_{px}$$

$$F_{px} \geq 0.35 Z I w_{px} = 0.14 w_{px}$$

Assuming that the weights of the walls are negligible compared with the diaphragm weight at each level, the weight tributary to each diaphragm is identical with the structure dead load at that level and, hence

$$w_{px} = w_x$$

The diaphragm force at each level is shown in Table 3–5.

Level	F_x	ΣF_i	w_x	Σw_i	$\Sigma F_i / \Sigma w_i$	w_{px}	F_{px}
Roof	95.33	95.33	700	700	0.14 Minimum	700	98
3rd Floor	80.14	175.47	850	1550	0.14 Minimum	850	119
2nd Floor	44.53	220.00	850	2400	0.14 Minimum	850	119

Table 3–5 Diaphragm force

2a and 2b. DYNAMIC METHOD

The base shear and story forces may be obtained by the method of response spectrum analysis given in SEAOC Appendix 1F or in other reference texts[4,5] For a two-dimensional structure, the total number of node points equals the number of stories. Each node is located at a floor level and has one degree of freedom in the horizontal direction.

FIRST MODE

The natural period of the first mode is given as

$$T = 1.5 \text{ seconds}$$
$$> 0.65 \text{ seconds}$$

Hence, from the problem statement, the spectral velocity is

$$S_v = 12 \text{ inches per second}$$

and the corresponding spectral acceleration is

$$S_a = 2\pi S_v/T$$
$$= 2\pi \times 12/1.5$$
$$= 50.27 \text{ inches per second}^2$$
$$= 0.13g$$

From Table 3-6, where the weight at each floor level is used for convenience, the effective weight is given by

$$W^E = (\Sigma W_i\phi_i)^2/\Sigma W_i\phi_i^2$$
$$= 1488^2/1,141$$
$$= 1940 \text{ kips}$$

The base shear is given by

$$V = W^E S_a/g$$
$$= 1940 \times 0.13g/g$$
$$= 252 \text{ kips}$$

The lateral force at each node is given by

$$F_i = (W_i\phi_i/\Sigma W_i\phi_i)V$$
$$= (W_i\phi_i/1488)252$$
$$= 0.169 W_i\phi_i$$

and the force at each level and the shear at each story V_i is shown in Table 3-6

Level	W_i	ϕ_i	$W_i\phi_i$	$W_i\phi_i^2$	F_i	V_i
Roof	700	1.000	700	700	118.6	–
3rd Floor	850	0.675	574	387	97.2	118.6
2nd Floor	850	0.252	214	54	36.2	215.8
1st Floor	–	–	–	–	–	252.0
Total	–	–	1488	1141	252.0	–

Table 3-6 Story shear force: first mode

SECOND MODE

The natural period of the second mode is given as

$$T = 0.64 \text{ seconds}$$
$$< 0.65 \text{ seconds}$$

Hence, from the problem statement, the spectral acceleration is

$$S_a = 0.3g$$

From Table 3-7 the effective weight is given by

$$W^E = 619^2/854$$
$$= 449 \text{ kips}$$

The base shear is given by

$$V = 449 \times 0.3g/g$$
$$= 134.7 \text{ kips}$$

The lateral force at each node is given by

$$F_i = (W_i\phi_i/619)134.7$$
$$= 0.2176W_i\phi_i$$

and the force at each level and the shear at each story V_i is shown in Table 3-7.

Level	W_i	ϕ_i	$W_i\phi_i$	$W_i\phi_i^2$	F_i	V_i
Roof	700	1.000	700	700	152.3	-
3rd Floor	850	0.250	212	53	46.1	152.3
2nd Floor	850	-0.345	-293	101	-63.7	198.4
1st Floor	-	-	-	-	-	134.7
Total	-	-	619	854	134.7	-

Table 3-7 Story shear force: second mode

2c. COMBINED STORY FORCE

As a percentage of the total structural weight, the sum of the effective weights for the first two modes is given by

$$100(W_1^E + W_2^E)/W = 100(1940 + 449)/2400$$
$$= 99.5 \text{ percent}$$
$$> 90 \text{ percent}$$

Hence, combining the first two modes ensures that a minimum of 90 percent of the structural mass participates in the determination of the response parameters and UBC Section 1629.5.1 is satisfied. The combined force at each level for the two modes may be obtained by using the square-root-of-the-sum-of-the-squares method as detailed in SEAOC commentary Section 1F.5.b. This is acceptable for two-dimensional structures when the ratio of the periods of any higher mode to any lower mode is 0.75 or less. The combined force at level i is given by

$F_{ci} = (F_{1i}^2 + F_{2i}^2)^{1/2}$

where F_{1i} = lateral force at level i for the first mode

F_{2i} = lateral force at level i for the second mode

The combined force at each level is shown in Table 3–8.

Level	F_{1i}	F_{2i}	F_{ci}
Roof	118.6	152.3	193
3rd Floor	97.2	46.1	108
2nd Floor	36.2	-63.7	73
Base	252.0	134.7	286

Table 3–8 Combined vertical force distribution

3.2.3 Dynamic lateral response procedure

The time–history analysis technique[2], which determines the system response from numerical integration over short time increments, must be used for all structures located on soil type S_4. Response spectra analysis may be employed for all other conditions.

References

1. Sabol, T.A. *Dynamic analysis or not.* Structural Engineers Association of Southern California design seminar. Los Angeles, CA, February 1992.

2. Structural Engineering Association of California. *Recommended lateral force requirements and commentary.* Sacramento, CA, 1990.

3. Chopra, A.K. *Dynamics of structures.* Earthquake Engineering Research Institute, Berkeley, CA, 1980.

4. Paz, M. *Structural dynamics.* Van Nostrand Reinhold, New York, 1991.

5. Corps of Engineers, *Seismic design guidelines for essential buildings. NAVFAC, Technical Manual P-355.1, Washington, D.C., 1986.*

6. American Plywood Association, *Diaphragms.* APA Design/Construction Guide, Tacoma, WA, 1989, pp. 9–23.

4

Seismic design of steel structures

4.1 GENERAL REQUIREMENTS

4.1.1 Member design

In accordance with UBC Section 1603.6 steel structures shall be designed for the following loading combination:

DL + Floor LL + Seismic

In areas where the snow load exceeds 30 pounds per square foot, the following combination is also applicable:

DL + Floor LL + Snow + Seismic

The snow load may be reduced by up to 75 percent when approved by the building official.

In seismic zones 3 and 4, UBC Section 2211.5 requires columns to be designed for the additional loading combinations:

$1.0P_{DL} + 0.7P_{LL} + 0.375R_WP_E$ ... compression

and $0.85P_{DL} \pm 0.375R_WP_E$ ... tension

where P_{DL} = axial dead load
P_{LL} = axial live load
P_E = axial load due to earthquake
R_W = response modification factor

For working stress design, UBC Section 1603.5 permits allowable stresses to be increased by one third when considering seismic loading.

4.1.2 Member strength

Member strength is defined in UBC Section 2211.4 as:

M_s = flexural strength
= ZF_y
V_s = shear strength
= $0.55F_y dt$
P_{sc} = axial compression
= $1.7F_a A$
P_{st} = axial tension
= $F_y A$

where
Z = plastic modulus of the section
F_y = yield stress of the steel
d = depth of the member
t = thickness of the member
F_a = axial design strength
A = cross-sectional area of the member.

The strength of connectors is further specified as:

$F_y A$ ... for full-penetration welds
1.7 allowable ... for partial penetration welds
1.7 allowable ... for fillet welds
1.7 allowable ... for bolts

These values are applicable to the strength requirements of UBC Sections 2211 and 2212.

Example 4–1 (Structural Engineering Examination 1989, Section B, weight 8.0 points)

GIVEN: A portal frame with a truss girder is shown in Figure 4–1. Applied loads are as follows:
P(DL + LL) = 10 kips
V(Seismic) = 20 kips

CRITERIA: Materials:
- A36 steel
- A325-SC bolts, ¾-inch diameter

Assumptions:
- Each column resists half of the applied lateral force.
- Truss girder is rigid for lateral deflection calculation.
- All chord members are laterally braced at panel points only.
- Effective length factors for the columns are K_x = 1.8, K_y = 1.0.

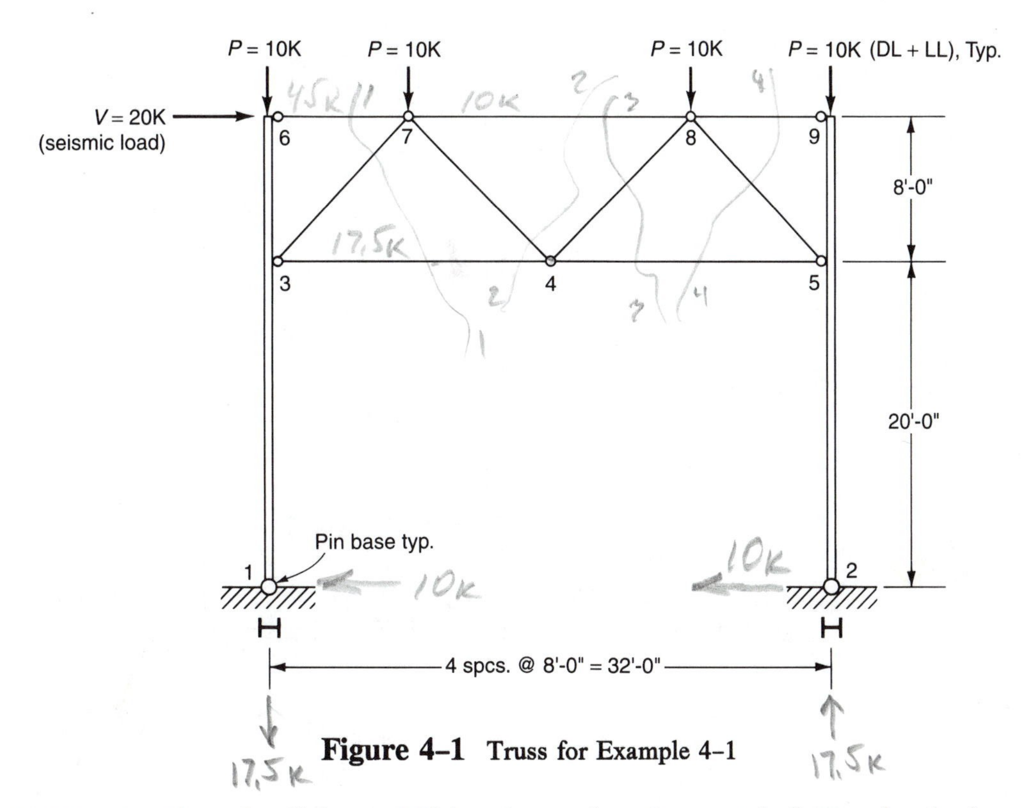

Figure 4–1 Truss for Example 4–1

REQUIRED: 1. Size the lightest W14 column for the portal frame for both stress and deflection, using dead, live and lateral loads. The allowable deflection shall be 0.005 times the story height. Show all stress calculations and draw the shear and moment diagrams for the column.

2. Size the lightest WT5 top chord for the truss girder.
3. Size the lightest 6" double channel bottom chord for the truss girder.
4. With only the forces given in Figure 4–2 compute the number of bolts required at the connection between the vertical double angle and the gusset plate. The forces are for combined dead, live and lateral loads.

SOLUTION

MEMBER FORCES

(i) DUE TO LATERAL LOAD

The horizontal reactions at the supports are given by

$$H_1 = H_2 = 20/2 = 10 \text{ kips}$$

The vertical reactions at the supports are obtained by taking moments about each support in turn and are given by

$$V_1 = V_2 = 20 \times 28/32 = 17.5 \text{ kips}$$

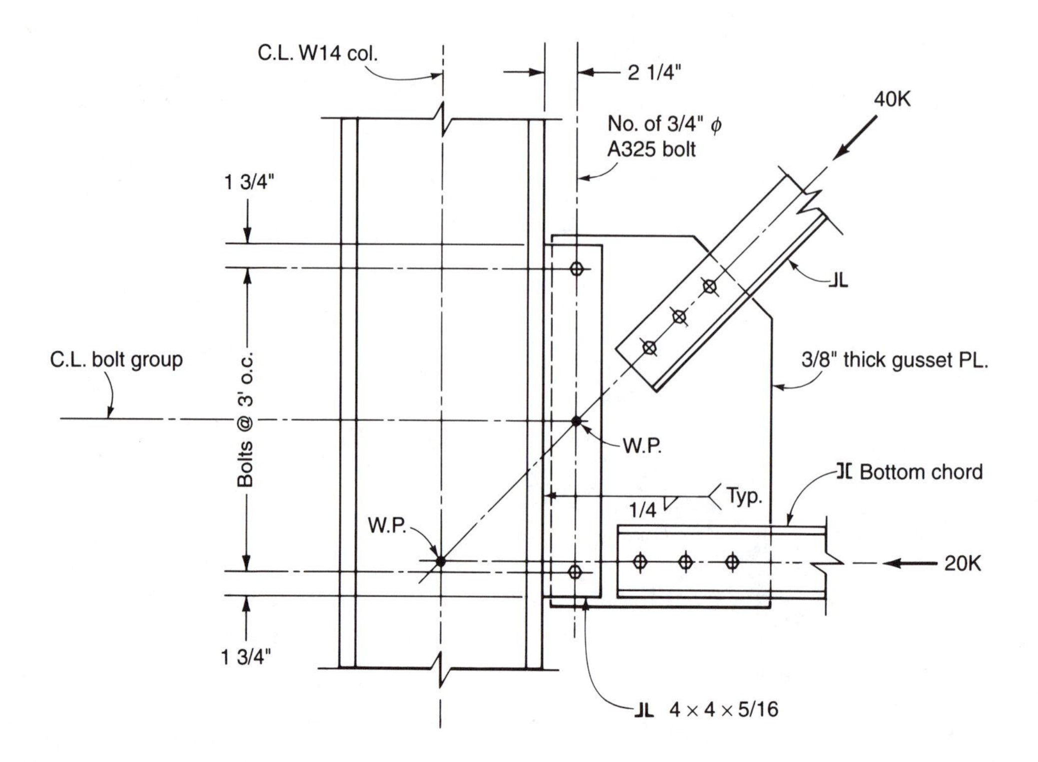

Figure 4–2 Gusset plate connection for Example 4–1

Applying the method of sections to panel 3–4–7–6, and taking moments about joint 3, gives the force in the top chord member 67 as

$$P_{67} = (8V + 20H_1)/8$$
$$= 45 \text{ kips compression}$$

Taking moments about joint 7 gives the force in the bottom chord member 34 as

$$P_{34} = (28H_1 - 8V_1)/8$$
$$= 17.5 \text{ kips tension}$$

Applying the method of sections to panel 3–4–8–7, and taking moments about joint 4, gives the force in the top chord member 78 as

$$P_{78} = (8V + 20H_1 - 16V_1)/8$$
$$= 10 \text{ kips compression}$$

Similarly $P_{89} = 25$ kips tension

The forces acting on column 2–5–9, and the corresponding shear force and bending moment diagrams, are shown in Figure 4–3. The maximum moment occurs at joint 5 and is given by

$$M_5 = 20H_2$$
$$= 20 \times 10$$
$$= 200 \text{ kip feet}$$

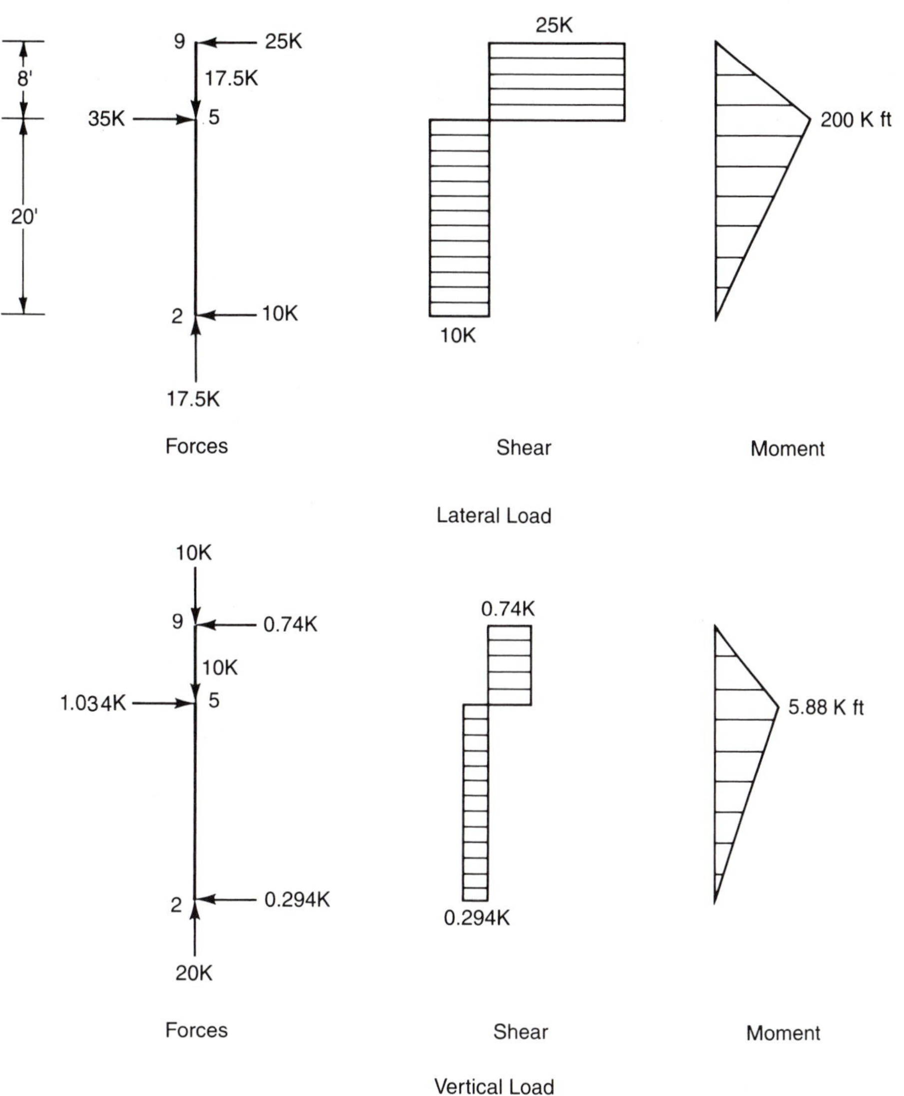

Figure 4-3 Force diagrams for Example 4-1

(ii) DUE TO VERTICAL LOADS

The two-hinged frame is one degree redundant, and it is convenient to consider the horizontal reaction R at the hinges as the external redundant. The cut-back structure is produced by removing this redundant, and the applied loads and a unit virtual load replacing R are applied, in turn, to the cut-back structure, as shown in Figure 4-4. The redundant reaction is obtained from the virtual work method[1,2] as

$$R = -(\Sigma PuL/AE + \int Mmds/EI)/(\Sigma u^2L/AE + \int m^2ds/EI)$$

where P and u are the member forces in the cut-back structure due to the applied loads and unit virtual load, respectively, and M and m are the bending moments at any point in a member of the cut-back structure due to the applied loads and the unit virtual load, respectively. Since M is zero for all members

$$\int Mmds/EI = 0$$

The summations are obtained in Table 4–1 from the relevant values of P and u and by using an initial estimate of the cross sectional areas of the members. Only the top chord and bottom chord members need be included in the Table since the force u is zero for all other members. Tensile forces are considered positive.

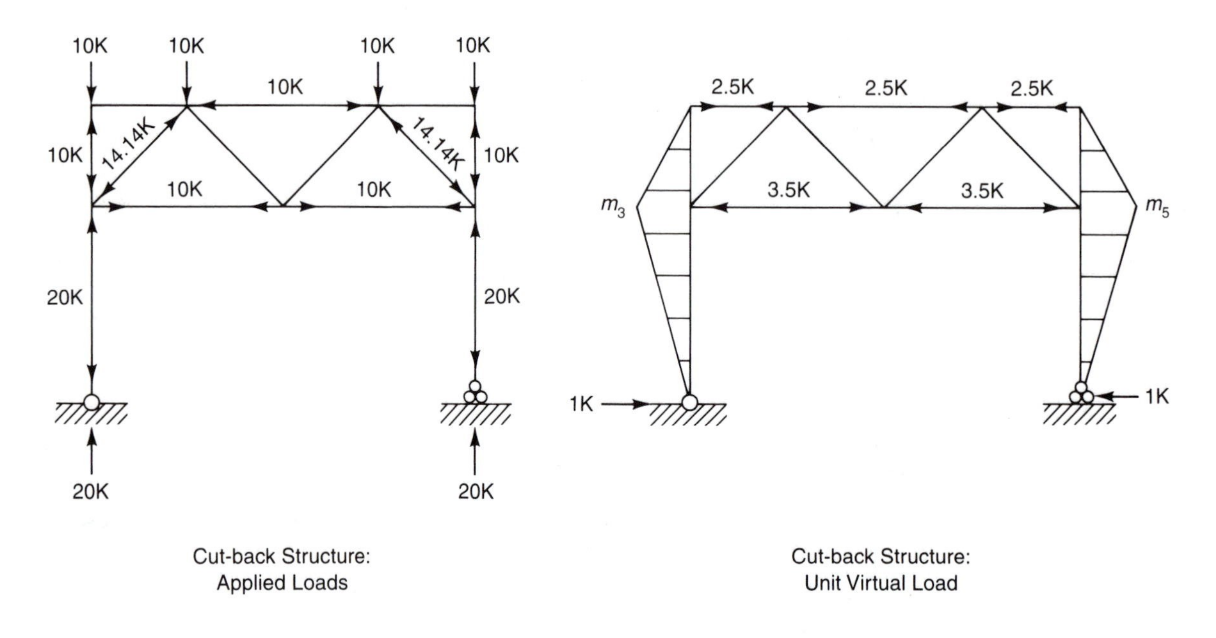

Figure 4–4 Forces in cut-back structure

The allowable lateral displacement at the top of the frame is

	δ	$= 0.005H$
where	H	= story height of 28 feet
then,	δ	$= 0.005 \times 28 \times 12$
		= 1.68 inches

Since the truss girder is assumed to be rigid for lateral displacement, no rotation occurs at the top of a column which acts as a vertical cantilever as shown in Figure 4–5. Then the minimum allowable moment of inertia for each column, so as not to exceed the allowable displacement, is given by

	I	$= WL^3/3E\delta$
where	I	= required moment of inertia of one column
	W	= applied lateral force of 10 kips on one column
	L	= column height of 240 inches
Then	I	$= 10 \times (240)^3/(3 \times 29{,}000 \times 1.68)$
		$= 946 \text{ inches}^4$

The lightest W14 column, which satisfies this requirement, is a W14 × 90 which has a moment of inertia of 999 inches^4.

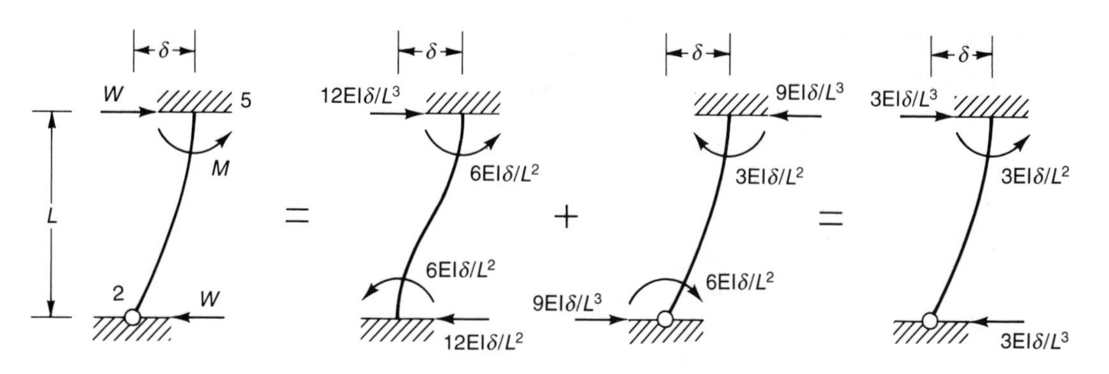

Figure 4–5 Column forces

The term $\int m^2ds/EI$ is applicable to the columns only with

$$m_3 = m_5 = 1 \times 20 \times 12$$
$$= 240 \text{ kip inches}$$

and $$I = 999 \text{ in}^4\text{, as calculated}$$

Using the volume integration technique,

$$\int m^2ds/EI = 2 \times (1/3)m_3^2(L_{13} + L_{36})/EI$$
$$= 2 \times (1/3) \times 240^2(20 + 8)/999E$$
$$= 1076/E$$

Then the redundant reaction is given by

$$R = 359/(146 + 1076)$$
$$= 0.294 \text{ kips}$$

Member	P	u	L	A	PuL/A	u^2L/A	P+uR
34	10	-3.5	16	4.8	-117	41	9.00
45	10	-3.5	16	4.8	-117	41	9.00
67	0	2.5	8	3.2	0	16	0.74
78	-10	2.5	16	3.2	-125	32	-9.30
89	0	2.5	8	3.2	0	16	0.74
Total	-	-	-	-	-359	146	-

Table 4–1 Determination of forces in Example 4–1

The final forces in the actual structure are obtained from the expression (P + uR) and are shown in the Table. The forces acting on column 2–5–9, and the corresponding shear force and bending moment diagrams, are shown in Figure 4–3. The maximum moment occurs at joint 5 and is given by

$$M_5 = 20R$$
$$= 20 \times 0.294$$
$$= 5.88 \text{ kip feet}$$

1. COLUMN MEMBER 2-5-9

The properties of a W14 × 90 column are:

$$
\begin{aligned}
F_y &= 36 \text{ kips per square inch} \\
F_y' &= 40.4 \text{ kips per square inch} \\
I_x &= 999 \text{ inches}^4 \\
S_x &= 143 \text{ inches}^3 \\
A &= 26.5 \text{ inches}^2 \\
r_x &= 6.14 \text{ inches} \\
r_y &= 3.70 \text{ inches} \\
L_c &= 15.3 \text{ feet} \\
L_u &= 34.0 \text{ feet (for } C_b = 1)
\end{aligned}
$$

The design of the column is governed by the combination of vertical load plus lateral load and the column is subjected to combined axial compression and bending. The forces acting on the column are given by

$$
\begin{aligned}
P &= 17.5 + 20 \\
&= 37.5 \text{ kips} \\
M &= 200 + 5.88 \\
&= 205.88 \text{ kip feet}
\end{aligned}
$$

The compressive stress produced in the column flange by the bending moment is

$$
\begin{aligned}
f_b &= M/S_x \\
&= 205.88 \times 12/143 \\
&= 17.3 \text{ kips per square inch}
\end{aligned}
$$

The section is compact since

$$F_y' > F_y$$

and the allowable bending stress is dependent on the value of the maximum unbraced length of the compression flange which is assumed to be

$$l = 20 \text{ feet}$$

The bending coefficient is given by UBC Section 2251 F1.3 as

$$C_b = 1.75 + 1.05(M_1/M_2) + 0.3(M_1/M_2)^2$$

where M_1 = smallest bending moment in the column = 0

M_2 = largest bending moment in the column = 202.88 kip feet

Hence $C_b = 1.75$

> 1

and the actual value of L_u the maximum unbraced length of the compression flange at which the allowable bending stress may be taken as $0.6F_y$, is greater than 34 feet.

Hence $L_c < l < L_u$

and the allowable bending stress is given by

$$F_b = 0.6F_y = 21.6 \text{ kips per square inch}$$

The slenderness ratios about the Y–axis and the X–axis are given by

$$K_y\, l/r_y = 1 \times 20 \times 12/3.7 = 64.86$$

and

$$K_x\, l/r_x = 1.8 \times 20 \times 12/6.14 = 70.36 > 64.86$$

Hence, the slenderness ratio about the X–axis governs and the allowable compressive stress is obtained from AISC Table C–36 as

$$F_a = 16.39 \text{ kips per square inch}$$

The actual compressive stress in the column, due to dead load plus live load, is given by

$$f_a = P/A = 37.5/26.5 = 1.42 \text{ kips per square inch} < 0.15F_a$$

Hence in checking the adequacy of the column in combined axial compression and bending, in accordance with UBC Section 2251 H1, Formula (H1–3) is applicable. Formula (H1–3), after allowing for the one–third increase in permissible stresses when designing for seismic loads, is given by

$$f_a/F_a + f_b/F_b \leq 1 \times 1.33$$

The left hand side of the Formula is evaluated as

$$1.42/16.39 + 17.3/21.6 = 0.09 + 0.80 = 0.89 < 1.33$$

In addition, for seismic zones 3 and 4 and in accordance with UBC Section 2211.5, the column must satisfy the Formula

$$1.0P_{DL} + 0.7P_{LL} + 0.375R_W P_E \leq 1.7F_a A$$

Assuming the given vertical loads are equally due to dead load and live load, and obtaining the value of R_W from Table 16–N for a moment–resisting frame, the left hand side is evaluated as

$$1 \times 10 + 0.7 \times 10 + (3 \times 12/8)17.5 = 95.75$$

The right hand side is evaluated as

$$1.7 \times 16.39 \times 26.5 = 738 > 95.75$$

Hence, the W14 × 90 section is satisfactory.

2. TOP CHORD MEMBER 67

Member 67 is the most highly loaded top chord member with a compressive force, due to vertical load plus lateral load, of

$$P_{67} = 45 - 0.74$$
$$= 44.26 \text{ kips}$$

In accordance with UBC Section 1603.5, a one–third increase is allowed in permissible stresses when designing for seismic forces. The equivalent design force in compression, based on the basic allowable stress is, then

$$P_E = P_{67}/1.33$$
$$= 44.26/1.33$$
$$= 33.28 \text{ kips}$$

Member 67 is braced laterally at its ends and the effective length, with respect to the Y–axis is

$$Kl = 1 \times 8$$
$$= 8 \text{ feet}$$

The effective length, with respect to the X–axis, is specified by AISC Section C–C2 as equal to the actual length and is given by

$$Kl = 1 \times 8$$
$$= 8 \text{ feet}$$

From AISC Table 3–103, a WT 5 × 11 is the lightest suitable section and has an allowable axial load of

$$P_A = 50 \text{ kips}$$
$$> 33.28$$

and complies with UBC Section 2251 B7 since the Kl/r values about both axes are less than 200.

2. TOP CHORD MEMBER 78

Member 78 is more lightly loaded than member 67 but, as it has a greater effective length, must also be checked. The compressive force due to vertical load plus lateral load is

$$P_{78} = 10 + 9.3$$
$$= 19.3 \text{ kips}$$

and the equivalent design force is

$$P_E = P_{78}/1.33$$
$$= 14.5 \text{ kips}$$

The effective length of member 78 with respect to both the X and Y–axes is

$$Kl = 1 \times 16$$
$$= 16 \text{ feet}$$

From AISC Table 3–103, a WT 5 × 11 with an effective length of 16 feet has an allowable load of

$$P_A = 23 \text{ kips}$$
$$> 14.5$$

Hence, the WT 5 × 11 is adequate.

2. TOP CHORD MEMBER 89

The tensile force due to vertical load plus lateral load is

$$P_{89} = 25 + 0.74$$
$$= 25.74 \text{ kips}$$

and the equivalent design force is

$$P_E = P_{89}/1.33$$
$$= 19.4 \text{ kips}$$

By inspection, the WT 5 × 11 is adequate. Hence, the WT 5 × 11 is adequate for all top chord members.

3. BOTTOM CHORD MEMBER 34

Member 34 is subjected to a tensile force, due to vertical load plus lateral load, of

$$P_{34} = 17.5 + 9$$
$$= 26.5 \text{ kips tension}$$

The equivalent design force, allowing for the seismic load, is

$$P_E = P_{34}/1.33$$
$$= 26.5/1.33$$
$$= 19.9 \text{ kips tension}$$

The allowable tensile stress is governed by the gross area and is given by UBC Section 2251 D1 as

$$F_t = 0.6 F_y$$
$$= 21.6 \text{ kips per square inch}$$

The required sectional area is given by

$$A_g = P_E/F_t$$
$$= 19.9/21.6$$
$$= 0.92 \text{ square inches}$$

From AISC Table 1-40 a C6 × 8.2 channel is the lightest suitable section and provides a sectional area of

$$A = 2 \times 2.4$$
$$= 4.8 \text{ square inches}$$
$$> 0.92$$

Hence, a double C6 × 8.2 member is adequate for the bottom chord tensile forces. However, when the seismic force reverses, member 34 is subjected to a compressive force of

$$P_{34} = 17.5 - 9$$
$$= 8.5 \text{ kips compression}$$

The equivalent design force, allowing for the seismic load, is

$$P_E = P_{34}/1.33$$
$$= 8.5/1.33$$
$$= 6.4 \text{ kips compression}$$

The relevant properties of a single C6 × 8.2 channel are

$$A = 2.4 \text{ inches}^2$$
$$I_y = 0.693 \text{ inches}^4$$
$$\bar{x} = 0.511 \text{ inches}$$
$$r_y = 0.537 \text{ inches}$$
$$r_x = 2.34 \text{ inches}$$

The relevant properties of a double C6 × 8.2, back-to-back with a ⅜-inch gusset plate are obtained as

$$I_y = 2[0.693 + 2.4(0.511 + 0.375/2)^2]$$
$$= 3.73 \text{ inches}^4$$
$$r_y = [3.73/(2 \times 2.4)]^{0.5}$$
$$= 0.88 \text{ inches}$$
$$r_x = 2.34 \text{ inches}$$

The effective length of member 34 with respect to both the X and Y-axis is

$$Kl = 1 \times 16$$
$$= 16 \text{ feet}$$

The slenderness ratio about the Y-axis governs and is given by

$$Kl/r_y = 16 \times 12/0.88$$
$$= 218$$

This is less than the limit of 300 specified in UBC Section 2251 B7 for tension members, but exceeds the limit of 200 specified for compression members. However, the limit is exceeded by only 9 percent and the compressive stress in the member is low. Also, as specified in UBC Section 2251 B7, members which have been designed as tension members, but experience some compression loading need not satisfy the compression slenderness limit. The allowable compressive stress in member 34 is given by UBC Formula (E2-2) as

$$F_a = 12\pi^2E/23(Kl/r)^2$$
$$= 12 \times 3.14^2 \times 29{,}000/)(23 \times 218^2)$$
$$= 3.14 \text{ kips per square inch.}$$

The actual stress in member 34 due to the combined loading is

$$f_a = P_E/A$$
$$= 6.4/4.8$$
$$= 1.33 \text{ kips per square inch}$$
$$< 3.14$$

Hence, the double C6 × 8.2 section is satisfactory for the bottom chord members.

4. BOLTED GUSSET PLATE

Assuming a total of seven bolts, the applied loading on the bolt group is given by:

Vertical shear, V = 0.707×40
= 28.3 kips

Horizontal shear, H = $0.707 \times 40 + 20$
= 48.3 kips

Eccentricity moment, M = $20(14/2 + 2.25)$
= 185 kip inches

The geometrical properties of the bolt group are obtained by applying the unit area method. Then, the inertia of the bolt group about its centroid is

$$I = 2(3^2 + 6^2 + 9^2)$$
$$= 252 \text{ inches}^4$$

The modulus of an outer bolt is

$$S = 252/9$$
$$= 28 \text{ inches}^3$$

The co-existent forces on an outer bolt are:

Vertical force dues to vertical shear = V/7
= 28.3/7
= 4.04 kips

Horizontal force due to horizontal shear = H/7
= 48.3/7
= 6.90 kips

Horizontal force due to eccentricity moment = M/S
= 185/28
= 6.61 kips

Total horizontal force = 6.90 + 6.61
= 13.51 kips

The resultant force on an outer bolt is given by

$$R = (4.04^2 + 13.41^2)^{1/2}$$
$$= 14.10 \text{ kips}$$

Allowing for the one-third increase in permissible stress when designing for seismic loads, the allowable double shear value of a ¾-inch diameter A325-SC bolt is obtained from AISC Table 1-D as

$$P_s = 15 \times 1.33$$
$$= 20 \text{ kips}$$

This exceeds the applied force of 14.10 kips and is satisfactory.

Example 4–2 (Structural Engineering Examination 1988, Section A, weight 5.0 points)

GIVEN: The concrete roof slab is supported by 3 steel tube columns, each of TS 10 × 10 × ½ as shown in Figure 4–6.

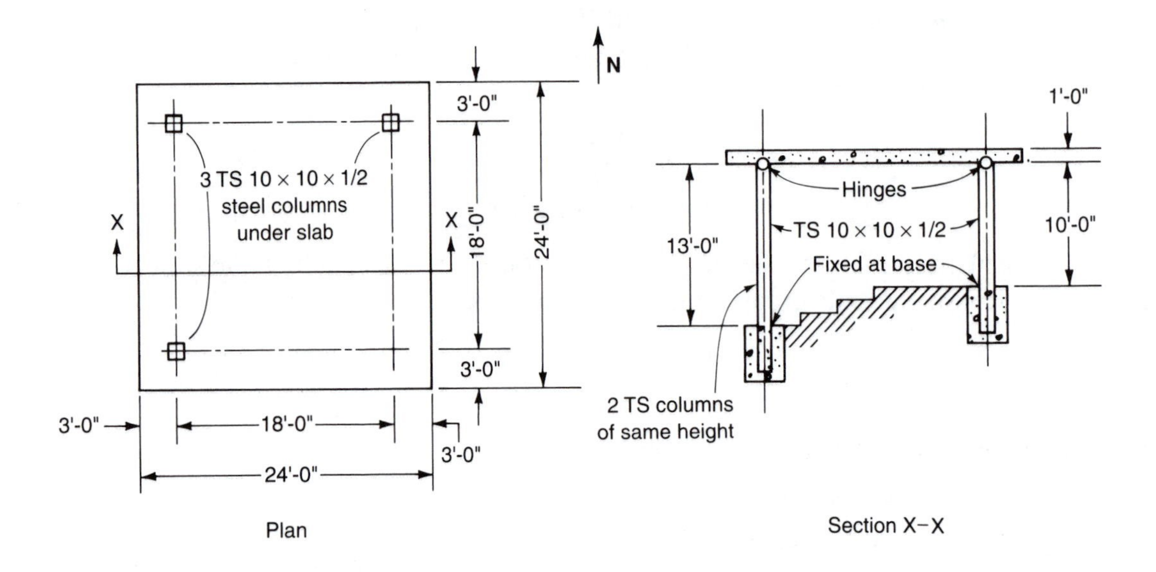

Figure 4–6 Details for Example 4–2

CRITERIA: Seismic zone 4, $R_W = 4$, $C = 2.75$
Steel tube column $F_y = 46$ kips per square inch
Wt. of concrete = 150 pounds per cubic foot
Assumptions:
Concrete slab is adequate.
Weight of steel columns may be neglected.

REQUIRED:
1. Determine the maximum reactions at the base of each column due to seismic forces in East - West direction only.
2. Determine the adequacy of all steel columns.

SOLUTION

DEAD LOAD

The dead load of the roof slab is

$$W = 0.15 \times 1 \times 24^2$$
$$= 86.4 \text{ kips}$$

BASE SHEAR

From UBC Section 1628.2 the base shear is given by Formula (28–1) as

$$V = (ZIC/R_W)W$$

where Z = 0.4 for zone 4 from Table 16–I
I = 1.0 from Table 16–K for a non-essential facility
C = 2.75 as given
R_W = 4 as given
and W = 86.4 as calculated
Then V = $(0.4 \times 1 \times 2.75/4)86.4$
= 0.275×86.4
= 23.76 kips

CENTER OF MASS

By inspection, the center of mass lies at the center of the concrete roof slab and its distance from column 2, as indicated in Figure 4–7, is

$\bar{y}$ = 9 feet

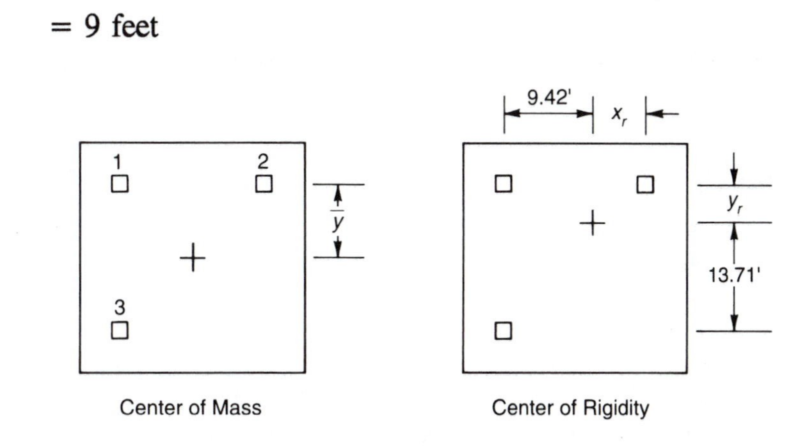

Figure 4–7 Torsional properties

COLUMN RIGIDITIES

The rigidity of each column is identical in the North–South direction and in the East–West direction and the rigidities are given by

$R_1 = R_3$ = $3EI/L^3$
= $3EI/13^3$
= $3EI/2197$
and R_2 = $3EI/L^3$
= $3EI/10^3$
= $3EI/1000$

The relative rigidities of the columns are

$R_1 = R_3$ = 2197/2197
= 1
and R_2 = 2197/1000
= 2.197

CENTER OF RIGIDITY

For seismic loads in the North–South direction, the distance of the center of rigidity from column 2 is obtained by taking moments about 2 and, as indicated in Figure 4–7, is given by

$$\begin{aligned} x_r &= (R_1 + R_3) \times 18/\Sigma R \\ &= (1 + 1) \times 18/(1 + 1 + 2.197) \\ &= 36/4.197 \\ &= 8.58 \text{ feet} \end{aligned}$$

For seismic loads in the East–West direction, the distance of the center of rigidity from column 2 is

$$\begin{aligned} y_r &= R_3 \times 18/\Sigma R \\ &= 1 \times 18/4.197 \\ &= 4.29 \text{ feet} \end{aligned}$$

The distances of column 3 from the center of rigidity are shown in Figure 4–7.

APPLIED TORSION

For seismic loads in the East–West direction, the eccentricity is

$$\begin{aligned} e_y &= \bar{y} - y_r \\ &= 9 - 4.29 \\ &= 4.71 \text{ feet} \end{aligned}$$

To comply with UBC, Section 1628.5, accidental eccentricity must be considered and this amounts to five percent of the roof slab dimension perpendicular to the direction of the seismic force. Neglecting torsional irregularity effects, the accidental eccentricity is given by

$$\begin{aligned} e_a &= \pm 0.05 \times 24 \\ &= \pm 1.2 \text{ feet} \end{aligned}$$

The net eccentricity is, then

$$\begin{aligned} e &= 4.71 \pm 1.2 \\ &= 5.91 \text{ or } 3.51 \text{ and the critical value is} \\ e &= 5.91 \text{ feet} \end{aligned}$$

The torsional moment acting about the center of rigidity is

$$\begin{aligned} T &= Ve \\ &= 23.76 \times 5.91 \\ &= 140.42 \text{ kip feet} \end{aligned}$$

FORCES DUE TO UNIT ROTATION

Imposing a unit clockwise rotation on the roof slab produces forces in the columns which are given by the expression

$$F = rR$$

and these forces are shown in Figure 4–8

MOMENT TO PRODUCE UNIT ROTATION
In order to produce a unit rotation of the roof slab, a moment must be applied to the center of rigidity and this is given by

$$\begin{aligned} M &= \Sigma rF \\ &= 18.84 \times 18 + 13.71 \times 18 \\ &= 586 \text{ kip feet} \end{aligned}$$

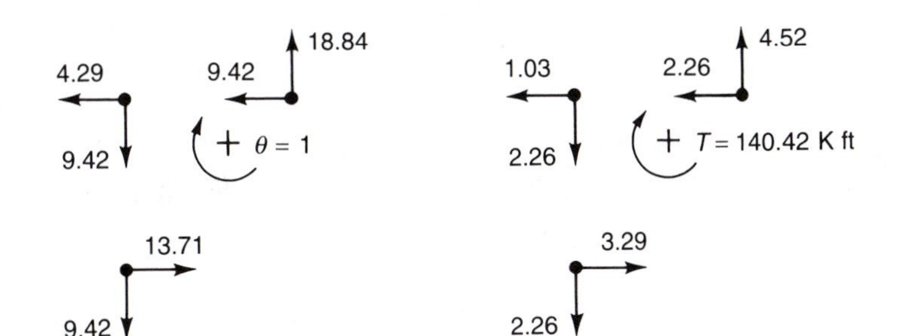

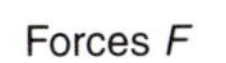

Forces F_T

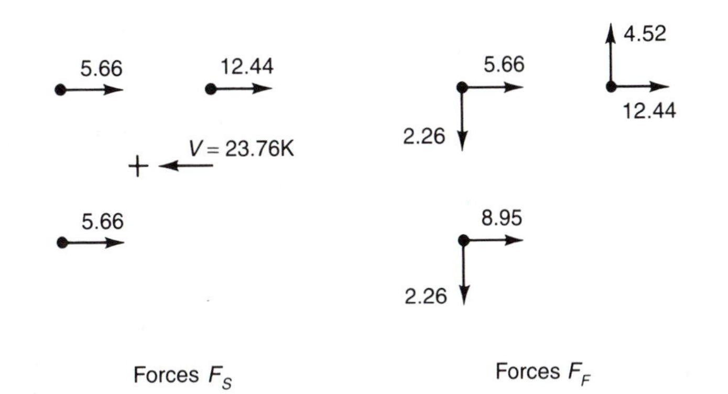

Figure 4–8 Column forces

FORCES DUE TO APPLIED TORSION
The torsion T, applied to the roof slab, produces forces in the columns which are given by the expression

$$\begin{aligned} F_T &= FT/M \\ &= F \times 140.42/586 \\ &= 0.24F \text{ kips, and these forces are shown in Figure 4–8} \end{aligned}$$

FORCES DUE TO LATERAL SEISMIC FORCE
For seismic loads in the East–West direction, the in–plane shear forces produced in the columns are given by

$$F_S = VR_x/\Sigma R_x$$

where R_x is the rigidity of a column in the East–West direction. Then,

$$F_S = 23.76R_x/(1 + 2.197 + 1)$$
$$= 23.76R_x/4.197$$
$$= 5.66R_x \text{ kips}$$

and the forces are shown in Figure 4-8

1.COLUMN REACTIONS DUE TO SEISMIC LOADS

The total shear forces produced in the columns are

$$F_F = F_S + F_T$$

and, to comply with UBC Section 1603.3, the torsional shear force, F_T, is neglected when of opposite sense to the in-plane shear force, F_S. The total shear forces are shown in Figure 4-8 and the X and Y components may be resolved as indicated in Figure 4-9.

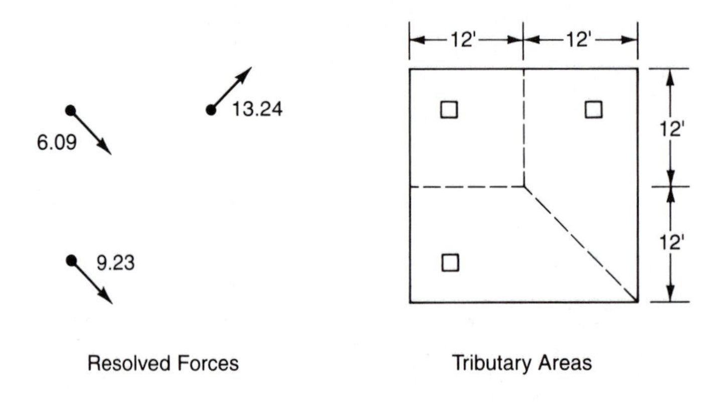

Figure 4-9 Details for Example 4-2

Because of the hinge at the top of each column, no moments are produced in the roof slab by seismic loads, and hence, no axial forces are produced in the columns. Moments are produced, by seismic loads, at the bottom of the columns. For column 1, the moments produced about the x-axis and the y-axis are

$$M_{1x} = 2.26 \times 13 \times 12$$
$$= 353 \text{ kip inches}$$
$$M_{1y} = 5.66 \times 13 \times 12$$
$$= 883 \text{ kip inches}$$

For column 2, the corresponding moments are

$$M_{2x} = 4.52 \times 10 \times 12$$
$$= 542 \text{ kip inches}$$
$$M_{2y} = 12.44 \times 10 \times 12$$
$$= 1493 \text{ kip inches}$$

For column 3, the corresponding moments are

$$M_{3x} = 2.26 \times 13 \times 12$$
$$= 353 \text{ kip inches}$$
$$M_{3y} = 8.95 \times 13 \times 12$$
$$= 1396 \text{ kip inches}$$

2. ADEQUACY OF THE COLUMNS

AXIAL LOADS

The column tributary areas are shown in Figure 4-9 and the column axial loads are given by

$$P_1 = W/4$$
$$= 86.4/4$$
$$= 21.6 \text{ kips}$$
$$P_2 = P_3 = 3W/8$$
$$= 32.4 \text{ kips}$$

COLUMN 3

The properties of the 10" x 10" x ½" structural tubing are obtained from AISC Table 1-94 as

$$A = 18.4 \text{ inches}^2$$
$$S = 54.2 \text{ inches}^3$$
$$r = 3.84 \text{ inches}$$

To establish the allowable flexural stress in the tubing, compliance with UBC Section 2251 B5.1 must be determined. The limiting width-thickness ratio for a compact section is

$$b/t \leq 190/(F_y)^{1/2}$$
$$\leq 190/(46)^{1/2}$$
$$\leq 28$$

The actual width-thickness ratio of the section is

$$b/t = 10/0.5$$
$$= 20$$

which is satisfactory. In addition, the laterally unsupported length is limited by UBC Section 2251 F3.1 to

$$l_b \leq 1200\, b/F_y$$
$$\leq 1200 \times 10/46$$
$$\leq 260 \text{ inches}$$

The actual unbraced length of the column is

$$l_b = 13 \times 12$$
$$= 156 \text{ inches}$$

which is satisfactory. Hence, the section is compact and the allowable bending stress is

$$F_b = 0.66F = 0.66 \times 46 = 30.4 \text{ kips per square inch}$$

The actual bending stresses in the column are

$$f_{bx} = M_{3x}/S = 353/54.2 = 6.50 \text{ kips per square inch}$$

and

$$f_{by} = M_{3y}/S = 1396/54.2 = 25.76 \text{ kips per square inch}$$

The effective length factor for a column fixed at its base and free to rotate and translate at its top is obtained from AISC Table C-C2.1 item (e) as

$$K = 2.1$$

and the effective slenderness ratio is

$$Kl/r = 2.1 \times 13 \times 12/3.84 = 85.3$$

The allowable compressive stress is then obtained from AISC Table C-50 as

$$F_a = 17.93 \text{ kips per square inch}$$

The actual compressive stress in the column is given by

$$f_a = P_3/A = 32.4/18.4 = 1.76 \text{ kips per square inch} < 0.15 \times F_a$$

Hence, in checking the adequacy of the column for combined axial compression and bending, in accordance with UBC Section 2251 H1, Formula (H1-3) is applicable and

$$f_a/F_a + f_{bx}/F_b + f_{by}/F_b = 1.76/17.93 + 6.5/30.4 + 25.76/30.4 = 0.098 + 0.214 + 0.847 = 1.16 < 1 \times 1.33$$

Hence, column 3 is adequate as the combined stresses do not exceed the one-third increase, allowed by UBC Section 1603.5, for seismic forces.

COLUMN 1

By inspection, column 1 is adequate

COLUMN 2

The actual bending stresses in column 2 are

$$f_{bx} = M_{2x}/S = 542/54.2 = 10.00 \text{ kips per square inch}$$

and

$$f_{by} = M_{2y}/S = 1493/54.2 = 27.55 \text{ kips per square inch}$$

The effective slenderness ratio for column 2 is

$$Kl/r = 2.1 \times 10 \times 12/3.84 = 65.6$$

and the allowable compressive stress is obtained from AISC Table C–50 as

$$F_a = 21.74 \text{ kips per square inch}$$

The actual compressive stress in the column is given by

$$f_a = P_2/A = 32.4/18.4 = 1.76 \text{ kips per square inch} < 0.15 \times F_a$$

Hence Formula (H1–3) is again applicable and

$$f_a/F_a + f_{bx}/F_b + f_{by}/F_b = 1.76/21.74 + 10/30.4 + 27.55/30.4 = 0.081 + 0.329 + 0.906 = 1.32 < 1 \times 1.33$$

Hence column 2 is adequate

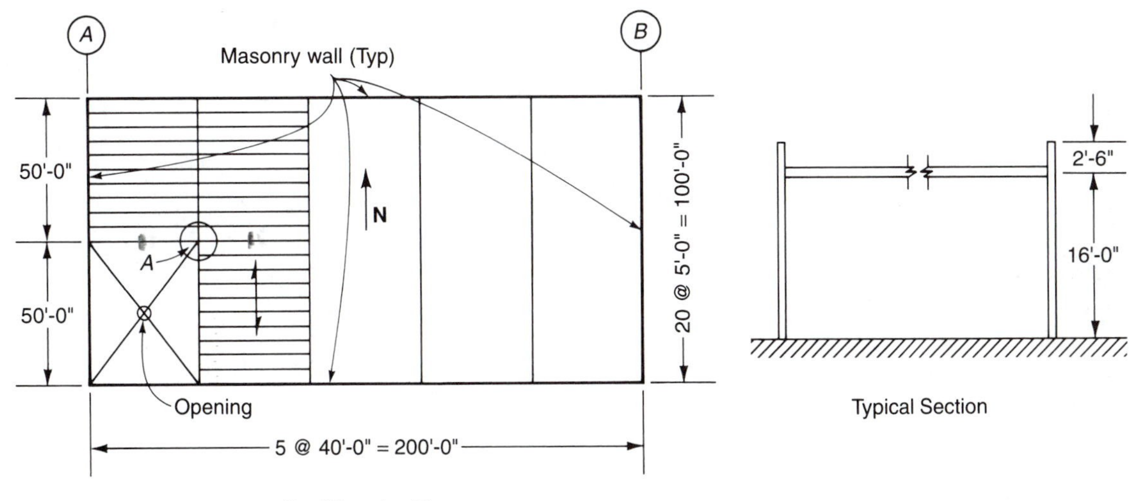

Figure 4–10 Building for Example 4–3

Example 4-3 (Structural Engineering Examination 1987, Section B, weight 3.0 points)

GIVEN: Figure 4–10 shows the roof framing plan of a one-story building with a metal deck roof and steel roof framing. All exterior walls are masonry shear walls. The building is located in zone 4.

CRITERIA: Roof DL = 20 pounds per square foot
Roof LL = 20 pounds per square foot
Wall Wt = 80 pounds per square foot
A36 steel, F_u = 58 kips per square inch
1½" metal deck diaphragm
E70XX Electrode

REQUIRED: Check the adequacy of the connection detail shown in Figure 4–11.

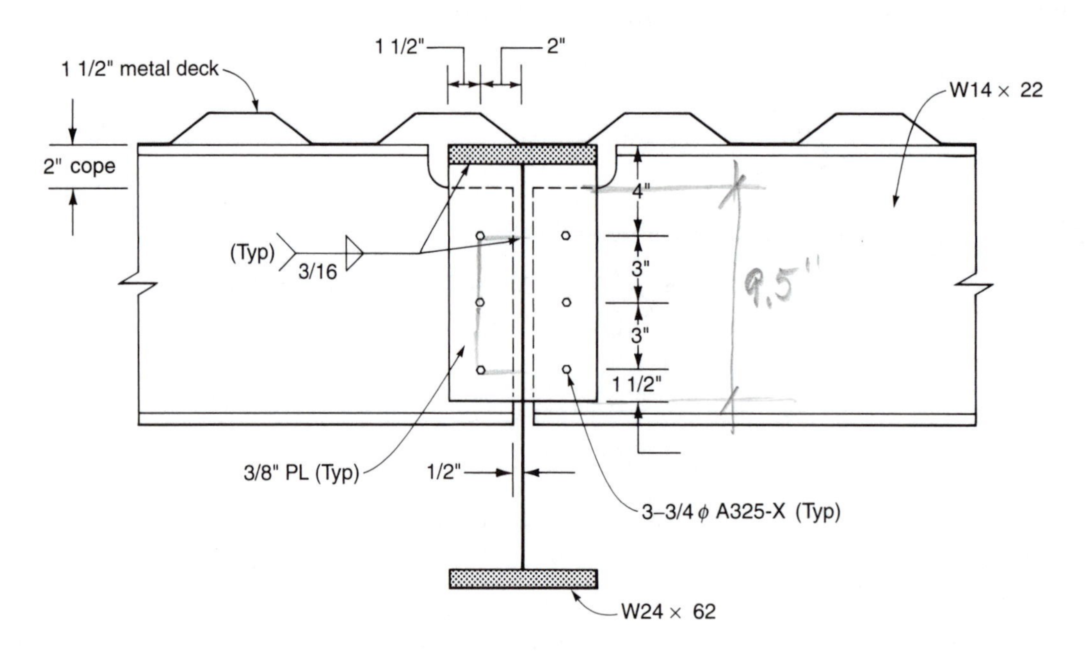

Figure 4–11 Detail A

SOLUTION

LATERAL FORCE

The seismic force acting on the structure is obtained from UBC Section 1628.2 2312 Formula (28–1) as

V = $(ZIC/R_W)W$

where Z = 0.4 for zone 4 from Table 16–I
I = 1.0 from Table 16–K for a standard occupancy structure
C = 2.75 from Section 1628.2 for an unspecified soil type
R_W = 6 from Table 16–N for a bearing wall system with masonry shear walls
W = structure dead load

Then $V = (0.4 \times 1 \times 2.75/6)W$
$= 0.183W$

NORTH-SOUTH SEISMIC

The lateral seismic loads acting at the roof diaphragm level are shown in Figure 4-12 and are given by

Roof, first bay $= 0.183 \times 0.02 \times 50$
$= 0.183$ kips per linear foot

Roof, remaining bays $= 0.183 \times 0.02 \times 100$
$= 0.366$ kips per linear foot

North wall $= 0.183 \times 0.08 \times 18.5^2/(2 \times 16)$
$= 0.157$ kips per linear foot

South wall $= 0.157$ kips per linear foot

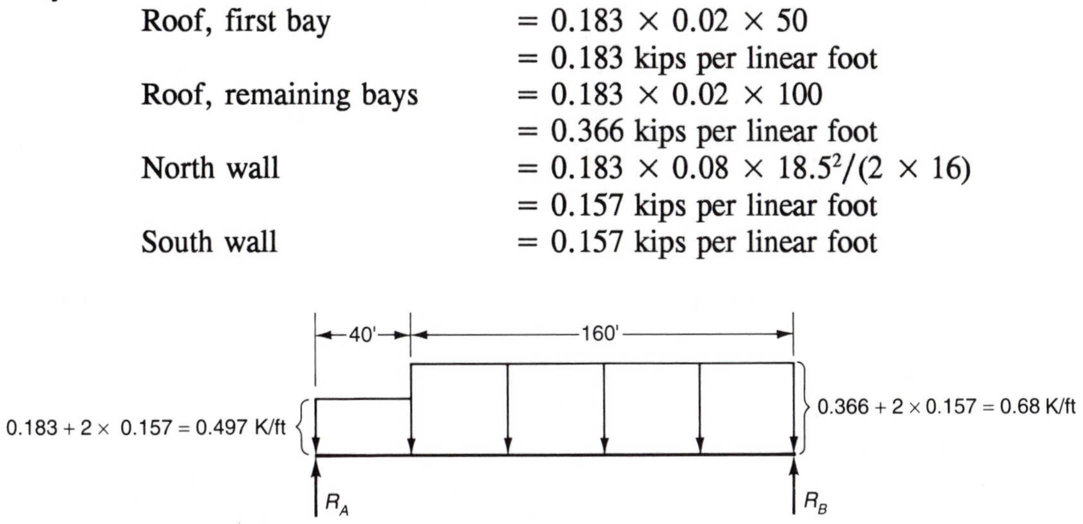

Figure 4-12 Diaphragm loading

The method of analysis employed[3] is to determine the overall diaphragm forces and then to correct for the local effects of the lateral loads at the opening. Analyzing the overall diaphragm, the shear force acting on the West wall is given by

$$R_A = 0.68 \times 200/2 - 0.183 \times 40 \times 180/200$$
$$= 61.31 \text{ kips}$$

The resultant loads acting on the diaphragm at the opening are shown in Figure 4-13. The lateral loading consists of the seismic loads due to the roof plus the North wall, and is given by

$$w = 0.183 + 0.157$$
$$= 0.34 \text{ kips per linear foot}$$

The horizontal force acting on the connection at detail A is given by

$$F_h = 40R_A/50 - w \times 40^2/100$$
$$= 40 \times 61.31/50 - 0.34 \times 40^2/100$$
$$= 43.61$$

and this may be assumed to act at the centroid of the bolt group.

The vertical force acting on the connection is due to the roof dead load and is given by

$$F_v = 0.02 \times 40 \times 5/4$$
$$= 1.0 \text{ kip}$$

and this acts vertically through the bolts' centroid.

PLATE WELD

The flange thickness of the W24 × 62 is approximately 0.5 inches which gives the shear plate size as 3.5 × 11 inches. Assuming a one-inch corner snip is provided to the plate to clear the web fillet on the W24 × 62, the weld profile is as shown in Figure 4-13. The relevant weld properties for unit size of the weld are

Length of weld, L	$= 10.0 + 2.5$
	$= 12.5$ inches
Area, A	$= 1 \times L$
	$= 12.5$ inches^2 per inch
Centroid location, y	$= 10.0 \times 6.0/12.5$
	$= 4.80$ inches
Centroid location, x	$= 2.5 \times 2.25/12.5$
	$= 0.45$ inches
Inertia about x-axis, I_x	$= 10.0^3/12 + 10.0 \times 1.20^2 + 2.5 \times 4.80^2$
	$= 155.33$ inches^4 per inch
Inertia about y-axis, I_y	$= 2.5^3/12 + 2.5 \times 1.80^2 + 10.0 \times 0.45^2$
	$= 11.43$
Polar Inertia, I_o	$= I_x + I_y$
	$= 155.33 + 11.43$
	$= 166.76$

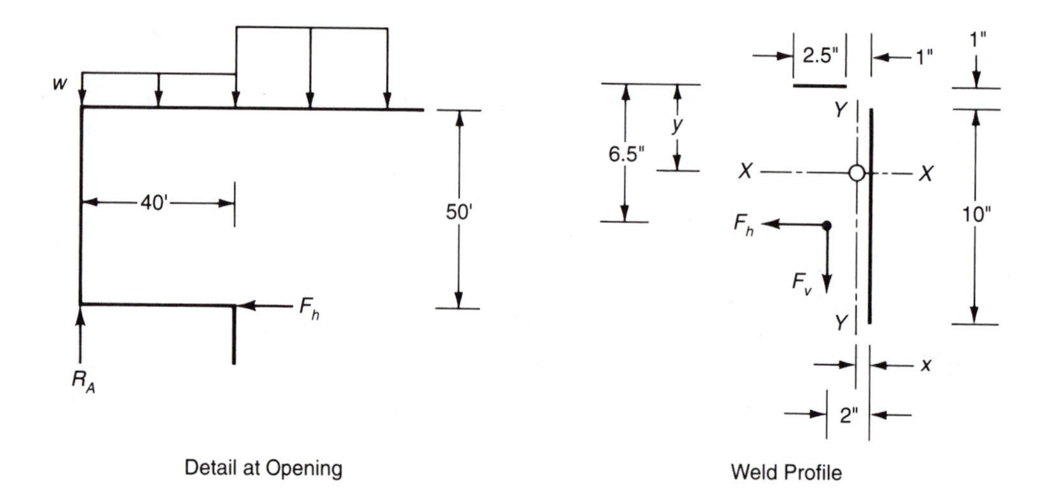

Figure 4-13 Details for Example 4-3

The moment acting about the centroid of the weld profile is

$$
\begin{aligned}
M_o &= F_h(6.5 - y) - F_v(2 - x) \\
&= 43.61(6.5 - 4.80) - 1(2 - 0.45) \\
&= 72.59 \text{ kip inch}
\end{aligned}
$$

The co-existent forces acting at the bottom of the weld profile in the x-direction and y-direction are given by

$$
\begin{aligned}
f_x &= F_h/A + M_o(11.0 - y)/I_o \\
&= 43.61/12.5 + 72.59(11.0 - 4.80)/166.76 \\
&= 6.19 \text{ kips per inch} \\
f_y &= F_v/A + M_o x/I_o \\
&= 1/12.5 + 72.59 \times 0.45/166.76 \\
&= 0.28 \text{ kips per inch}
\end{aligned}
$$

The resultant force at the bottom of the weld profile is

$$
\begin{aligned}
f_r &= (f_x^2 + f_y^2)^{1/2} \\
&= (6.19^2 + 0.28^2)^{1/2} \\
&= 6.19 \text{ kips per inch}
\end{aligned}
$$

The allowable force on the double 3/16-inch E70XX grade fillet weld is obtained from UBC Table J2.5 and, in accordance with UBC Section 1631.2.9 this may not be increased by one-third for seismic loads because of the re-entrant corner. The allowable force is, then,

$$
\begin{aligned}
q &= 2 \times 3/16 \times 0.707 \times 0.3F_u \\
&= 2 \times 3/16 \times 0.707 \times 0.3 \times 70 \\
&= 2 \times 3 \times 0.928 \\
&= 5.57 \text{ kips per inch} \\
&< f_r
\end{aligned}
$$

Hence, the weld provided is inadequate

DESIGN FORCE

The resultant force acting on the connection is given by

$$
\begin{aligned}
F_r &= (F_h^2 + F_v^2)^{1/2} \\
&= (43.61^2 + 1^2)^{1/2} \\
&= 43.62 \text{ kips}
\end{aligned}
$$

PLATE CAPACITY

The gross plate area is given by

$$
\begin{aligned}
A_g &= Bt \\
&= 11.5 \times 0.375 \\
&= 4.31 \text{ square inches}
\end{aligned}
$$

Based on the gross area, the allowable plate strength is defined in UBC Section 2251 D1 and is given by

$$
\begin{aligned}
P_t &= 0.6F_yA_g \\
&= 0.6 \times 36 \times 4.31 \\
&= 93.15 \text{ kips}
\end{aligned}
$$

The maximum allowable value of the effective net area is defined by UBC Section 2251 B3 as

$$
\begin{aligned}
A_e &= 0.85A_g \\
&= 0.85 \times 4.31 \\
&= 3.66 \text{ square inches}
\end{aligned}
$$

The actual effective net area is defined by UBC Section 2251 B3 and is given by

$A_e = t(B-nd_e)$

where $n = 3$ = number of bolts of diameter d_b in one vertical row

d_e = effective width of hole as given by Section 2251 B2 and Table J3.1

$= d_b + 0.125$

$= 0.75 + 0.125$

$= 0.875$ inches

Then $A_e = 0.375\,(11.5 - 3 \times 0.875)$

$= 3.33$ square inches

< 3.66

Hence the applicable net area is

$A_e = 3.33$ square inches

Based on the effective net area, the allowable plate strength is defined by UBC Section 2251 D1 and is given by

$P_t = 0.5\,F_uA_e$

$= 0.5 \times 58 \times 3.33$

$= 96.57$ kips

> 93.15

Hence, the allowable plate capacity, as governed by the gross area is

$P_t = 93.15$ kips

This exceeds the horizontal design force and is satisfactory.

BOLT SHEAR

The allowable single shear force on a ¾-inch diameter A325-X bolt is given in AISC Table 1-D as 13.3 kips. The allowable force on three bolts, conforming to the minimum edge distance of 1.25 inches given in UBC Table J3.5 is

$P_s = 3 \times 13.3$

$= 39.9$ kips

This is less than the resultant design force and is unsatisfactory.

BOLT BEARING

Bearing on the beam web governs and AISC Table I-F defines the bearing capacity of the three bolts, conforming to minimum edge distance and spacing criteria, as

$P_b = 52.2nt_w$

$= 52.2 \times 3 \times 0.23$

$= 36.02$ kips

This is less than the resultant design force and is unsatisfactory.

WEB TEAR OUT

The web thickness of the W14 × 22 is

$$t_w = 0.23 \text{ inches}$$

The net effective shear area resisting tearing failure is given by

$$\begin{aligned} A_v &= t_w(2 \times 1.5 - d_e) \\ &= 0.23\ (3 - 0.875) \\ &= 0.49 \text{ square inches} \end{aligned}$$

The net effective tension area resisting tearing failure is given by

$$\begin{aligned} A_t &= t_w(6 - 2d_e) \\ &= 0.23\ (6 - 2 \times 0.875) \\ &= 0.98 \text{ square inches} \end{aligned}$$

The block shear strength is determined from UBC Section 2251 J4 as

$$\begin{aligned} P_r &= 0.3F_uA_v + 0.5F_uA_t \\ &= 0.3 \times 58 \times 0.49 + 0.5 \times 58 \times 0.98 \\ &= 36.85 \\ &< F_h \end{aligned}$$

Hence, the web tear out capacity of the beam is inadequate.

4.2 BRACED FRAMES

4.2.1 Concentrically braced frames

Special concentrically braced frames provide a more ductile system than ordinary concentrically braced frames. Because of this, UBC Table 16–N gives a values for the response modification factor R_W of 9 for a special concentrically braced frame and 8 for an ordinary braced frame. More stringent design requirements are imposed on special concentrically braced frames which are restricted to all steel construction.

To reduce the possibility of local buckling, UBC Section 2211.9.2 requires compression members in special concentrically braced frames to be compact sections. The outside diameter to wall thickness ratio of circular hollow sections in limited to a value of

$$D/t = 1300/F_y$$

The outside width to wall thickness ratio of rectangular tubes is limited to a value of

$$b/t = 110/(F_y)^{1/2}$$

The width to thickness ratio of angle sections is limited to a value of

$$b/t = 52/(F_y)^{1/2}$$

For built-up members, a minimum of two stitches is required and bolted stitches may not be located within the central quarter of the brace. The slenderness ratio of the individual members between stitches may not exceed 40 percent of the governing slenderness ratio of the build-up member.

In special concentrically braced frames utilizing chevron bracing, UBC Section 2211.9.4 requires the beam to be continuous between columns and to be designed to carry all gravity loads without support from the bracing. The top and bottom flanges of the beam at the point of intersection of chevron braces must be laterally supported. This may be achieved[4] by designing for a lateral force of

$$P_{bl} = 0.015F_y b_f t_f$$

where F_y = beam yield strength
b_f = beam flange width
t_f = beam flange thickness

In addition, to provide improved post buckling behavior and resistance to beam deformation the beam shall be designed to resist the unbalanced brace forces given by the additional loading combinations:

$$1.2D + 0.5L + P_{b(min)}$$

and $0.9D - P_{b(max)}$

where D = tributary dead load
L = tributary live load
$P_{b(min)} = P_{st}$
$P_{b(max)} = 0.3P_{sc}$

UBC Section 2211.9.3, for braces that can buckle in the plane of the gusset plate, requires the end connections to have a design strength not less than that of the brace gross section. For braces that can buckle out-of-plane of the gusset plate, the gusset plate shall be designed to carry the compressive strength of the brace without buckling. Also, the brace shall terminate on the gusset plate a minimum of twice the gusset plate thickness from a line about which the gusset plate can bend unrestrained by the column or beam. This is shown in Figure 4-14.

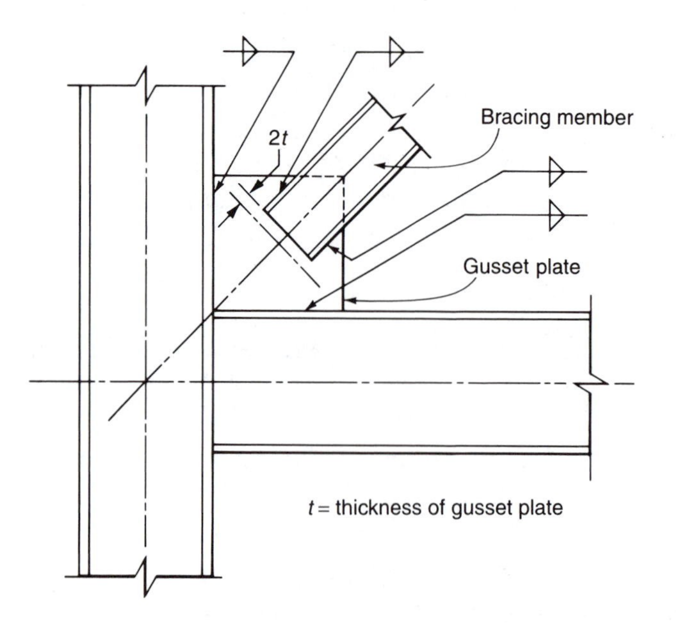

Figure 4-14 Gusset plate requirements

Example 4–4 (Structural Engineering Examination 1988, Section B, weight 5.0 points)

GIVEN: Plan and section of an office building as shown in Figure 4–15.

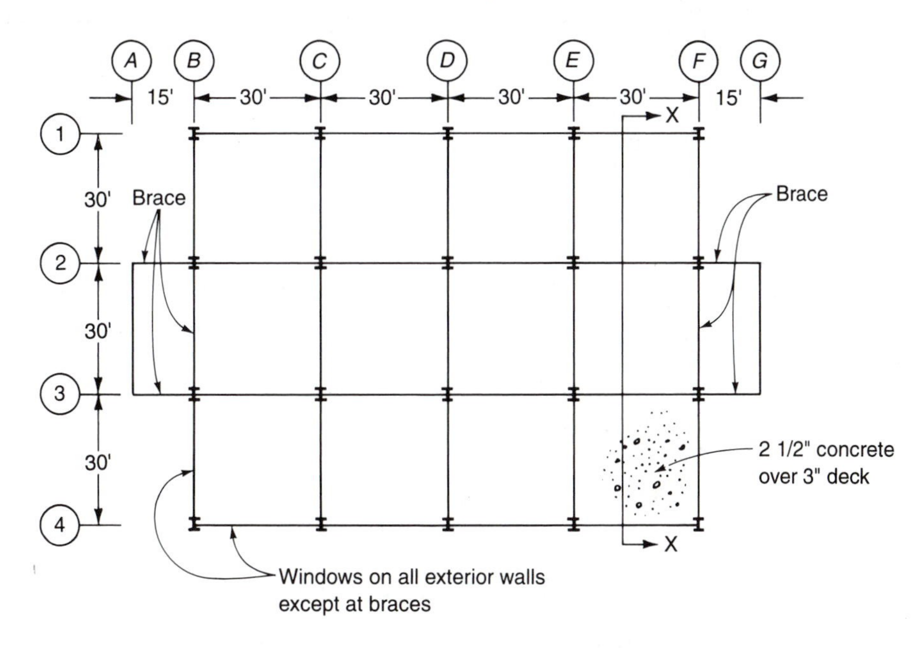

Figure 4–15 Building for Example 4–4

CRITERIA: Uniform Building Code seismic zone 3, building period = 0.56 seconds, S = 1.50, I = 1.00, R_W = 6. Weight of each floor and roof is 1100 kips each. Non-moment-resisting frame with braces in bays shown. The stiffness of each bracing along B and F is twice that of those along 2 and 3.

REQUIRED:
1. Determine the seismic force at each level.
2. Determine the maximum seismic forces in members of the frame on line F. Neglect vertical loads.
3. Check the adequacy of member Z indicated in Figure 4–16. Show all calculations.

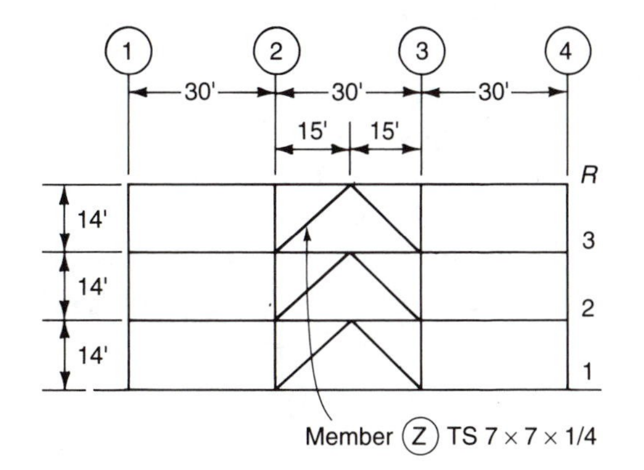

Figure 4–16 Section X–X

SOLUTION

DEAD LOAD

The seismic dead load, assuming no storage occupancies, partition loads, snow loads, or permanent equipment loads, is given by

$W = 3 \times 1100$
$= 3300$ kips

BASE SHEAR

The force coefficient is given by UBC Formula (28–2) as

$C = 1.25S/T^{2/3}$

where $S = 1.5$ for the site coefficient, as given

and $T = 0.56$ for the structure period, as given

Thus $C = 1.25 \times 1.5/(0.56)^{2/3}$
$= 2.77$
> 2.75 hence use the upper limit of
$C = 2.75$

The seismic base shear is given by UBC Formula (28–1) as

$V = (ZIC/R_W)W$

where $Z = 0.3$ for zone 3 from Table 16–I
$I = 1.0$ for standard occupancy from Table 16–K
$R_W = 6$ for the response factor, as given
$W = 3300$ kips for the total structure dead load, as calculated
$C = 2.75$ for the force coefficient, as calculated

Then $V = (0.3 \times 1 \times 2.75/6)3300$
$= 0.1375 \times 3300$
$= 454$ kips

1. BASE SHEAR DISTRIBUTION

The base shear is distributed over the height of the structure[5] in accordance with UBC Formula (28–8)

$F_x = (V - F_t)\, w_x h_x / \Sigma w_i h_i$

where F_x = the lateral force at level x
V = the base shear of 454 kips, as calculated
F_t = the additional concentrated force at roof level
$= 0$, as T is less than 0.7 seconds
w_x = the structure dead load at level x
h_x = the height in feet above the base to level x
$\Sigma w_i h_i$ = the sum of the product $w_x \times h_x$ at all levels
$= 92{,}400$ kips feet, from Table 4–2

Then $F_x = Vw_x h_x/\Sigma w_i h_i$
$= 454 w_x h_x/92{,}400$
$= 0.00491 w_x h_x$ kips

In Table 4–2,

F_x = the lateral force at level x

and V_x = the shear at story x

Level	w_x	h_x	w_xh_x	F_x	V_x
Roof	1100	42	46,200	227	–
3rd floor	1100	28	30,800	151	227
2nd floor	1100	14	15,400	76	378
1st floor	–	–	–	–	454
Total	3300	–	92,400	454	–

Table 4–2 Vertical distribution of forces

2. MEMBER FORCES

The roof and floor diaphragms may be considered rigid and, in determining the distribution of the base shear to the braced frames, torsional effects must be considered. Since the structure is symmetrical in plan, the center of mass and the center of rigidity coincide. However, to comply with UBC Section 1628.5 accidental eccentricity shall be allowed for by assuming that the center of mass is displaced a distance equal to five percent of the building dimension perpendicular to the direction of the seismic force. The maximum forces are produced in the frame on line F by a North–South seismic load when the center of mass is displaced to the East. Torsional irregularity does not exist and, in accordance with UBC Section 1628.6, further amplification of the torsional effects is not required. Then, the accidental eccentricity is given by

$$e = 0.05 \times \text{East–West dimension}$$
$$= 0.05 \times 150$$
$$= 7.5 \text{ feet}$$

The polar moment of inertia of the frames is given by the sum of the products of the relative rigidity of each frame and its squared distance from the center of rigidity. Thus,

$$\Sigma r^2R = 2 \times 2 \times 60^2 + 4 \times 1 \times 15^2$$
$$= 15{,}300 \text{ square feet}$$

The torsion acting on the structure at level x is given by

$$T_x = F_xe$$
$$= 7.5F_x \text{ kip feet}$$

The torsional shear force acting on the frame on line F at level x is

$$F_T = T_xr_FR_F/\Sigma r^2R$$

where r_F = the distance, 60 feet, of the frame on Line F from the center of rigidity

R_F = the relative rigidity of the frame on line F = 2 as given

Σr^2R = the polar moment of inertia = 15,300 as calculated

T_x = the applied torsion = $7.5F_x$ as calculated

Then $F_T = 2 \times 60 \times 7.5F_x/15{,}300$
$= 0.0588F_x$ kips

The in-plane shear force acting on the frame on line F at level x is

$F_S = F_xR_F/\Sigma R_y$

where F_x = the lateral force at level x
R_F = the relative rigidity of the frame on line F = 2 as given
ΣR_y = sum of braced frame rigidities for frames in the North–South direction
$= 2 + 2$
$= 4$

Then $F_S = 2F_x/4$
$= 0.5F_x$ kips

The total force in the braced frame on line F is given by

$F_F = F_S + F_T$
$= 0.5F_x + 0.0588F_x$
$= 0.5588F_x$

In Table 4–3,

F_F = the lateral force at the indicated level
V = the shear at the indicated story
M = the moment at the indicated story
$= \Sigma VH$

where H = the height of the indicated story and the summation ΣVH extends from the indicated story to all stories above.

The forces in the members of the frame on line F are obtained by resolving forces at the joints and assuming that the lateral force F_F is applied symmetrically on either side of the frame. The force in a beam is equal to half the corresponding story shear. The force in a diagonal equals the beam force times $\cos\theta$, where θ is the angle of inclination of the diagonal. The force in a column equals the corresponding story moment divided by the frame width. These values are given in Table 4–3 and indicated in Figure 4–17

Level	F_F	V	VH	M	V/2	$(V/2)\cos\theta$	M/B
Roof	127	–	–	–	–	–	–
3rd floor	84	127	1778	–	63.5	87	–
2nd floor	43	211	2954	1778	105.5	144	59
1st floor	–	254	3556	4732	127.0	174	158
Foundation	–	–	–	8288	–	–	276

Table 4–3 Member forces

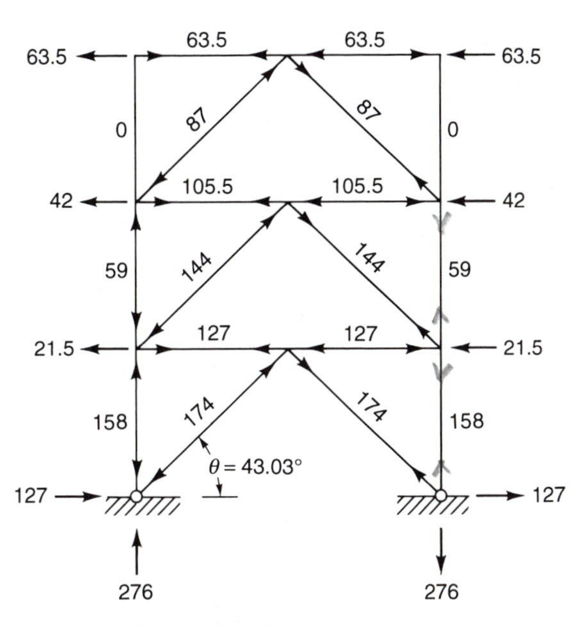

Figure 4–17 Member forces

3. CHECK MEMBER Z

MEMBER PROPERTIES
The section properties of structural tubing size 7 × 7 × ¼ are given in AISC Table 3–41

$$A = 6.59 \text{ square inches}$$
$$r = 2.74 \text{ inches}$$
$$F_y = 46 \text{ kips per square inch}$$

The unbraced length of member Z, using center line dimensions, is

$$\begin{aligned} l &= H/\sin\theta \\ &= 14/\sin(43.03)^\circ \\ &= 20.52 \text{ feet} \end{aligned}$$

The effective length factor for the brace, assuming hinged ends, is given by AISC Table C–C2.1 item (d) as

$$K = 1.0$$

BRACING CONFIGURATION
In accordance with UBC Section 2211.8.4 chevron bracing in an ordinary concentrically braced frame shall be designed for 1.5 times the prescribed seismic force. The required design force is, then

$$\begin{aligned} P_R &= 1.5 \times 87 \\ &= 131 \text{ kips} \end{aligned}$$

STRESS INCREASE
In accordance with UBC Section 1603.5, a one-third increase is allowed in stresses when designing for seismic forces. The equivalent design force, based on the allowable stress, is

$$P_E = P_R/1.33 = 131/1.33 = 98 \text{ kips}$$

The equivalent member stress, based on the gross sectional area, is

$$f = P_E/A = 98/6.59 = 14.9 \text{ kips per square inch}$$

and this stress may be either compressive or tensile.

TENSILE CAPACITY

Assuming that the brace connections are welded, the gross area of the member governs the tension capacity and the allowable tensile stress is given by UBC Section 2251 D1 as

$$F_t = 0.6F_y = 0.6 \times 46 = 27.6 \text{ kips per square inch} > 14.9$$

and the brace is adequate for the applied tensile force.

SLENDERNESS

As the building is located in seismic zone 3, the slenderness ratio of the brace is limited, by UBC Section 2211.8.2 to a maximum value of

$$l/r = 720/(F_y)^{1/2} = 720/(46)^{1/2} = 106$$

The actual slenderness ratio is

$$l/r = 20.52 \times 12/2.74 = 90 < 106$$

and the slenderness ratio of the brace is satisfactory.

LOCAL BUCKLING

The width to thickness ratio of square structural tubing is limited, by UBC Section 2251.7 Table B5.1, to a maximum value of

$$b/t = 238/(F_y)^{1/2} = 238/(46)^{1/2} = 35$$

The actual width to thickness ratio is

$$b/t = 7/0.25 = 28 < 35$$

and the width to thickness ratio of the brace is satisfactory.

DISTRIBUTION OF LATERAL FORCE

UBC Section 2211.8.2 stipulates that neither the sum of the horizontal components of the compressive member forces nor the sum of the horizontal components of the tensile member forces, along a line of bracing, shall exceed seventy percent of the applied lateral force. In each story of the building, both braces resist the lateral force equally and the braces are, therefore, satisfactory.

COMPRESSIVE CAPACITY

The effective slenderness ratio is given by

$$Kl/r = 1 \times 90 = 90$$

The axial compressive stress permitted by UBC Section 2251 E2 is obtained from AISC Table C-50 as

$$F_a = 16.94 \text{ kips per square inch}$$

The stress reduction factor for a bracing member resisting seismic forces is determined from UBC Formula (11-5) as

$$\beta = 1/[1 + ([Kl/r]/2C_c)]$$

where $C_c = 111.6$ as given by UBC Division IX Table 4

Hence, $\beta = 1/[1 + 90/(2 \times 111.6)] = 0.71$

Then, the allowable stress for a bracing member resisting seismic forces in compression is determined from UBC Formula (11-4) as

$$F_{as} = \beta F_a = 0.71 \times 16.94 = 12.07 < 14.9$$

and the brace is inadequate for the applied compressive force.

Example 4-5 (Structural Engineering Examination 1990, Section B, weight 7.0 points)

GIVEN: The elevation of a metal stud wall with one "X" braced frame "abcd" to resist seismic load is shown in Figure 4-18. Seismic load is as follows:

- Seismic Load = 300 pounds per linear foot

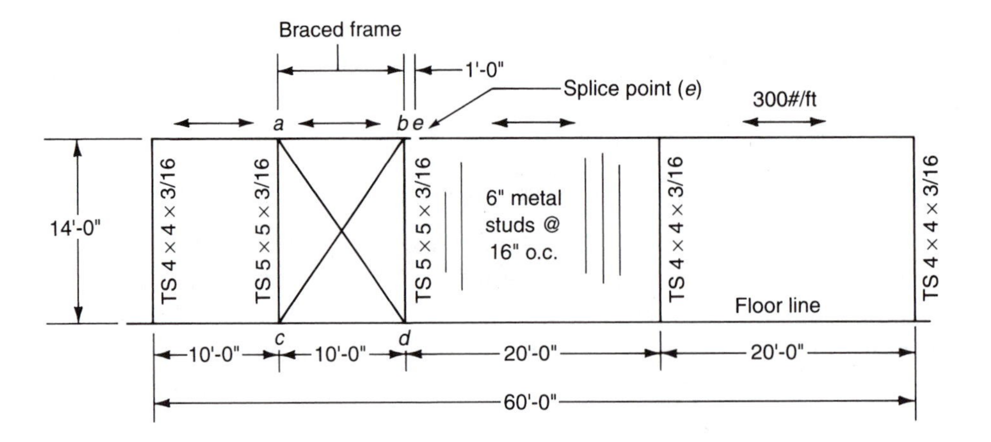

Figure 4–18 Elevation of stud wall

CRITERIA: Seismic zone 4, R_W = 8

Materials:

- Structural tubes F_y = 46 kips per square inch
- Steel beams A36 F_y = 36 kips per square inch
- A325-SC bolts
- E70XX electrodes

Assumptions:

- The seismic load of 300 pounds per linear foot (reversible) is adequately transferred through a nailer on the top flange of the W14 × 22 beam.
- The braced frame is adequately anchored to the concrete foundation.
- Neglect gravity loads on the columns.
- All joints are pin connected.

REQUIRED:

1. Determine the drag strut force at points "a", "b", and "e".
2. Design and detail the beam to beam bolted connection at point "e" for drag forces only. Use ¾-inch diameter bolts. Complete the detail shown in Figure 4-19.
3. Check the adequacy of the diagonal "ad" assuming an axial force of 16 kips.
4. Using the loads from Figure 4–20 and assuming that the maximum force that can be transferred to the brace by the system is 60 kips:
   a. Determine the forces necessary to design the connection at point "b" on Figure 4–20.
   b. Design and detail the welded connection at "b". Complete the detail shown in Figure 4-19. Calculations are required for fillet welds L_1 for length of gusset plate shown.

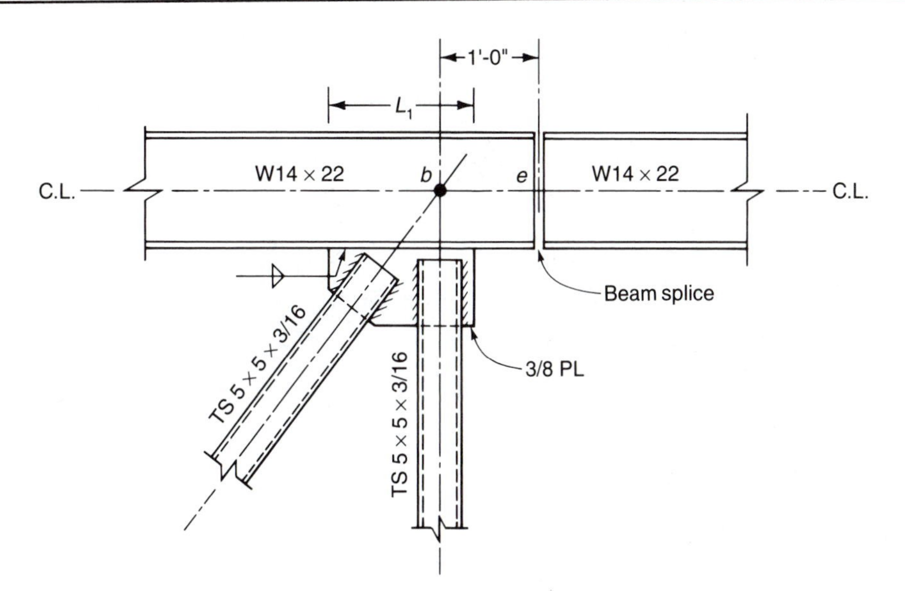

Figure 4–19 Frame detail

SOLUTION

1. DRAG FORCES

The shear distribution, net shear and drag force distribution along the wall are shown in Figure 4–21.

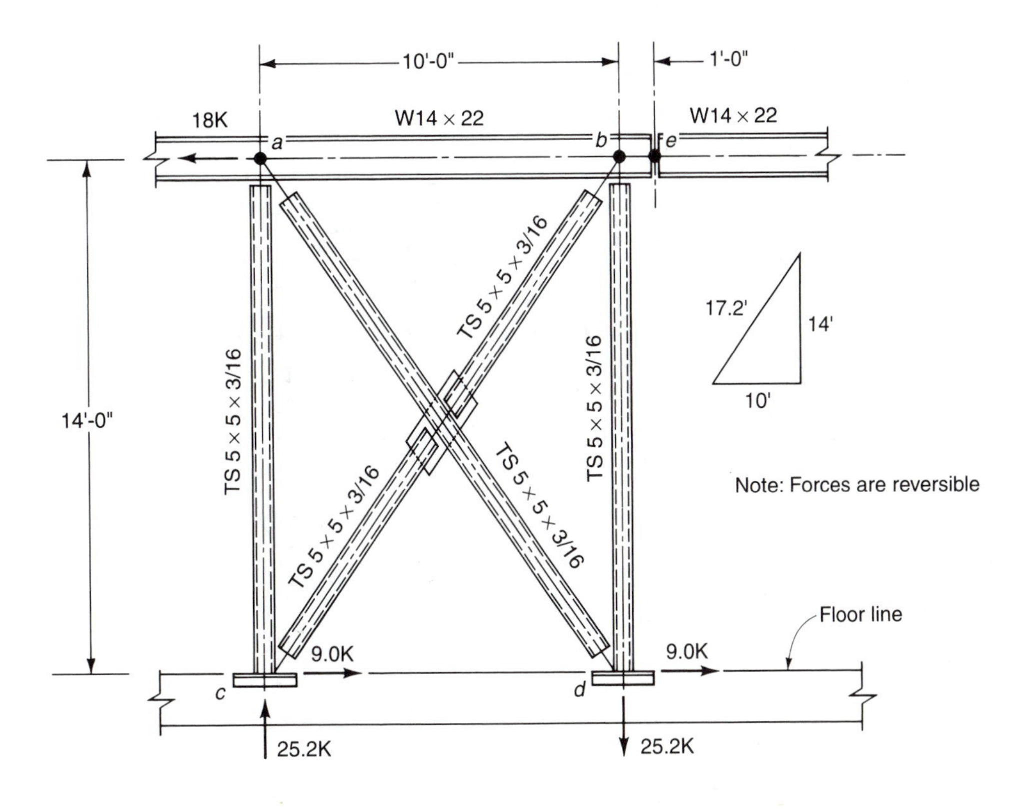

Figure 4–20 Frame forces

The shear along the length of the nailer is given as

$$q_N = 300 \text{ pounds per linear foot}$$

The total seismic force applied to the wall is

$$\begin{aligned} V &= L_N q_N \\ &= 60 \times 300/1{,}000 \\ &= 18 \text{ kips} \end{aligned}$$

The shear developed by each brace is

$$\begin{aligned} Q_B &= V/2 \\ &= 18/2 \\ &= 9 \text{ kips} \end{aligned}$$

The drag forces at "a", "b" and "e" are

$$\begin{aligned} F_a &= 10q_N/1{,}000 \\ &= 10 \times 300/1{,}000 \\ &= 3 \text{ and } 6 \text{ kips} \\ F_b &= 40q_N/1{,}000 \\ &= 40 \times 300/1{,}000 \\ &= 12 \text{ and } 3 \text{ kips} \\ F_e &= 39q_N/1{,}000 \\ &= 39 \times 300/1{,}000 \\ &= 11.7 \text{ kips} \end{aligned}$$

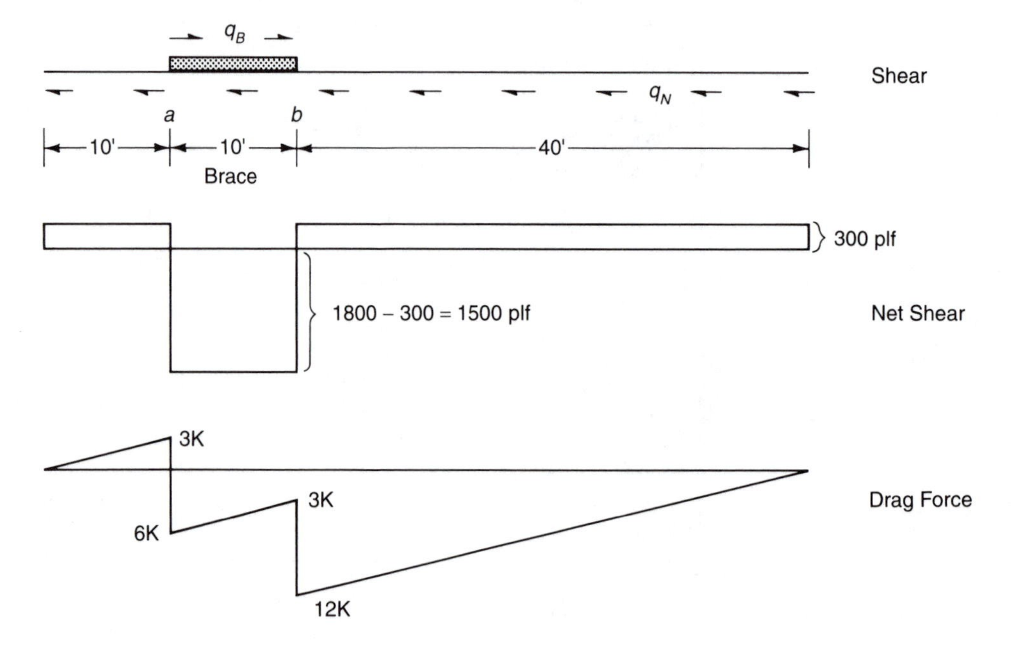

Figure 4–21 Drag force distribution

2. BEAM SPLICE
The splice design load is

$$F_e = 11.7 \text{ kips}$$

Allowing for the one-third increase in permissible stress for seismic loads, in accordance with the UBC Section 1603.5, the equivalent design force is

$$\begin{aligned} P_E &= F_e/1.33 \\ &= 11.7/1.33 \\ &= 8.8 \text{ kips} \end{aligned}$$

The W14 × 22 beam has a clear web depth between fillets of twelve inches and the splice may be fabricated with four ¾-inch A325-SC bolts and a ¼ × 7 × 6 inch splice plate, as shown in Figure 4-22.

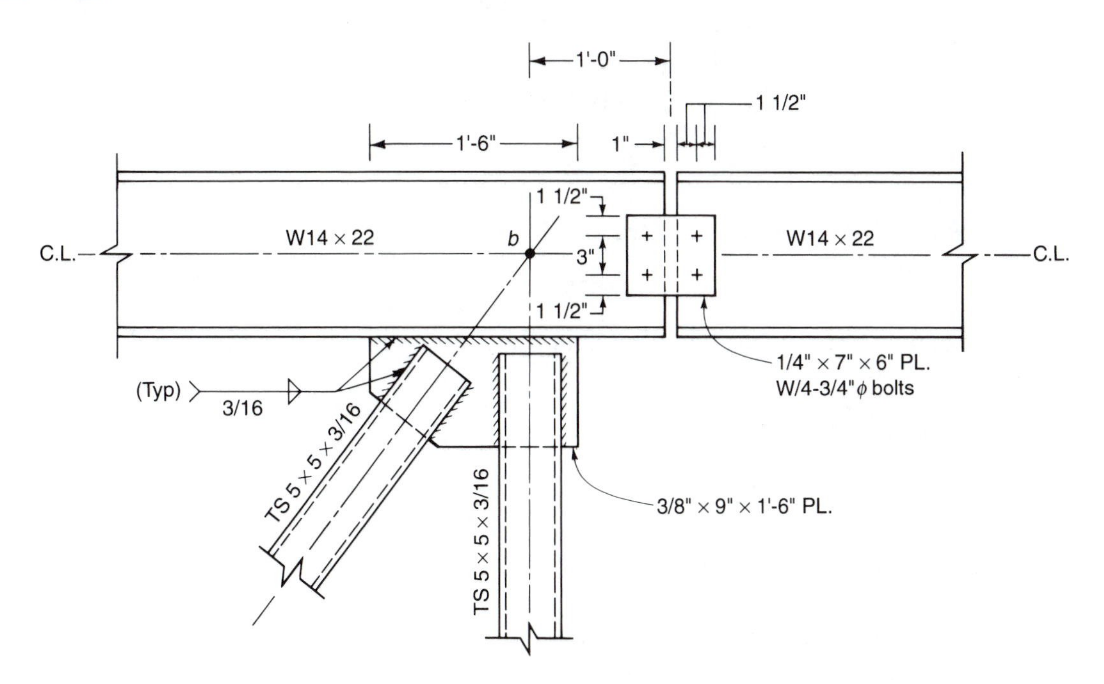

Figure 4-22 Connection details

PLATE STRENGTH IN TENSION
The gross plate area is given by

$$\begin{aligned} A_g &= Bt \\ &= 6 \times 1/4 \\ &= 1.5 \text{ square inches} \end{aligned}$$

based on the gross area, the allowable plate strength is defined by UBC Section 2251 D1 and is given by

$$P_t = 0.6F_yA_g$$
$$= 0.6 \times 36 \times 1.5$$
$$= 32.4 \text{ kips}$$

The maximum allowable value of the effective net area is defined by Section 2251 B3 as

$$A_e = 0.85A_y$$
$$= 0.85 \times 1.5$$
$$= 1.275 \text{ square inches}$$

The actual effective net area is defined by UBC Section 2251 B3 and is given by

$$A_e = t(B - nd_e)$$

where $n = 2$ = the number of bolts, of diameter d_b, in one vertical row

d_e = effective width of hole as given by Section 2251 B2 and Table J3.1

$$= d_b + 0.125$$
$$= 0.875 \text{ inches}$$

Then

$$A_e = 0.25\,(6 - 2 \times 0.875)$$
$$= 1.06 \text{ square inches}$$
$$< 1.275$$

Hence, the applicable effective net area is

$$A_e = 1.06 \text{ square inches}$$

based on the effective net area, the allowable plate strength is defined by UBC Section 2251 D1 and is given by

$$P_t = 0.5\,F_uA_e$$
$$= 0.5 \times 58 \times 1.06$$
$$= 30.8 \text{ kips}$$

This exceeds the equivalent design force and is satisfactory.

PLATE STRENGTH IN COMPRESSION

The effective slenderness ratio is given by

$$Kl/r = 1 \times 6.5 \times (12)^{1/2}/t$$
$$= 90$$

The axial compressive stress permitted by UBC Section 2251 E2 is obtained from AISC Table C-36 as

$$F_a = 14.2 \text{ kips per square inch}$$

Then, the allowable compressive strength of the plate is

$$P_c = F_a \times A_g$$
$$= 14.2 \times 1.5$$
$$= 21.3$$

This exceeds the equivalent design force and is satisfactory.

WEB TEAR OUT

The web thickness of the W14 × 22 is

$$t_w = 0.23 \text{ inches}$$

The net effective shear area resisting tearing failure is given by

$$\begin{aligned} A_v &= t_w(2 \times 1.5 - d_e) \\ &= 0.23\ (3 - 0.875) \\ &= 0.49 \text{ square inches} \end{aligned}$$

The net effective tension area resisting tearing failure is given by

$$\begin{aligned} A_t &= t_w(3 - d_e) \\ &= 0.23\ (3 - 0.875) \\ &= 0.49 \text{ square inches} \end{aligned}$$

The block shear strength is determined from UBC Section 2251 J4 as

$$\begin{aligned} P_r &= 0.3F_uA_v + 0.5F_uA_t \\ &= 0.3 \times 58 \times 0.49 + 0.5 \times 58 \times 0.49 \\ &= 22.74 \end{aligned}$$

This exceeds the equivalent design force and is satisfactory.

BOLT SHEAR

The allowable single shear force on a ¾-inch diameter A325-SC bolt is given in AISC Table 1-D as 7.51 kips. The allowable load on two bolts, conforming to the minimum edge distance and spacing criteria, is

$$\begin{aligned} P_s &= 2 \times 7.51 \\ &= 15.02 \text{ kips} \end{aligned}$$

This exceeds the equivalent design force and the splice is adequate as shown in Figure 4-22.

3. BRACE CAPACITY

Design the brace in accordance with the requirements for an ordinary concentrically braced frame. The relevant properties of the 5 × 5 × 3/16 structural tubing are

Yield stress, F_y = 46 kips per square inch
Area, A = 3.52 square inches
Radius of gyration, r = 1.95 inches

STRESS INCREASE

Allowing for the one-third increase in permissible stress for seismic loads, in accordance with UBC Section 1603.5 the equivalent design force is

$$\begin{aligned} P_E &= 16/1.33 \\ &= 12.0 \text{ kips} \end{aligned}$$

The equivalent member stress, based on the gross sectional area is

$$\begin{aligned} f &= P_E/A \\ &= 12/3.52 \\ &= 3.41 \text{ kips per square inch} \end{aligned}$$

and this stress may be either tensile or compressive.

TENSILE CAPACITY
Since the brace connections are welded, the gross area of the member governs the tensile capacity and the allowable tensile stress is given by UBC Section 2251 D1 as

$$\begin{aligned} F_t &= 0.6F_y \\ &= 0.6 \times 46 \\ &= 27.6 \text{ kips per square inch} \\ &> 3.41 \end{aligned}$$

and the brace is adequate for the applied tensile force.

SLENDERNESS
Since the building is located in seismic zone 4, the slenderness ratio of the brace is limited, by UBC Section 2211.8.2 to a maximum value of

$$\begin{aligned} l/r &= 720/(F_y)^{1/2} \\ &= 720/(46)^{1/2} \\ &= 106 \end{aligned}$$

The distance between lateral supports, for each brace, may be assumed to be half the full theoretical length[6] of the brace, for both in-plane and out-of-plane effects. The actual slenderness ratio is, then

$$\begin{aligned} l/r &= 17.2 \times 12/(2 \times 1.95) \\ &= 52.9 \\ &< 106 \end{aligned}$$

and the slenderness ratio of the brace is satisfactory.

LOCAL BUCKLING
The width to thickness ratio of square structural tubing is limited, by UBC Section 2251.7 Table B5.1 to a maximum value of

$$\begin{aligned} b/t &= 238/(F_y)^{1/2} \\ &= 238/(46)^{1/2} \\ &= 35 \end{aligned}$$

The actual width to thickness ratio is

$$\begin{aligned} b/t &= 5 \times 16/3 \\ &= 26.7 \end{aligned}$$

and the width to thickness ratio of the brace is satisfactory.

DISTRIBUTION OF LATERAL FORCE

UBC Section 2211.8.2 stipulates that neither the sum of the horizontal components of the compressive member forces nor the sum of the horizontal components of the tensile member forces, along a line of bracing, shall exceed seventy percent of the applied lateral force. Both braces resist the lateral force equally and are, therefore, satisfactory.

COMPRESSIVE CAPACITY

The effective length factor for a pin-ended member is given by AISC Table C-C2.1 item (d) as

$$K = 1$$

The effective slenderness ratio of the brace, using center line dimensions, is given by

$$Kl/r = 1 \times 52.9$$
$$= 52.9$$

The axial compressive stress permitted by UBC Section 2251 E2 is obtained from AISC Table C-50 as

$$F_a = 23.90 \text{ kips per square inch.}$$

The stress reduction factor for a bracing member resisting seismic forces is determined from UBC Formula (11-5) as

$$\beta = 1/[1 + ([kl/r]/2C_c)]$$

where $C_c = 111.6$ as given by UBC Division IX Table 4

Hence $\beta = 1/[1 \times 52.9/(2 \times 111.6)]$

$$= 0.808$$

Then, allowable stress for a bracing member resisting seismic forces in compression is determined from UBC Formula (11-4) as

$$F_{as} = \beta F_a$$
$$= 0.808 \times 23.9$$
$$= 19.31$$
$$> 3.41$$

Hence, the brace satisfies all the requirements of the UBC code.

4a. CONNECTION FORCES

The force applied to the brace is given by

$$P = 9 \times 17.2/10$$
$$= 15.48 \text{ kips}$$

and this stress may be either tensile or compressive. The brace connection shall be designed, in accordance with UBC Section 2211.8.3 for the lesser of the following three values:

(i) The tensile strength of the brace,

$$\begin{aligned} P_{st} &= AF_y \text{ from UBC Section 2211.4} \\ &= 3.52 \times 46 \\ &= 162 \text{ kips} \end{aligned}$$

(ii) The applied force times 0.375RW, where R_W is the applicable response factor for a building frame system, given in UBC Table 16–N item 2.4. Hence,

$$\begin{aligned} P_b &= 0.375R_WP \\ &= 0.375 \times 8 \times 15.48 \\ &= 46.44 \end{aligned}$$

(iii) The maximum force that can be transferred to the brace by the system

$$\begin{aligned} P_m &= 60 \times 17.2/(2 \times 10) \\ &= 51.6 \text{ kips} \end{aligned}$$

Hence, the design load value for the brace is

$$P_b = 46.44 \text{ kips}$$

By resolving forces at the joint, the design loads at the connection are obtained as shown in Figure 4–23 and are

$$\begin{aligned} P_c &= \text{design load value for the column} \\ &= 46.44 \times 14/17.2 \\ &= 37.8 \text{ kips} \\ P_g &= \text{design load value for the beam} \\ &= 46.44 \times 10/17.2 \\ &= 27.0 \text{ kips} \end{aligned}$$

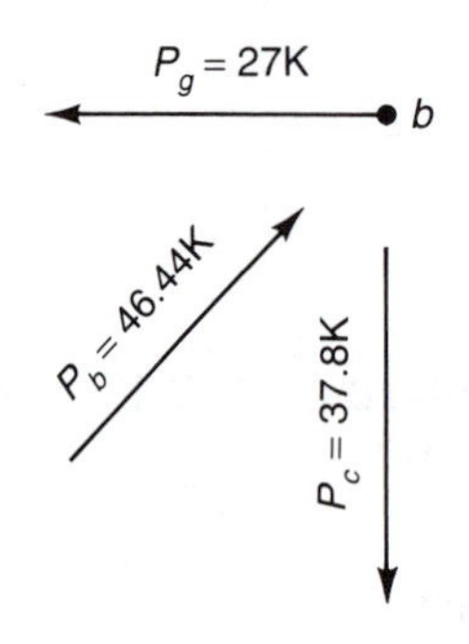

Figure 4–23 Member forces

4b. CONNECTION DESIGN

To adequately connect the brace and the column to the W14 × 22 without eccentricity, as shown in Figure 4–22, requires a ⅜ × 9 × 18 inch gusset plate. Use a five inch length of weld to the brace and a five inch length of weld to the column. In accordance with UBC Table J2.4 of Division IX the minimum size of weld required for the 5/16 beam flange and the ⅜–inch gusset plate is 3/16–inch. The strength capacity of the 3/16–inch E70XX grade fillet weld is given by UBC Section 2211.4 as

$$
\begin{aligned}
q_u &= 1.7q_{(allowable)} \\
&= 1.7 \times 3/16 \times 0.707 \times 0.3F_u \\
&= 1.7 \times 3/16 \times 0.707 \times 0.3 \times 70 \\
&= 1.7 \times 3 \times 0.928 \\
&= 4.73 \text{ kips per inch}
\end{aligned}
$$

The strength of the five inch weld to the column is given by

$$
\begin{aligned}
P_w &= 4q_u \times 5 \\
&= 4 \times 4.73 \times 5 \\
&= 94.6 \text{ kips} \\
&> P_c
\end{aligned}
$$

Hence, the column connection is satisfactory.
The strength of the five inch weld to the brace is given by

$$
\begin{aligned}
P_w &= 4q_u \times 5 \\
&= 94.6 \text{ kips} \\
&> P_b
\end{aligned}
$$

Hence, the brace connection is satisfactory.
The geometrical properties of the gusset plate weld are obtained by assuming unit size of weld.

$$
\begin{aligned}
\text{Weld length, } L &= 18 \text{ inches} \\
\text{Weld area, } A &= 1 \times 18 \\
&= 18 \text{ inches}^2 \text{ per inch} \\
\text{Plastic modulus, } Z &= 1 \times L^2/4 \\
&= 1 \times 18^2/4 \\
&= 81 \text{ inches}^3 \text{ per inch}
\end{aligned}
$$

The depth of the W14 × 22 beam is

$$d = 13.74 \text{ inches}$$

and the applied loading on the weld is obtained from Figures 4–22 and 4–23 as

$$
\begin{aligned}
\text{Horizontal shear, } H &= P_g \\
&= 27 \text{ kips} \\
\text{Eccentricity moment, } M &= P_g \times d/2 \\
&= 27 \times 13.74/2 \\
&= 185.5 \text{ kip inches}
\end{aligned}
$$

The co–existent forces acting at the extreme ends of the unit weld are

$$
\begin{aligned}
\text{Horizontal force} &= H/A \\
&= 27/18 \\
&= 1.5 \text{ kips per inch} \\
\text{Vertical force} &= M/Z \\
&= 185.5/81 \\
&= 2.29 \text{ kips per inch}
\end{aligned}
$$

The resultant force at the ends of the weld is

$$R = (1.5^2 + 2.29^2)^{1/2} = 2.74 \text{ kips per inch}$$

The strength capacity of the double 3/16-inch fillet weld is

$$q_d = 2q_u = 2 \times 4.73 = 9.46 \text{ kips per inch} > R$$

Hence, the gusset plate weld is adequate.
The gusset plate shear capacity is given by UBC Section 2211.4 as

$$V_s = 0.55\ F_y dt = 0.55 \times 36 \times 18 \times 3/8 = 133.7 \text{ kips} > P_g$$

The gusset plate moment capacity is

$$M_s = ZF_y = (td^2/4)F_y = 36 \times 3 \times 18^2/(8 \times 4) = 1{,}094 \text{ kip inches} > M$$

Hence, the gusset plate is satisfactory.

4.2.2 Eccentrically braced frames

An eccentrically braced frame provides a high stiffness in the elastic range which is similar to that of a concentrically braced frame. During a major earthquake, the link beam is designed to deform inelastically and provide a non-linear energy absorbing capacity similar to a special moment-resisting frame. The other framing elements are designed to remain elastic and to be sufficiently strong to cause the link beam to yield[7].

Link beam requirements

- To ensure stability of the link, compact sections shall be used with a flange width-thickness ratio not exceeding

$$b_f/2t_f = 52/(F_y)^{1/2}$$

- When the link beam strength is controlled by shear, the web capacity is fully utilized in shear and the strengths in flexure and in axial compression are determined using the beam flanges only. Shear is assumed to control[8] when the link beam length is less than

$$e = 2.2M_s/V_s$$

- When the axial stress in the link beam is

$$f_a \geq 0.15f_y$$

the reduced strength in flexure is given by

$$M_{rs} = Z(F_y - f_a)$$

- For a total frame drift of $0.375R_W$ times the drift determined for the prescribed seismic forces, plastic hinges may be assumed to have formed at each end of the link beam. The elements of the frame are considered rigid and the link beam rotation angle θ is calculated as shown in Figure 4-24. To limit the inelastic deformation of the frame, the link rotation angle is limited to the following values:

$$\theta \leq 0.060 \text{ radian for } e \leq 1.6M_s/V_s$$
$$\theta \leq 0.015 \text{ radian for } e \leq 3.0M_s/V_s$$

Linear interpolation may be used for intermediate link lengths.

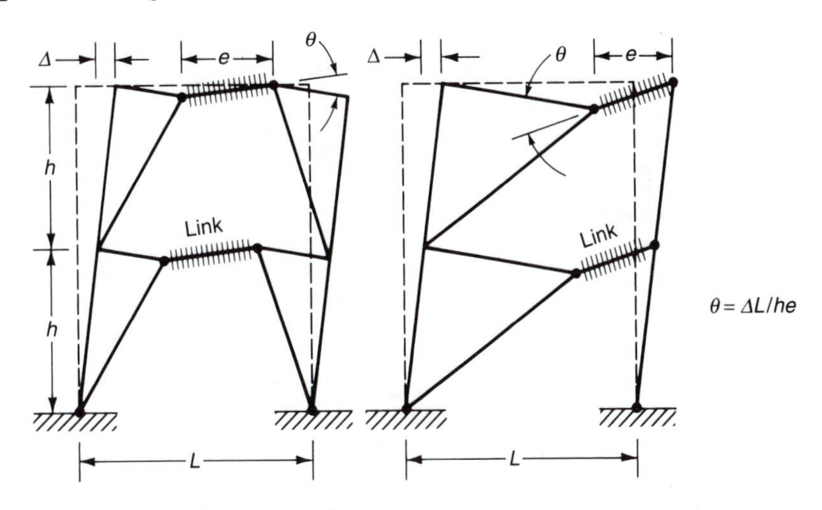

Figure 4-24 Link beam rotation

- Doubler plates on the web of the link beam are not allowed as they are ineffective during inelastic deformation.
- Holes are not allowed in the web of the link beam as these affect the inelastic deformation of the link beam web.
- To provide the same factor of safety as other structural systems, under the prescribed seismic forces the link beam web shear is limited to

$$f_v d t_w \leq 0.8V_s$$

- To prevent web instability under cyclic loading, full-depth web stiffeners shall be provided on both sides of the link beam web at the brace end of the link. As shown in Figure 4-25, the stiffeners shall have a combined width of

$$2b_{st} \geq b_f - 2t_w$$

and a thickness of

$t_{st} \geq 0.75t_w$
$\geq$ ⅜-inch

where b_f = link flange with
t_w = link web thickness

An additional stiffener is required at a distance of b_f from the brace when the link beam length exceeds $1.6\ M_s/V_s$

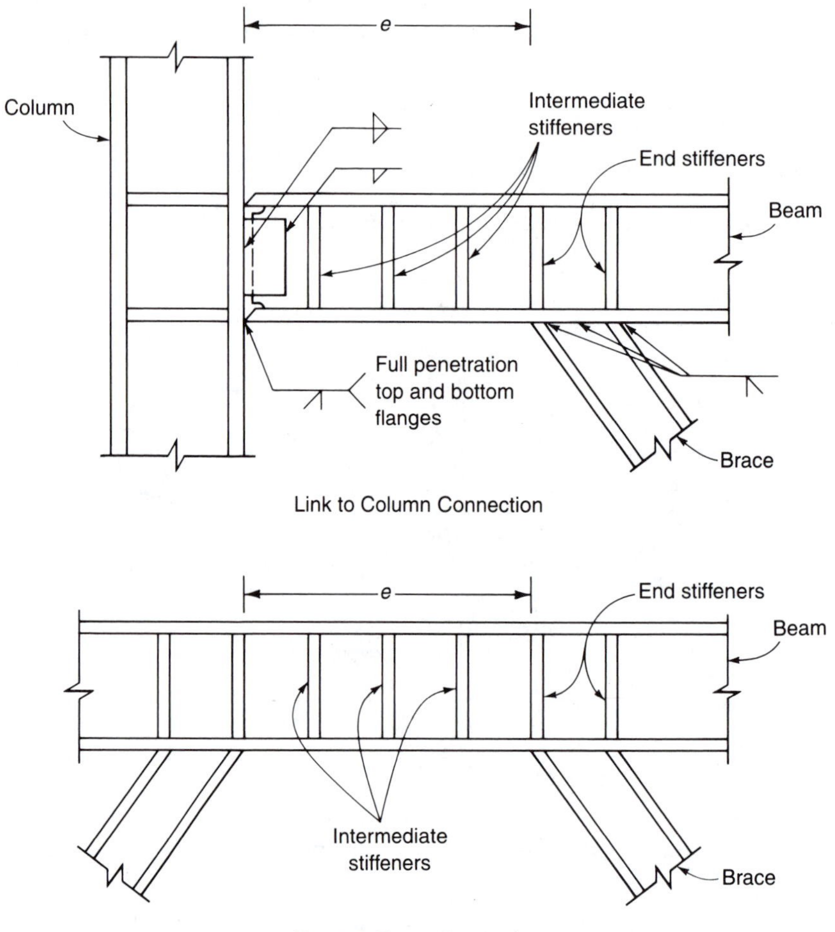

Figure 4–25 Link beam details

- To ensure full development of the shear capacity of the link beam web, intermediate stiffeners are required when

$e < 1.6M_s/V_s$
or $f_v > 0.45F_y$
where $f_v = 2M_{rs}/edt_w$

Single sided, full depth web stiffeners are permitted provided the link beam depth is less than 24 inches.

- The required stiffener spacing s is given by

$$s \leq 38t_w - d/5 \text{ for } \theta = 0.06 \text{ radian}$$
$$s \leq 56t_w - d/5 \text{ for } \theta \leq 0.03 \text{ radian}$$

Linear interpolation may be used for intermediate link rotations.

- The weld between the stiffener and the web is required to develop the full strength of the stiffener as shown in Figure 4-26. The weld force is given by

$$P_w \geq A_{st}F_y$$

where A_{st} = area of stiffener
$= b_{st}t_{st}$

The weld between the stiffener and the flange is necessary to develop the rigidity of the stiffener and restrain flange buckling. The weld force is given by

$$P_w \geq A_{st}F_y/4$$

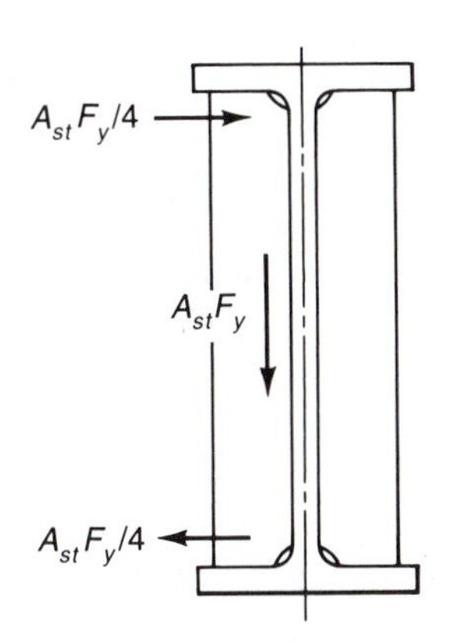

Figure 4-26 Stiffener weld forces

Link beam-to-column connection requirements

- Due to the large inelastic rotations generated, the link beam length shall not exceed

$$e = 1.6M_s/V_s$$

- Because of the high shear requirements the link flanges shall have complete penetration welds to the column. Where the link beam strength is controlled by shear, the web shall be welded to the column to develop the full strength of the link beam web.
- Where the link beam is connected to the column web, the link flanges shall have complete penetration welds to connection plates and the web connection shall be welded to develop the full strength of the link beam web. In addition, the link beam rotation angle shall not exceed

$$\theta = 0.015 \text{ radian.}$$

Brace and beam outside of link requirements

- To ensure the integrity of the frame, the required strength of the brace and of the beam outside of the link shall be at least 1.5 times the forces corresponding to the link beam strength.

Column strength requirements

- To ensure that yielding occurs in the link beam, columns shall be designed to remain elastic when subjected to forces corresponding to 1.25 times the forces due to the link beam strength.

Bracing requirements

- Bracing is necessary to the top and bottom flanges of the beam to prevent instability and restrain the link from twisting out of plane. Lateral support shall be provided at the ends of the link with a design strength of

$$P_{bl} = 0.06F_yb_ft_f$$

Intermediate bracing shall be provided at intervals of

$$s_{bl} = 76b_f/(F_y)^{1/2}$$

Intermediate bracing shall have a design strength of one percent of the link beam strength.

Beam-to-column connection

- A pinned connection is adequate provided that restraint is provided to resist accidental eccentricities. The connection shall have a capacity to resist a torsional moment of

$$M_T = 0.01F_yb_ft_fd$$

Example 4-6 (Eccentrically braced frame)
The unfactored lateral seismic forces acting on the bottom story of a four-story frame are shown in Figure 4-27. Determine a suitable W12 section for the beam and a suitable W12 section for the brace. Determine the maximum allowable drift due to the prescribed seismic forces.
The brace may be considered pinned at each end and the beam may be considered pinned at the column. The link beam is connected to the column web.
Gravity loading and axial force in the link beam may be neglected. All stories in the frame are the same height.

Solution

PRESCRIBED FORCES
The length of the brace is given by

$$\begin{aligned} L_B &= (a^2 + h^2)^{1/2} \\ &= (10^2 + 15^2)^{1/2} \\ &= 18.0 \text{ feet} \end{aligned}$$

The horizontal component of the force in the second story brace is

$$\begin{aligned} H_{B2} &= R_{B2}a/L_B \\ &= 180 \times 10/18 \\ &= 100 \text{ kips} \end{aligned}$$

The vertical reaction at the foundations is

$$
\begin{aligned}
V_0 &= (H_{B2} + F_{x2})h/L \\
&= (100 + 10)15/15 \\
&= 110 \text{ kips}
\end{aligned}
$$

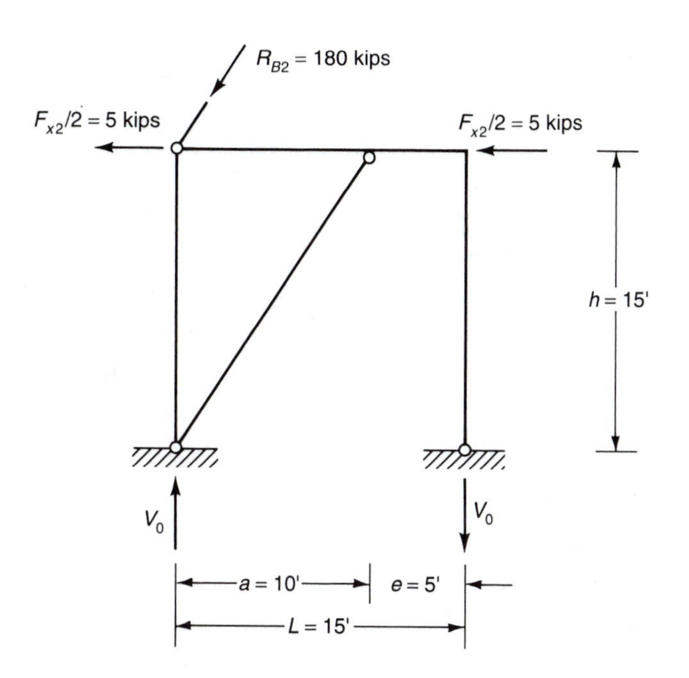

Figure 4–27 Details for Example 4-6

The shear force applied to the link beam is given by

$$
\begin{aligned}
V_{LB} &= V_0 \\
&= 110 \text{ kips}
\end{aligned}
$$

Assuming that a point of contraflexure occurs at the midpoint of the link beam, the moment produced at each end of the link beam is

$$
\begin{aligned}
M_{LB} &= V_{LB}e/2 \\
&= 110 \times 60/2 \\
&= 3300 \text{ kip inches}
\end{aligned}
$$

The bending moment and shear force diagrams for the beam and the link beam are shown in Figure 4–28. From the diagrams, the shear force applied to the beam at the pinned support at the column is given by

$$
\begin{aligned}
V_C &= M_{LB}/a \\
&= 3300/120 \\
&= 27.5 \text{ kips}
\end{aligned}
$$

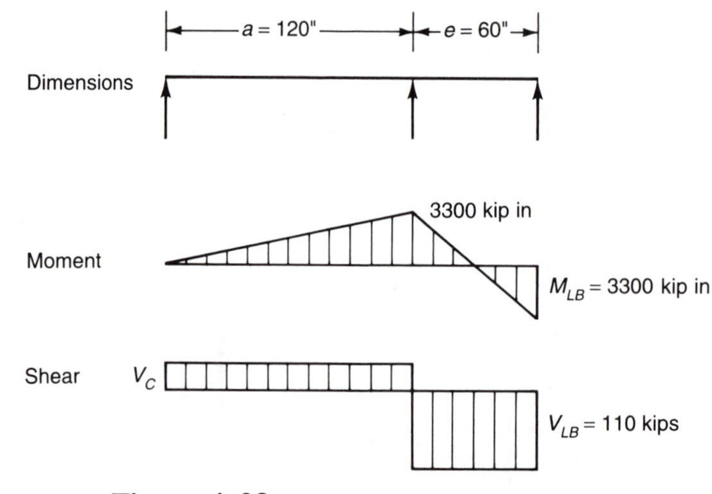

Figure 4–28 Shear and moment diagrams

The vertical component of the force in the brace is, then

$$\begin{aligned} V_B &= V_{LB} + V_C \\ &= 110 + 27.5 \\ &= 137.5 \text{ kips} \end{aligned}$$

The force in the brace is given by

$$\begin{aligned} P_B &= V_B L_B / h \\ &= 137.5 \times 18/15 \\ &= 165 \text{ kips} \end{aligned}$$

LINK BEAM SECTION

The shear in the link beam is limited by UBC Section 2211.10.5 to

$$\begin{aligned} f_v A_v &= 0.8 V_s \\ f_v &= \text{shear stress in the web} \\ A_v &= \text{net shear area} \\ V_s &= \text{shear strength of the web} \\ &= 0.55 F_y A_v \ldots \text{ from UBC Section 2211.4} \\ F_y &= \text{minimum yield stress} \\ &= 36 \text{ kips per square inch} \end{aligned}$$

Then

$$\begin{aligned} f_v &= 0.8 \times 0.55 \times 36 \\ &= 15.84 \text{ kips per square inch} \end{aligned}$$

The minimum allowable web area is, then

$$\begin{aligned} dt_w &= V_{LB} / f_v \\ &= 110/15.84 \\ &= 6.94 \text{ square inches} \end{aligned}$$

The minimum sized W12 section which is suitable is a W12 × 96 with a web area of

$$\begin{aligned} dt_w &= 6.99 \\ &> 6.94 \ldots \text{ satisfactory} \end{aligned}$$

CHECK COMPACT FLANGE
The maximum flange width to thickness ratio allowed by UBC Section 2211.10.2 is

$$(b_f/2t_f)_{max} = 52/(F_y)^{1/2}$$
$$= 52/6$$
$$= 8.67$$

The width to thickness ratio provided by the W12 × 96 section is

$$b_f/2t_f = 6.76$$
$$< 8.67 \text{ ... satisfactory}$$

CHECK COMPACT WEB
The maximum web depth to thickness ratio allowed by UBC Section 2251 B5 is

$$(d/t_w)_{max} = 257/(F_{y)})^{1/2}$$
$$= 257/6$$
$$= 42.83$$

The depth to thickness ratio provided by the W12 × 96 section is

$$d/t_w = 23.1$$
$$< 42.83 \text{ ... satisfactory}$$

CHECK LENGTH OF LINK
The flexural strength of the W12 × 96 section is defined in UBC Section 2211.4 as

$$M_s = ZF_y$$
$$= 147 \times 36$$
$$= 5292.0 \text{ kip inches}$$

The shear strength of the W12 × 96 section is defined in UBC Section 2211.4 as

$$V_s = 0.55F_y dt_W$$
$$= 0.55 \times 36 \times 6.99$$
$$= 138.4 \text{ kips}$$

The maximum permissible link beam length for a link beam-to-column connection is specified in UBC Section 2211.10.12 as

$$e_{max} = 1.6M_s/V_s$$
$$= 1.6 \times 5292.0/138.4$$
$$= 61.18 \text{ inches}$$

The link beam length provided is

$$e = 60 \text{ inches}$$
$$< 61.18 \text{ ... satisfactory}$$

Hence, the W12 × 96 section is adequate for the link beam.

MAXIMUM ALLOWABLE DRIFT

The story drift of $0.375R_W$ times the drift due to the prescribed seismic forces is given by

$x = 0.375\Delta R_W$

where Δ = elastic story drift due to the prescribed seismic forces

R_W = response modification factor

= 10 from UBC Table 16–N

Then $x = 0.375\Delta \times 10$

$= 3.75\Delta$

From Figure 4–24, the story drift is determined by

$x = eh\theta/L$

where θ = link rotation angle at a drift of x

= 0.015 radian maximum from UBC Section 2211.10.12

Hence $x = 60 \times 180 \times 0.015/180$

= 0.90 inches

The maximum allowable drift due to the prescribed seismic forces is

$\Delta = x/3.75$

$= 0.90/3.75$

= 0.24 inches

LINK BEAM STRENGTH

The reduced flexural strength of the link beam is specified by UBC Section 2211.10.3 as

$M_{rs} = Z(F_y - f_a)$

where f_a = 0 as given in the problem statement

Then $M_{rs} = ZF_y$

$= 147 \times 36$

= 5292 kip inches

The shear strength of the link beam is specified by UBC Section 2211.4 as

$V_s = 0.55F_y dt_w$

$= 0.55 \times 36 \times 6.99$

= 138.4 kips

The shear strength of the link beam as limited by the flexural strength is given by

$V_{rs} = 2M_{rs}/e$

$= 2 \times 5292/60$

= 176.4 kips

$> V_s$

Hence the controlling shear strength is

V_s = 138.4 kips

The excess link beam flexural capacity is given by

$$\begin{aligned}\phi_f &= M_{rs}/M_{LB}\\ &= 5292/3300\\ &= 1.60\end{aligned}$$

The excess link beam shear capacity is

$$\begin{aligned}\phi_s &= V_s/V_{LB}\\ &= 138.4/110\\ &= 1.26\\ &< \phi_f\end{aligned}$$

Hence the shear strength of the link beam controls and the overstrength factor is

$$\phi_{LB} = 1.26$$

DESIGN OF THE BRACE

In accordance with UBC Section 2211.10.13, the brace shall be designed for 1.5 times the force corresponding to the link beam strength. The design axial force in the brace is, then

$$\begin{aligned}P_{BD} &= 1.5\phi_{LB}P_B\\ &= 1.5 \times 1.26 \times 165\\ &= 312 \text{ kips}\end{aligned}$$

In accordance with UBC Section 2211.4.2 an increase of 1.7 is allowed in the permissible stress when designing for seismic forces. The equivalent design axial force, based on the basic allowable stress is then

$$\begin{aligned}P &= P_{BD}/1.7\\ &= 312/1.7\\ &= 183 \text{ kips}\end{aligned}$$

From AISC Section C–C2 the effective length is given by

$$\begin{aligned}KL_B &= 1.0 \times 18\\ &= 18 \text{ feet}\end{aligned}$$

From AISC Table 3–28, a W12 × 53 is the lightest suitable section and has an allowable axial load of

$$\begin{aligned}P_A &= 227 \text{ kips}\\ &> P \ldots \text{ satisfactory.}\end{aligned}$$

4.3 SPECIAL MOMENT–RESISTING FRAMES

4.3.1 Beam details

In seismic zones 3 and 4, UBC Section 2211.7 stipulates special detailing requirements to ensure the ductile behavior and energy absorption capacity of special moment–resisting frames.

- To limit local flange buckling, compact sections shall be used with a flange width–thickness ratio not exceeding

$$b_f/2t_f = 52/(F_y)^{1/2}$$

- To prevent stress concentrations resulting in a brittle mode of failure, abrupt changes in flange area are not permitted in plastic hinge regions.
- Bracing is necessary to the top and bottom flanges of the beam to prevent instability. Bracing is required at all concentrated loads where a hinge may form and at a maximum spacing of

$$l_{cr} = 96r_y$$

4.3.2 Column details

- In situations where the strong–column/weak–beam concept is not applicable, compact sections shall be used with a flange width–thickness ratio not exceeding

$$b_f/2t_f = 52/(F_y)^{1/2}$$

- Under normal circumstances, a strong–column/weak–beam concept should be adopted to ensure frame stability. This may be achieved by ensuring that

$$\Sigma Z_c(F_{yc} - f_a)/\Sigma Z_b F_{yb} > 1.0$$

or $$\Sigma Z_c(F_{yc} - f_a)/1.25\Sigma M_{pz} > 1.0$$

where f_a = axial stress in the column
> 0
Z_c = plastic section modulus of a column
Z_b = plastic section modulus of a beam
F_{yc} = yield stress of a column
F_{yb} = yield stress of a beam
M_{pz} = beam moment due to DL + LL + 1.85 seismic
$\leq 0.8M_s$

- Where a strong–column/weak–beam arrangement is impractical the column is required to:
  - (i) have an axial stress less than $0.4F_y$
  - or (ii) be located in a story with a shear strength 50 percent greater than that of the story above
  - or (iii) not contribute to the story shear strength
- Under normal circumstances, column flanges should be braced at the beam–to–column connection to restrain out–of–plane buckling. Where the column remains elastic, bracing is required at the level of the beam top flange. Otherwise, bracing must be provided at the levels of both the top and bottom flanges of the beam. The bracing shall have a design strength of

$$P_{bl} = 0.01F_y b_f t_f$$

and shall limit the lateral displacement of the column flanges to 0.2 inch.

- Where bracing at the beam-to-column connection is impractical, the column shall be designed as pin ended and have a slenderness ratio not exceeding 60. The column shall be designed in accordance with the requirements of UBC Section 2251 HI. The axial stress f_a shall be the least value determined from the loading combinations:

$$1.0P_{DL} + 0.7P_{LL} + 0.375R_W P_E$$
$$1.0P_{DL} + 0.7P_{LL} + 1.25P_G$$
$$1.0P_{DL} + 0.7P_{LL} + 1.00P_{PZ}$$

where P_{DL} = axial dead load
P_{LL} = axial live load
P_E = axial load due to prescribed seismic forces
P_G = axial load due to flexural strength of the girders
P_{PZ} = axial load due to panel zone shear strength
R_W = response modification factor

The flexural stress transverse to the column, f_{by}, shall include the effect of the bracing force and the second order P-delta effects.

4.3.3 Girder-to-column connection

- Where the joint contributes to the lateral-force-resisting system it shall be designed to develop the lesser of the flexural capacity of the beam or the moment corresponding to the panel zone shear strength. When the joint does not form part of the lateral-force-resisting system the elements at the joint, using an allowable stress increase of 1.7, shall be designed for a displacement of $0.375R_W$ times the displacement due to the prescribed seismic forces and for the second order P-delta effects. The flexural capacity of a beam is defined in UBC Section 2211.4.2 as

$$M_s = ZF_y$$

The panel zone shear strength is defined by UBC Formula (11-1) as

$$V = 0.55F_y d_c t(1 + 3b_c t_{cf}^2/d_b d_c t)$$

where b_c = column flange width
t_{cf} = column flange thickness
d_c = column depth
d_b = beam depth
t = total thickness of panel zone including doubler plates

The shear force developed in the panel zone is derived[9] as shown in Figure 4-29 and is given by AISC Formula (C-E6-1) as

$$V_{PZ} = M_1/0.95d_1 + M_2/0.95d_2 - (M_1 + M_2)/H$$

where H = distance between column midheights
≈ distance between points of contraflexure
M_1 = bending moment in beam 1
M_2 = bending moment in beam 2
$0.95d_1 \approx (d-t_f)$ for beam 1
$0.95d_2 \approx (d-t_f)$ for beam 2

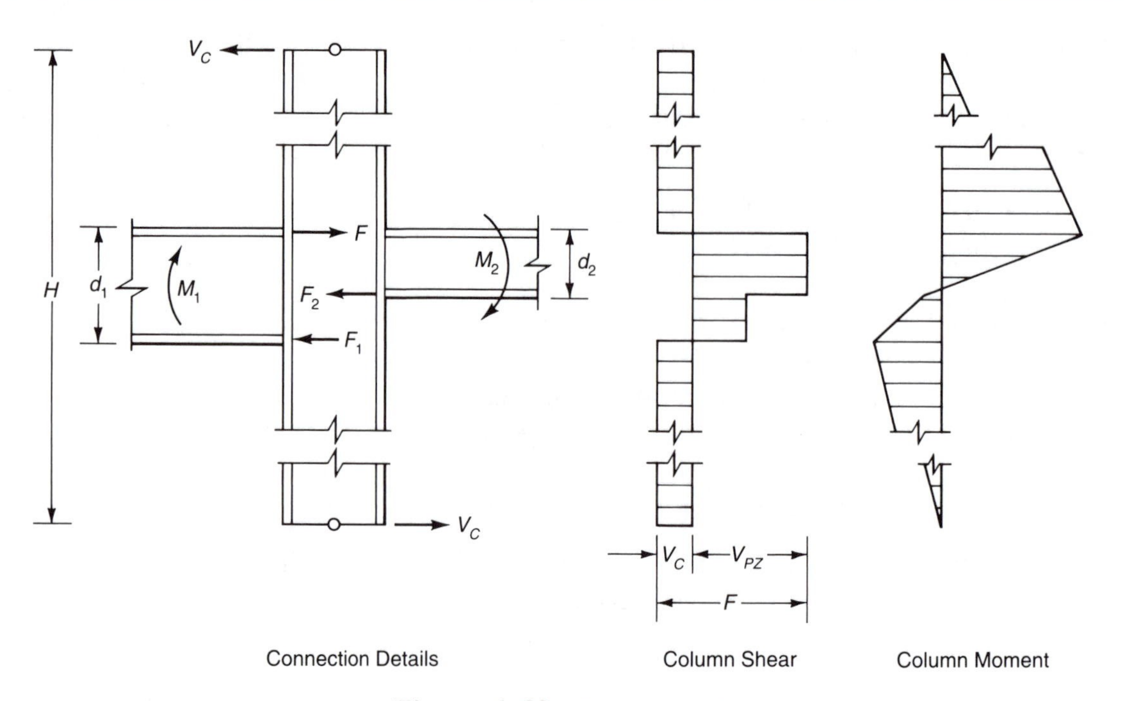

Figure 4–29 Shear in panel zone

The force developed in the beam flanges is

$$F = M_1/0.95d_1 + M_2/0.95d_2$$

The shear developed in the column outside the panel zone is

$$V_C = (M_1 + M_2)/H$$

- To develop the flexural capacity of the beam the flanges shall have full-penetration butt welds to the column. Where the flexural strength of the beam flanges is given by

$$b_f t_f (d - t_f) F_y > 0.7 Z F_y$$

the web connection may be made by slip-critical high strength bolts as shown in Figure 4-30. The required shear capacity shall be determined from the loading combination:

$$DL + \text{Floor LL} + 0.375R_W \times \text{Seismic Load}$$

Where the flexural strength of the beam flanges is less than 70 percent of the strength of the full section, the connection shall also be provided with supplementary welds to develop 20 percent of the flexural strength of the beam web. This is shown in Figure 4-31. When the panel zone shear strength controls the design of the joint, the required shear capacity of the connection shall be determined as shown in Figure 4-32. The beam shear is given by

$$V_B = 1.85 \times 2M_E/L$$
$$\leq 0.8 \times 2M_s/L$$

where M_E = beam moment due to the prescribed seismic forces
M_s = flexural capacity of the beam

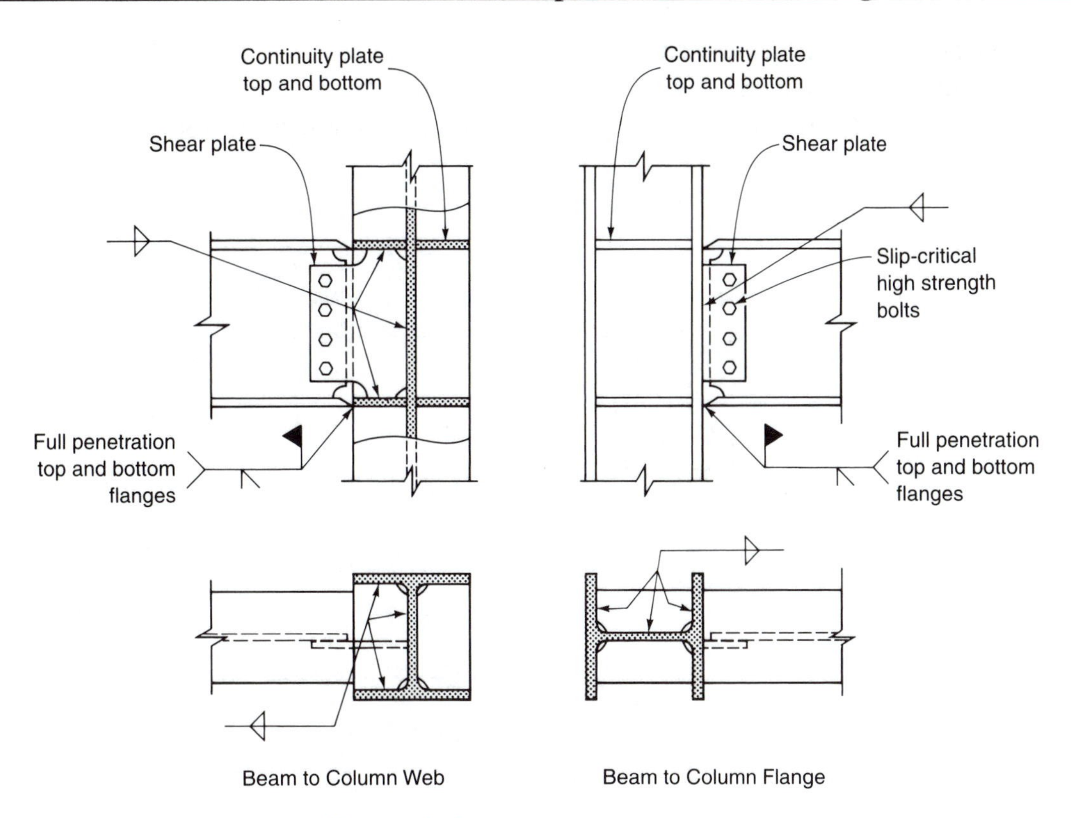

Figure 4–30 Beam-column joint

- The required shear capacity of the panel zone is given by

$$V_{PZ} = 1.85F - V_C$$
$$\leq 0.8F_s$$

where F = force developed in the flanges due to (dead + live + seismic) moments

V_C = shear developed in the column outside the panel zone due to (dead + live + seismic) moments

F_s = force developed in the beam flanges due to the flexural capacity of the beams

$$= M_{s1}/0.95d_1 + M_{s2}/0.95d_2$$

- To prevent shear buckling during cyclic loading, the panel zone thickness shall not be less than

$$t_z = (d_z + w_z)/90$$

where d_z = panel zone depth between continuity plates

w_z = panel zone width between column flanges.

The thickness of any doubler plate may be included in t_z provided it is connected to the column web with plug welds adequate to prevent local buckling of the plate. In addition, the doubler plate shall be spaced not more than 1/16–inch from the column web and welded top and bottom with at least a minimum fillet weld. The doubler plate shall be either butt or fillet welded to the column flanges to develop its shear strength. Doubler plates may extend between continuity plates or may extend one inch above and below the continuity plates as shown in Figure 4–33. When the doubler plate extends beyond the continuity plate, the welds to the column web must be adequate to transfer their pro rata share of the continuity plate force to the web.

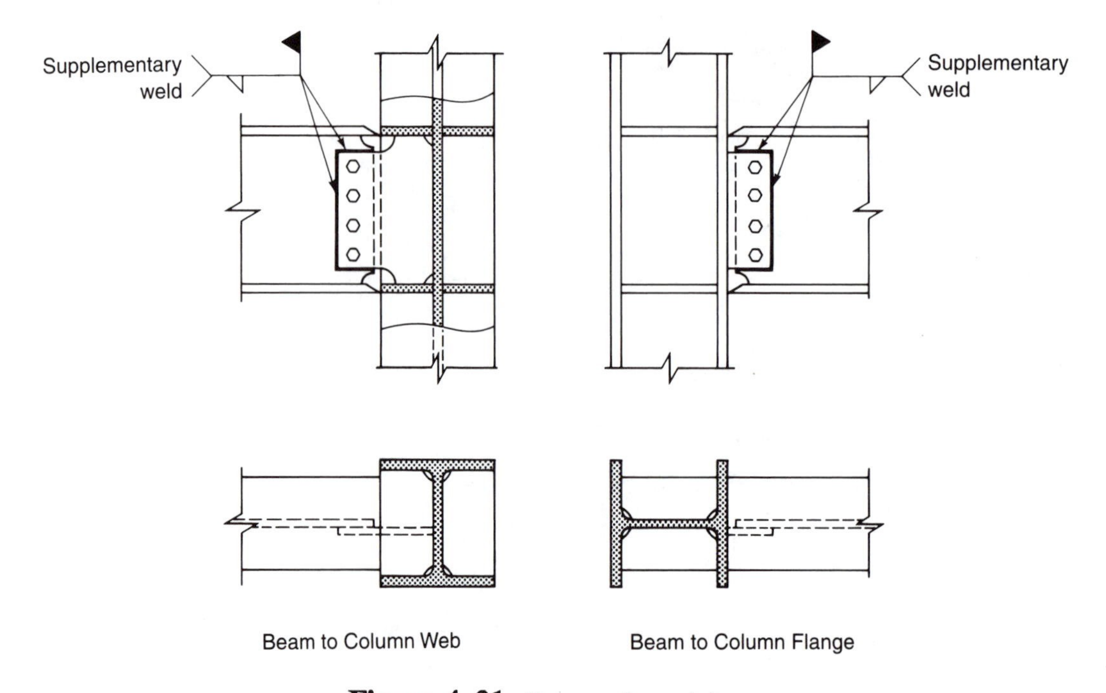

Figure 4–31 Beam–column joint

- Continuity plates are required, in accordance with UBC Division IX Sections K1.2, K1.6 and K1.8 when the column flange thickness is less than

$$t_f = 0.4(1.8P_{bf}/F_{yc})^{1/2}$$

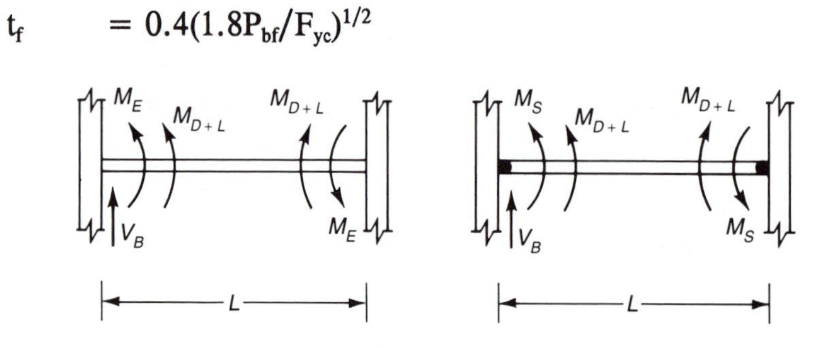

Figure 4–32 Beam shear force

or when the web depth clear of fillets is greater than

$$d_c = 4100t_{wc}(F_{yc})^{1/3}/P_{bf}$$

and the required cross-sectional area of the stiffener is given by

$$A_{st} = [P_{bf} - F_{yc}t_{wc}(t_b + 5k)]/F_{yst}$$

where P_{bf} = beam flange force
= bt_bF_{yb}
F_{yb} = beam yield stress
F_{yc} = column yield stress
F_{yst} = stiffener yield stress
t_b = beam flange thickness
b = beam flange width
t_{wc} = column web thickness
k = column fillet depth

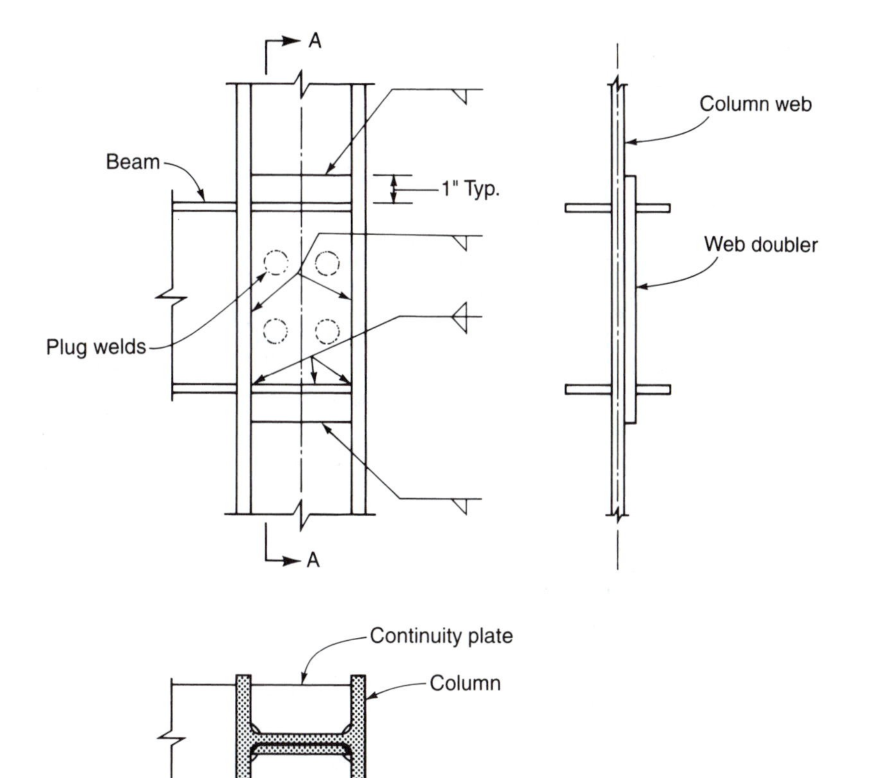

Figure 4–33 Detail of web doubler

The minimum stiffener thickness required is given by

$$t_{st} = t_b/2$$

The minimum stiffener width required is given by

$$b_{st} = b/3 = t_{wc}/2$$

The maximum width-thickness ratio is defined by UBC Division IX Section B5.1 as

$$b_{st}/t_{st} = 95/(F_{yst})^{1/2}$$

Seismic damage to special moment-resisting frames

Subsequent to the January 1994 Northridge earthquake in California, it was determined[10,11] that some damage had occurred at the girder-to-column connection. This consisted of cracking of the beam bottom flange weld which propagated into the adjacent column flange and web and into the beam bottom flange. This failure was accompanied in some instances by secondary cracking of the beam web shear plate and failure of the beam top flange weld. To improve the performance of the girder-to-column connection, the additional design details shown in Figure 4-34 have been recommended[10,12,13].

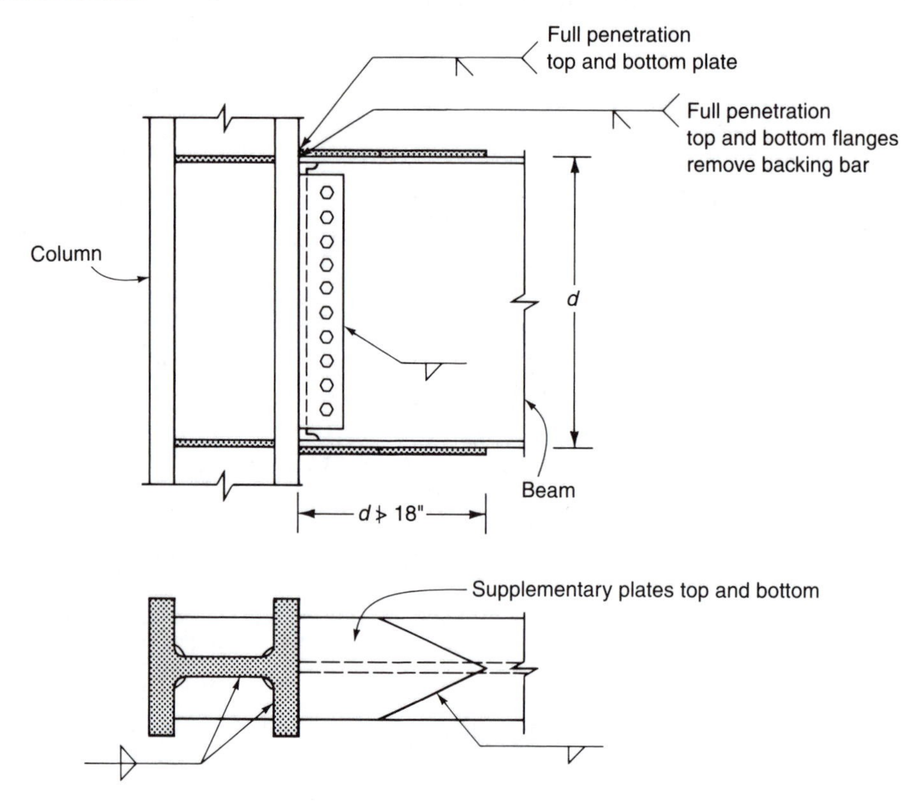

Figure 4–34 Strengthening details

Example 4–7 (Structural Engineering Examination 1988, Section B, weight 6.0 points)

GIVEN: Analyze the beam-column joint (panel zone) of the special moment-resisting frame shown in Figure 4-35.

CRITERIA: Seismic zone 3
All welds - E70XX Electrodes
All material ASTM A36
Assumptions:

- All requirements for compact sections are satisfied.
- Plastic hinges at this joint will not be the last to form.
- Beam-column shear connections are adequate.
- Design must adhere to UBC requirements for seismic design of steel.

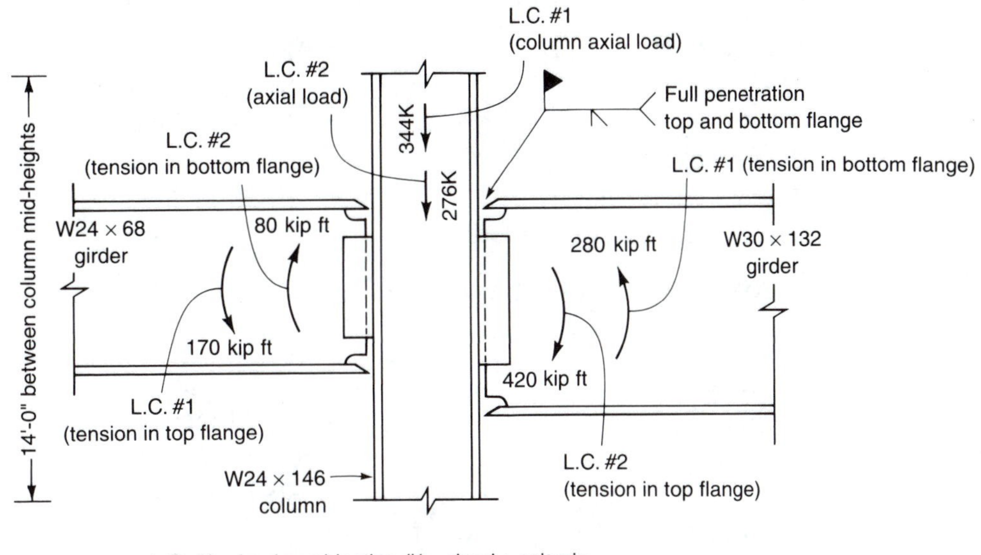

Figure 4–35 Details for Example 4–7

REQUIRED:
1. Complete the design of the beam–column connection, such that it satisfies seismic ductility requirements.
2. Complete a sketch showing the results of Part 1. Welding need not be shown or calculated.
3. Calculate the maximum permissible distance from the column face to a point of lateral support along each beam. Assume both beam deflected shapes are single curvature between the points considered.
4. Show by calculation that this beam–column represents a strong–column/weak–beam design.

SOLUTION

1. CONNECTION DESIGN

The relevant properties of the W24 × 146 column are

Plastic modulus, Z_c	= 418 inches^3
Yield stress, F_{yc}	= 36 kips per square inch
Depth, d_c	= 24.74 inches
Flange thickness, t_{cf}	= 1.09 inches
Flange width, b_c	= 12.9 inches
Web thickness, t	= 0.65 inches
Fillet depth, k	= 1.875 inches
Clear depth between fillets, T	= 21 inches
Area, A	= 43 square inches

The relevant section properties of the W30 × 132 beam 1 are

Plastic modulus, Z_1	= 437 inches^3
Yield stress, F_{yb}	= 36 kips per square inch
Depth, d_{b1}	= 30.31 inches
Flange thickness, t_{b1}	= 1.00 inches
Flange width, b_1	= 10.55 inches
Minimum radius of gyration, r_{y1}	= 2.25 inches

The relevant section properties of the W24 × 68 beam 2 are

Plastic modulus, Z_2	= 177 inches^3
Yield stress, F_{yb}	= 36 kips per square inch
Depth, d_{b2}	= 23.73 inches
Flange thickness, t_{b2}	= 0.585 inches
Flange width, b_2	= 8.97 inches
Minimum radius of gyration, r_{y2}	= 1.87 inches

APPLIED LOADS

From the problem statement, the gravity and seismic moments in beam 1 are

$$M_{V1} = (420 - 280)/2$$
$$= 70 \text{ kip feet}$$
$$M_{E1} = 420 - 70$$
$$= 350 \text{ kip feet}$$

The gravity and seismic moments in beam 2 are

$$M_{V2} = (170 - 80)/2$$
$$= 45 \text{ kip feet}$$
$$M_{E2} = 170 - 45$$
$$= 125 \text{ kip feet}$$

The applied forces at the joint, for seismic load acting from left to right, are shown in Figure 4-36.

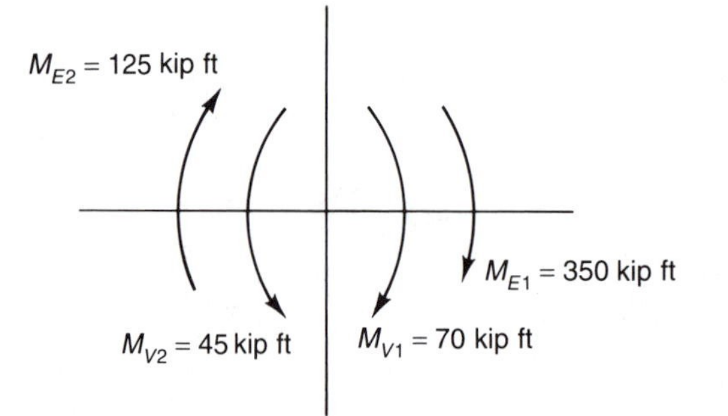

Figure 4-36 Applied forces at the joint

PANEL ZONE SHEAR

The panel zone design moments for seismic load acting from left to right, are obtained in accordance with UBC Section 2211.7.2.1 as

$$
\begin{aligned}
M_1 &= M_{V1} + 1.85\ M_{E1} \\
&= 70 + 1.85 \times 350 \\
&= 717.5 \text{ kip feet} \\
M_2 &= M_{V2} + 1.85\ M_{E2} \\
&= -45 + 1.85 \times 125 \\
&= 186.25 \text{ kip feet}
\end{aligned}
$$

The panel zone shear is obtained from AISC Formula (C-E6-1) as

$$V_{PZ} = M_1/0.95d_{b1} + M_2/0.95d_{b2} - (M_1 + M_2)/H$$

where H = distance between column midheights

= 14 feet

Hence

$$
\begin{aligned}
V_{PZ} &= 12 \times 717.5/(0.95 \times 30.31) + 12 \times 186.25/(0.95 \times 23.73) \\
&\quad -(717.5 + 186.25)/14 \\
&= 333.6 \text{ kips}
\end{aligned}
$$

Similarly, for seismic load acting from right to left, the panel zone design moments and the shear are

$$
\begin{aligned}
M_1 &= 717.5 - 2 \times 70 \\
&= 577.5 \text{ kip feet} \\
M_2 &= 186.25 + 2 \times 45 \\
&= 276.25 \text{ kip feet} \\
V_{PZ} &= 325.7 \text{ kips} \\
&< 333.6
\end{aligned}
$$

However, in accordance with UBC Section 2211.7.2.1, the panel zone design shear need not exceed the shear produced by 0.8 times the full plastic capacities of the beams. The plastic capacities of the beams are

$$
\begin{aligned}
M_{s1} &= F_{yb}Z_1 \\
&= 36 \times 437/12 \\
&= 1{,}311 \text{ kip feet} \\
M_{s2} &= F_{yb}Z_2 \\
&= 36 \times 177/12 \\
&= 531 \text{ kip feet}
\end{aligned}
$$

and these moments produce a panel zone shear in excess of that due to seismic load acting from left to right. Hence, the design panel zone shear is

$$V_{PZ} = 333.6 \text{ kips}$$

In determining the shear capacity of the unreinforced column web, the depth of the deeper beam is used as recommended in the SEAOC Commentary 4F.2.a, and the capacity is given by UBC Formula (11-1) as

$$V = 0.55F_{yc}d_ct[1 + 3b_ct_{cf}^2/d_{b1}d_ct]$$
$$= 0.55 \times 36 \times 24.74 \times 0.65\,[1 + 3 \times 12.9 \times 1.09^2/(30.31 \times 24.74 \times 0.65)]$$
$$= 348.44$$
$$> V_{PZ}$$

Hence the unreinforced web is adequate and no doubler plate is required.

The minimum panel zone thickness to prevent shear buckling failure is given by UBC Formula (11-2) as

$$t_z = (d_z + w_z)/90$$

where d_z = panel zone depth between continuity plates

$$\approx d_{b1} - t_{b1}$$
$$= 30.31 - 1$$
$$= 29.31 \text{ inches}$$

and w_z = panel zone width between flanges

$$= d_c - 2t_{cf}$$
$$= 24.74 - 2 \times 1.09$$
$$= 22.56 \text{ inches}$$

Hence $t_z = (29.31 + 22.56)/90$

$$= 0.58 \text{ inches}$$
$$< t$$

Thus, the column web thickness is adequate to prevent shear buckling failure.

STIFFENER REQUIREMENTS

The force in the beam flange for beam 1, for seismic zone 3, is defined in UBC Division IX Formula (K1-9) as

$$P_{bf} = A_{fb}F_{yb}$$
$$= t_{b1}b_1F_{yb}$$
$$= 1.00 \times 10.55 \times 36$$
$$= 380 \text{ kips}$$

The required stiffener area, to resist the concentrated flange force, is given by UBC Formula (K1-9) as

$$A_{st} \geq [P_{bf} - F_{yc}t(t_b + 5k)]/F_{yst}$$

Using calculator program A4.4 in Appendix III, the value is

$$A_{st} = 3.81 \text{ square inches}$$

Hence stiffeners, or continuity plates, are required and it is unnecessary to check Formulae (K1-8) and (K1-1) since the stiffeners are provided over the full depth of the column web. Using a pair of stiffener plates ½-inch × 4 inches provides a stiffener area of

$$A_{st} = 2 \times 4 \times 0.5$$
$$= 4.00 \text{ square inches}$$
$$> 3.81$$

The minimum stiffener thickness required is defined by UBC Division IX Section 2251 K8 as

$$t_{st} = t_{b1}/2$$
$$= 1.00/2$$
$$= 0.50 \text{ inches}$$

and this is provided.
The minimum stiffener width required is defined by UBC Division IX Section 2251 K8 as

$$b_{st} = b_1/3 - t/2$$
$$= 10.55/3 - 0.65/2$$
$$= 3.2 \text{ inches}$$
$$< 4$$

The maximum width–thickness ratio is defined by UBC Division IX Section B5.1 as

$$b_{st}/t_{st} = 95/(F_{yst})^{0.5}$$
$$= 95/(36)^{0.5}$$
$$= 15.8$$

The actual width–thickness ratio provided is

$$b_{st}/t_{st} = 4 \times 2/1$$
$$= 8$$
$$< 15.8$$

Hence, the stiffeners provided are adequate.

The force in the beam flange for beam 2 is

$$P_{bf} = t_{b2}b_2F_{yb}$$
$$= 0.585 \times 8.97 \times 36$$
$$= 189 \text{ kips}$$

The required stiffener area, to resist the concentrated flange force, is

$$A_{st} \geq [P_{bf} - F_{yc}t(t_b + 5k)]/F_{yst}$$

and this is evaluated by calculator program A4.4 as

$$A_{st} = -1.22$$

and stiffeners are not required.

To prevent buckling opposite the compression flange, UBC Formula (K1–8) requires the provision of stiffeners if the column web depth clear of fillets exceeds the value

$$T = 4100t^3(F_{yc})^{0.5}/P_{bf}$$

and this is evaluated by calculator program A4.4 as

$$T = 35.75 \text{ inches}$$

The actual clear web depth is

$$T = 21 \text{ inches}$$

and stiffeners are not required opposite the compression flange.

To prevent flange bending opposite the tension flange, UBC Formula (K1–1) and Section 2211.7.4 require the provision of stiffeners if the column flange thickness is less than the value

$$t_{cf} = 0.4(1.8P_{bf}/F_{yc})^{0.5}$$

and this is evaluated by calculator program A4.4 as

$$t_{cf} = 1.23 \text{ inches}$$

The actual flange thickness is

$$t_{cf} = 1.09 \text{ inches} < 1.23$$

and, hence, stiffeners are required opposite the tension flange and a pair of stiffeners ⅜-inch × 3 inches satisfy the requirements of UBC Division IX Section 2251 K8. Details of the connection are shown in Figure 4–37.

3. BEAM BRACING

In accordance with UBC Section 2211.7.8 both flanges of the beams shall be braced at a maximum distance, between column center lines, given by the expression

$$l = 96r_y$$

For beam 1, this length is given by

$$l = 96 \times 2.25/12 = 18 \text{ feet}$$

For beam 2, this length is given by

$$l = 96 \times 1.87/12 = 15 \text{ feet}$$

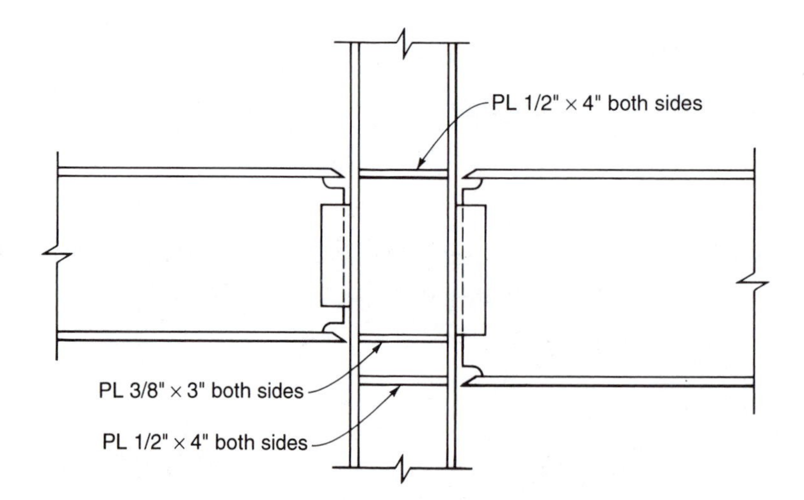

Figure 4–37 Girder–column connection

4. COLUMN-BEAM STRENGTH RATIO
From the problem statement, the maximum column load is

$$P_T = 344 \text{ kips}$$

The corresponding axial stress is

$$\begin{aligned} f_a &= P_T/A \\ &= 344/43 \\ &= 8 \text{ kips per square inch} \end{aligned}$$

To ensure that hinges will form in the beams rather than in the columns, when the panel zone shear strength does not control, UBC Formula (11-3.1) specifies that

$$\begin{aligned} 1 &< \Sigma Z_c(F_{yc} - f_a)/\Sigma Z_b F_{yb} \\ &< 2 \times 418(36-8)/36(437 + 177) \\ &< 1.06 \end{aligned}$$

Hence, the requirement for a strong-column/weak-beam design is satisfied.

Example 4–8 (Structural Engineering Examination 1992, Section B, weight 7 points)

GIVEN: The elevation, plan sections and section properties of an interior steel column as shown in Figure 4–38. The deflected shape of the column due to a seismic event about either axis as shown in Figure 4–39.

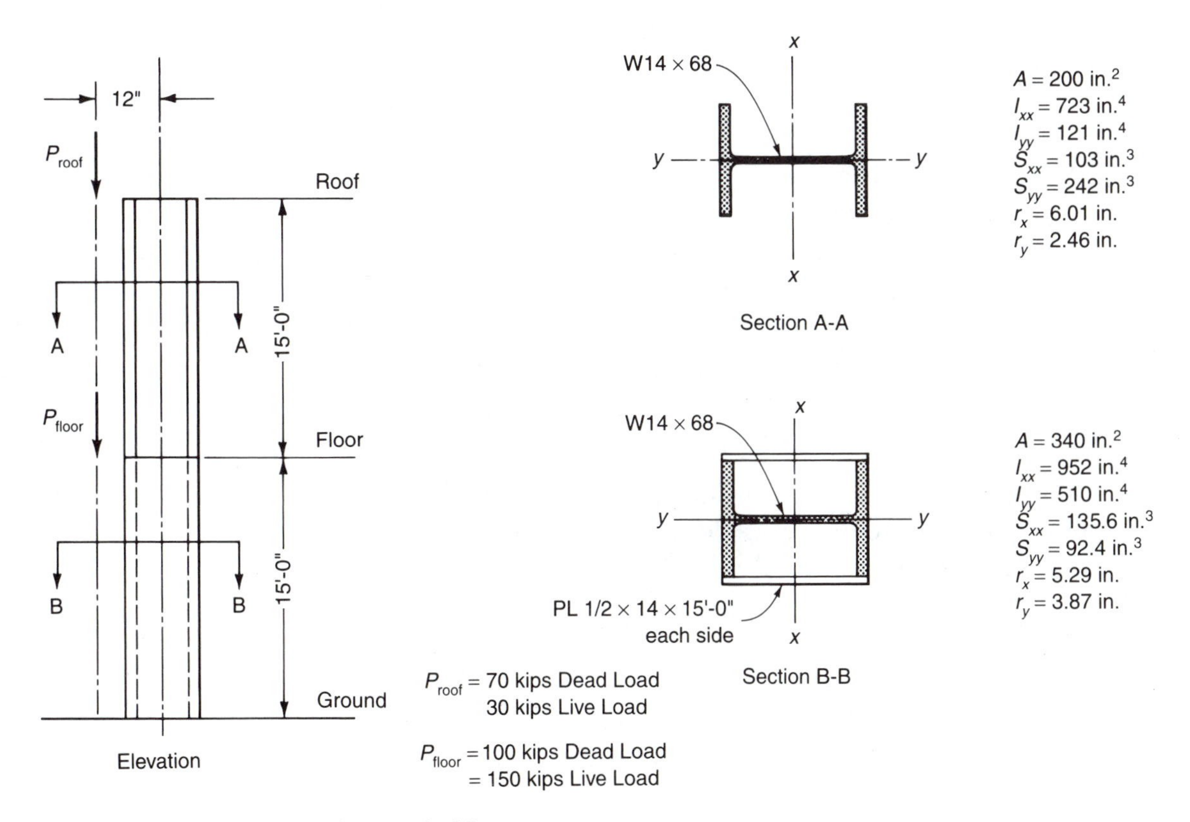

Figure 4–38 Details for Example 4–8

CRITERIA: Materials:

Steel A36, F_y = 36 kips per square inch.

Assumptions:

- Do not reduce the loads shown on Figure 1 for tributary area considerations.
- The column is not part of the lateral force resisting system.
- Cover plates are fully composite with column.
- Lateral support is provided at the floor and roof diaphragm in both directions.
- The column is continuous at the floor and pinned at the ground and roof.
- Assume the effective length factor K = 0.8 for the upper and lower portions of the column.

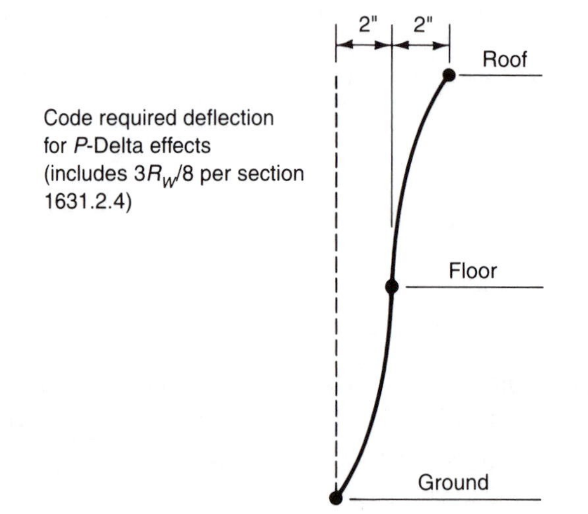

Figure 4–39 Column displacements

REQUIRED:

1. Determine the moment diagram for the column for eccentric loads from gravity only.
2. Determine the adequacy of the upper and lower portions of the column for gravity loads only.
3. Determine the adequacy of the upper and lower portions of the column including P–delta effects.

SOLUTION

1. MOMENT DIAGRAM FOR GRAVITY LOADING

Both the live load and the dead load at the floor and roof are applicable and the moment applied to the column at the roof level is

$$M_R = (70 + 30)12/12$$
$$= 100 \text{ kip feet}$$

The moment applied at the floor level is

$$M_F = (100 + 150)12/12$$
$$= 250 \text{ kip feet}$$

The resulting column moments are determined by the moment distribution procedure in Table 4–4. The convention adopted is that clockwise moments acting from the joint on the member are considered positive. There is no carry-over to the hinged ends and the cantilevers RO and FO, producing the applied moments, have zero stiffness.

Joint	R		F			G
Member	RO	RF	FR	FO	FG	GF
Relative EI/L	0		723	0	952	
Distribution factors	0	1	0.43	0	0.57	1
Carry-over factors	0	0.5→	←0.0	0	0.0→	←0.5
Applied moments	100			250		
Distribution		-100	-107.5		-142.5	
Carry-over			- 50.0			
Distribution			+21.5		28.5	
Final moments	100	-100	-136.0	250	-114.0	0

Table 4–4 Moment distribution for Example 4–8

The bending moment diagram is shown in Figure 4–40.

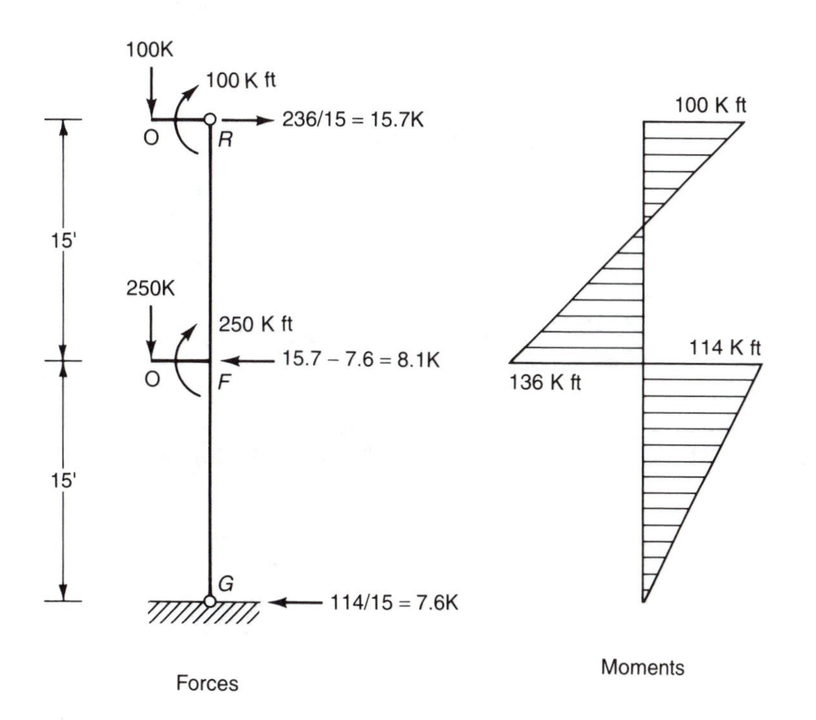

Figure 4–40 Column moments

2a. UPPER COLUMN WITH GRAVITY LOADING

The flexural stress produced in the upper column by the bending moment is

$$f_b = M/S_x$$
$$= 136 \times 12/103$$
$$= 15.85 \text{ kips per square inch}$$

For the W14 × 68 section, AISC Table 1-26 provides no value for F'_y indicating that

$$F'_y > 65 \text{ kips per square inch}$$
$$> F_y$$

Hence the section is compact, and the allowable bending stress is dependent on the value of the unbraced length which is given as

$$L_{RF} = 15 \text{ feet}$$

From AISC Table 2-65, the unbraced length at which the allowable bending stress changes are

$$L_c = 10.6 \text{ feet}$$

and $$L_u = 23.9 \text{ feet, for } C_b = 1$$

The actual value of the bending coefficient is obtained from UBC Division IX Section F1.3 as

$$C_b = 1.75 + 1.05(M_1/M_2) + 0.3(M_1/M_2)^2$$
$$> 1.0$$

where $$M_1 = +\ 100 \text{ kip feet}$$
$$M_2 = +\ 136 \text{ kip feet, for reverse curvature bending}$$

Then $$L_u > 23.9 \text{ feet}$$

and $$L_c < L_{RF} < L_u$$

and the allowable bending stress is given by

$$F_b = 0.6F_y$$
$$= 21.6 \text{ kips per square inch}$$

The slenderness ratios about the X-axis and the Y-axis are given by

$$KL_{RF}/r_x = 0.8 \times 15 \times 12/6.01$$
$$= 23.96$$

and $$KL_{RF}/r_y = 0.8 \times 15 \times 12/2.46$$
$$= 58.54$$
$$> 23.96$$

Hence the slenderness ratio about the Y-axis governs and the allowable axial stress is obtained from AISC Table C-36 as

$$F_a = 17.57 \text{ kips per square inch}$$

The actual compressive stress in the upper column due to dead load plus live load, is given by

$$
\begin{aligned}
f_a &= P/A \\
&= 100/20 \\
&= 5 \text{ kips per square inch} \\
&> 0.15F_a
\end{aligned}
$$

Hence, in checking the adequacy of the member in combined axial compression and bending in accordance with UBC Division IX, Formulas (H1-1) and (H1-2) are applicable . Formula (H1-1) is

$$f_a/F_a + C_m f_b/F_b(1-f_a/F'_e) \leq 1.0$$

where C_m = bending coefficient given by UBC Division IX Section H1

$$
\begin{aligned}
&= 0.6 - 0.4(M_1/M_2) \\
&= 0.6 - 0.4 \times 100/136 \\
&= 0.31
\end{aligned}
$$

and F'_e = factored Euler stress for a slenderness ratio KL_{RF}/r_x

= 260.18 kips per square inch, from AISC Table 8

Using calculator program A.4.2 in Appendix III, the left hand side of Formula (H1-1) is evaluated as

$$0.52 \quad < 1.0$$

Formula (H1-2) is

$$f_a/0.6F_y + f_b/F_b \leq 1.0$$

Using calculator program A.4.2 in Appendix III, the left hand side of Formula (H1-2) is evaluated as

$$0.97 \quad < 1.0$$

Hence both formulas are satisfied and the W14 × 68 section is adequate.

2b. LOWER COLUMN WITH GRAVITY LOADING

The flexural stress in the lower column is

$$
\begin{aligned}
f_b &= 114 \times 12/135.6 \\
&= 10.09 \text{ kips per square inch}
\end{aligned}
$$

Considering the composite section is behaving as an I-shaped section, as defined in UBC Division IX Section B5 and F1, the allowable bending stress, by inspection, is

$$
\begin{aligned}
F_b &= 0.6F_y \\
&= 21.6 \text{ kips per square inch.}
\end{aligned}
$$

The slenderness ratios about the X-axis and the Y-axis are

$$
\begin{aligned}
KL_{RF}/r_x &= 0.8 \times 15 \times 12/5.29 \\
&= 27.22 \\
KL_{RF}/r_y &= 0.8 \times 15 \times 12/3.87 \\
&= 37.21 \\
&> 27.22
\end{aligned}
$$

Hence the slenderness ratio about the Y–axis governs and the allowable axial stress is obtained from AISC Table C–36 as

F_a = 19.41 kips per square inch

The actual compressive stress is

f_a = 350/34
= 10.29 kips per square inch
> $0.15F_a$

The bending coefficient is

C_m = $0.6 - 0.4(M_1/M_2)$
= 0.6 - 0.4(0.0/114)
= 0.6

The factored Euler stress for a slenderness ratio of 27.22 is

F'_e = 201.68 kips per square inch

The left hand side of Formula (H1–1) is evaluated as

0.83 < 1.0

The left hand side of Formula H1–2) is evaluated as

0.94 < 1.0

Hence both formulas are satisfied and the section is adequate.

3. COLUMN WITH P–DELTA EFFECTS

In accordance with UBC Section 1628.9, P–delta effects shall be considered when the story drift ratio is

$\Delta'_R \geq 0.02/R_W$

The story drift ratio is given as

Δ_R = Δ/H
= $2/(15 \times 12 \times 0.375R_W)$
= $0.03/R_W$
> Δ'_R

Hence P–delta effects must be considered.

The most serious loading condition occurs when the seismic displacement occurs about the minor axis of the column. The roof dead load and the floor dead plus live load are applicable and the P–delta moments acting about the minor axis are obtained from Figure 4–41 as

M_{Uy} = moment in upper column about the Y–axis
= 70×2
= 140 kips

M_{Ly} = moment in lower column about the Y-axis
= 320 × 2
= 640 kip inches

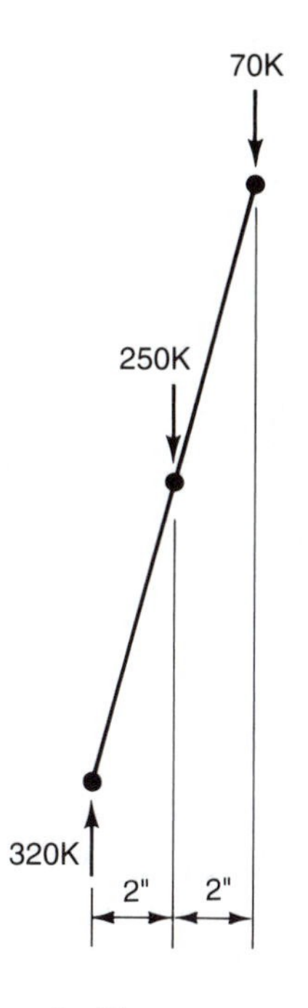

Figure 4–41 Column loads

The moments, due to gravity loading and eccentricity acting about the major axis, are determined by the moment distribution procedure in Table 4-5.

Joint	R		F			G
Member	RO	RF	FR	FO	FG	GF
Relative EI/L	0		723	0	952	
Distribution factors	0	1	0.43	0	0.57	1
Carry-over factors	0	0.5→	←0.0	0	0.0→	←0.5
Applied moments	70			250		
Distribution		-70	-107.5		-142.5	
Carry-over			- 35.0			
Distribution			+ 15.0		20.0	
Final moments	70	-70	-127.5	250	-122.5	0

Table 4–5 Moment distribution for Example 4-8

3a. UPPER COLUMN WITH P-DELTA EFFECTS
Formula (H1-2) is the most critical and the flexural stresses are

$$f_{bx} = 127.5 \times 12/103$$
$$= 14.85 \text{ kips per square inch}$$
$$= M_{Uy}/S_y$$
$$f_{by} = 140/24.2$$
$$= 5.79 \text{ kips per square inch}$$

The allowable flexural stresses are

$$F_{bx} = 0.6F_y \text{ from UBC Division IX Section F1.3}$$
$$= 21.6 \text{ kips per square inch}$$
$$F_{by} = 0.75F_y \text{ from UBC Division IX Formula (F2-1)}$$
$$= 27.0 \text{ kips per square inch}$$

The actual axial stress is

$$f_a = 70/20$$
$$= 3.5 \text{ kips per square inch}$$

In accordance with UBC Section 1631.2.4 allowable stresses may be increased by 1.7 times when P-delta effects are considered. Formula (H1-2) is, then

$$f_a/0.6F_y + f_{bx}/F_{bx} + f_{by}/F_{by} \leq 1.7$$
$$3.5/21.6 + 14.85/21.6 + 5.79/27.0 \leq 1.7$$
$$1.06 \leq 1.7$$

The formula is satisfied and the section is adequate.

3b. LOWER COLUMN WITH P-DELTA EFFECTS
The flexural stresses are

$$f_{bx} = 122.5 \times 12/135.6$$
$$= 10.84 \text{ kips per square inch}$$
$$f_{by} = M_{Ly}/S_y$$
$$= 640/92.4$$
$$= 6.93 \text{ kips per square inch}$$

The axial stress is

$$f_a = 320/34$$
$$= 9.41 \text{ kips per square inch}$$

The left hand side of Formula (H1-2) is

$$9.41/21.6 + 10.85/21.6 + 6.93/27 = 1.19$$
$$< 1.7$$

The formula is satisfied and the section is adequate.

References

1. Williams A. *The analysis of indeterminate structures.* Hart Publishing Company, Inc., New York, 1968, pp 9-41.

2. Tuma J.J. *Structural Analysis.* McGraw-Hill, New York, 1969, pp 162-179.

3. Applied Technology Council. *Guidelines for the design of horizontal wood diaphragms.* Berkeley, CA 1981, pp 13-14.

4. American Institute of Steel Construction. *Seismic provisions for structural steel buildings.* Chicago, 1992, p 12.

5. American Plywood Association. *Diaphragms.* APA Design/Construction Guide, Tacoma, WA, 1989, pp 9-23.

6. Becker, R. Naeim, F., Teal, E.J. *Seismic design practice for steel buildings.* Steel Committee of California, El Monte, 1988, p 26.

7. Ishler, M. *Seismic design practice for eccentrically braced frames.* Structural Steel Educational Council, Moraga, CA, 1992, p 43.

8. Structural Engineering Association of California. *Recommended lateral force requirements and commentary.* Sacramento, CA, 1990, p 111-C.

9. Lehmer, G.D. Detailing provisions for steel structures. *Proceedings of the 1988 seminar of the Structural Engineers Association of Southern California.* Los Angeles, CA, November 1988, p 21.

10. Sabol, T.A. Damage to ductile steel frames in the Northridge Earthquake. *Proceedings of the 1994 seminar of the Structural Engineers Association of Southern California.* Los Angeles, CA , March 1994.

11. Tide, R.H.R. Fracture of welded beam-to-column connections when subjected to seismic loading. *Proceedings of the AISC Special Task Committee on the Northridge Earthquake Meeting.* American Institute of Steel Construction, Chicago, IL, March 1994.

12. Engelhardt, M. Experimental evidence on welded flange-bolted web connections. *Proceedings of the AISC Special Task Committee on the Northridge Earthquake Meeting.* American Institute of Steel Construction, Chicago, IL, March 1994.

13. City of Los Angeles Building Bureau. *Requirements for connections in new steel frame structures.* Los Angeles, CA, 1994.

This page left blank intentionally.

5

Seismic design of concrete structures

5.1 ANCHORAGE TO CONCRETE

5.1.1 Design loads

The applicable loading combinations to be used in the strength design of concrete elements subjected to seismic loading is specified in UBC Section 1921.2.7 as

	U	$= 1.4(D + L + E)$
and	U	$= 0.9D \pm 1.4E$
where	D	= dead load
	L	= live load
	E	= seismic load
	U	= required strength

An additional load factor of 2 is required by UBC Section 1925.2 when concrete anchors are embedded in the tension zone of the concrete member and special inspection is provided. If special inspection is not provided, an additional load factor of 3 is required. When the anchors are not embedded in the tension zone, an additional load factor of 1.3 is required when special inspection is provided and an additional load factor of 2 when special inspection is not provided.

5.1.2 Design strength

The strength of an anchor bolt in tension is given by UBC Section 1925.3.2 as

$P_{ss} = 0.9A_b f'_s$

where A_b = bolt area

f'_s = bolt tensile strength

The strength of an anchor bolt in shear is

$V_{ss} = 0.75A_b f'_s$

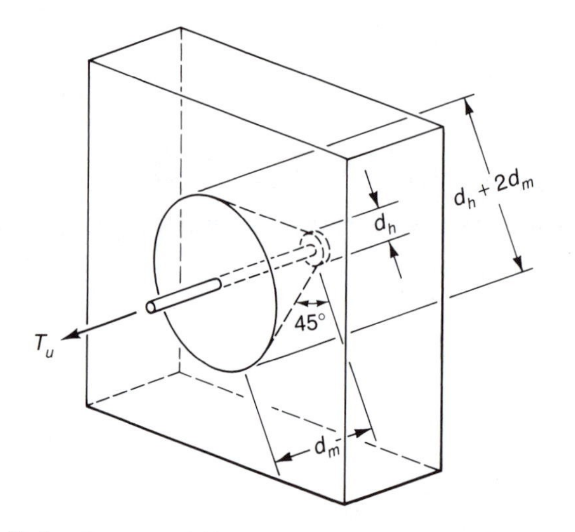

Figure 5-1 Concrete failure in tension for a single anchor

When the anchor bolts' spacing is not less than twice their embedment length, the concrete failure surface in tension is a truncated cone, as shown in Figure 5-1, with an area of

$A_s = 1.414\pi d_m(d_m + d_h)$

where d_m = embedment length

d_h = bolt head diameter

The concrete design capacity in tension is given by

$\phi P_c = \phi\lambda(2.8A_s)(f'_c)^{0.5}$

where ϕ = strength reduction factor

= 0.65

λ = concrete aggregate parameter

= 1.0 for normal-weight concrete

= 0.75 for all lightweight concrete

= 0.85 for sand-lightweight concrete

When the anchor bolts' spacing is less that twice their embedment length, the concrete failure surface in tension is a truncated pyramid, as shown in Figure 5-2, with an area of

$A_s = 2.828d_m(a + b + d_m)$

The area of the base of the truncated pyramid is

$A_t = ab$

The concrete design capacity in tension is given by

$$\phi P_c = \phi\lambda(2.8A_s + 4A_t)(f'_c)^{0.5}$$

When the edge distance is less than the embedment length, the value of ϕP_c shall be reduced proportionately.

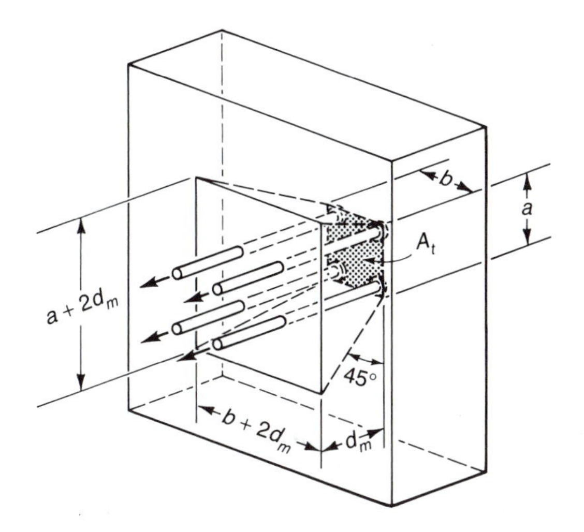

Figure 5–2 Concrete failure in tension for an anchor group

The concrete design capacity in shear, when the anchor is not less than ten diameters from the loaded edge of the member, is given by

$$\phi V_c = 800\phi\lambda A_b(f'_c)^{0.5}$$

The concrete design capacity in shear, when the anchor is less than ten diameters from the loaded edge of the member, is given by

$$\phi V_c = 2\phi\pi d_e^2\lambda(f'_c)^{0.5}$$

where d_e = edge distance of the anchor

The edge distance shall not be less than 3.33 diameters from the loaded edge and reinforcement shall be provided to carry the load when the edge distance is less than ten diameters.

For combined tension and shear, the following interaction equations shall be satisfied:

$$[(P_u/P_c)^2 + (V_u/V_c)^2]/\phi \leq 1$$

$$(P_u/P_{ss})^2 + (V_u/V_{ss})^2 \leq 1$$

where P_u = factored tensile load
V_u = factored shear load

In addition, the concrete design capacity in shear and tension shall each exceed the factored applied loads.

Example 5–1 (Structural Engineering Examination 1991, Section C, weight 8.0 points)

GIVEN: The bracket attached to a concrete wall as shown in Figure 5–3.

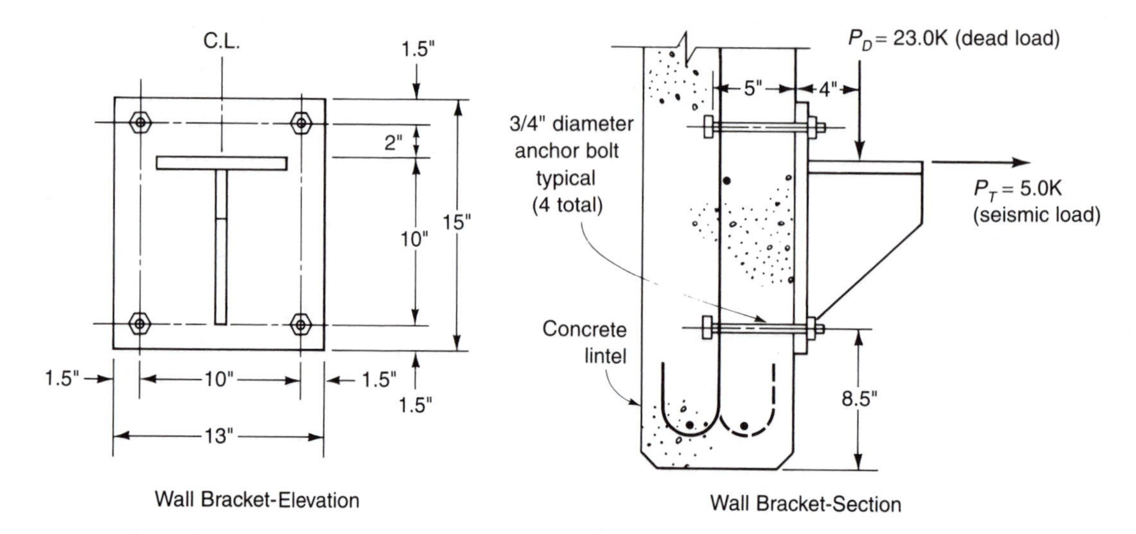

Figure 5–3 Details for Example 5–1

CRITERIA: Code: UBC, Strength Design Method, $\phi = 0.65$

Materials:

- Concrete: $f'_c = 4{,}000$ pounds per square inch, normal weight
- ¾-inches diameter A325 Bolts, head diameter = 1.31 inches

Assumptions:

- The steel bracket is infinitely rigid.
- The center of the rectangular compression block is at the center of the bottom bolts.
- The anchor bolts are embedded in the tension zone of the concrete.
- The concrete wall is adequate to carry the given loads.
- Special Inspection is provided.

REQUIRED:
1. Determine the maximum factored tension and shear forces on the bolts.
2. Determine if the connection is adequate for the given loads.

SOLUTION

1. FACTORED LOADS

Assuming that the location of the bracket is in seismic zone number 4, in accordance with UBC Section 1921.2.7, the required strength is

$$U = 1.4(D + E)$$

where

D = dead load = 23 kips

E = seismic load = 5 kips

An additional load factor of 2 is required by Section 1925.2 when the anchors are embedded in the tension zone of the concrete and special inspection is provided. Hence, the factored shear force acting on each bolt is given by

$$
\begin{aligned}
V_u &= 2 \times 1.4P_D/4 \\
&= 2 \times 1.4 \times 23/4 \\
&= 16.10 \text{ kips}
\end{aligned}
$$

Similarly, by taking moments about the bottom bolts, the factored tensile force in each top bolt is

$$
\begin{aligned}
P_u &= 2 \times 1.4(4P_D + 10P_T)/(2 \times 12) \\
&= 2 \times 1.4(4 \times 23 + 10 \times 5)/24 \\
&= 16.57 \text{ kips}
\end{aligned}
$$

2. ANCHOR CAPACITY

The bolt strength in tension is given by

$$P_{ss} = 0.9A_bf'_s$$

where A_b = bolt area = 0.442 square inches from AISC Table I-A.

f'_s = bolt tensile strength = 120 kips per square inch from AISC Table I-C.

Thus

$$
\begin{aligned}
P_{ss} &= 0.9 \times 0.442 \times 120 \\
&= 47.74 \text{ kips} \\
&> P_u
\end{aligned}
$$

The bolt strength in shear is given by

$$
\begin{aligned}
V_{ss} &= 0.75A_bf'_s \\
&= 0.75 \times 0.442 \times 120 \\
&= 39.78 \text{ kips} \\
&> V_u
\end{aligned}
$$

Hence the bolt strength is satisfactory in both tension and shear.

The ratio of anchor spacing to embedment length is

$$
\begin{aligned}
s/d_m &= 10/5 \\
&= 2
\end{aligned}
$$

Hence the failure surface in the concrete forms a truncated cone, radiating from the bolt head, with a surface area of

$$A_s = 1.414\pi(d_h + d_m)d_m$$

where d_h = bolt head diameter = 1.31 inches, as given

Thus

$$
\begin{aligned}
A_s &= 1.414\pi(1.31 + 5) \times 5 \\
&= 140.17 \text{ square inches}
\end{aligned}
$$

The ratio of the edge distance to the embedment length is

$$
\begin{aligned}
d_e/d_m &= (8.5 + 12)/5 \\
&= 4.1 \\
&> 1
\end{aligned}
$$

Hence the concrete capacity in tension for normal weight concrete is given by

$$\begin{aligned}\phi P_c &= \phi\lambda(2.8A_s)(f'_c)^{0.5}\\ &= 0.65 \times 1 \times (2.8 \times 140.17)(4000)^{0.5}/1,000\\ &= 16.14 \text{ kips}\\ &< P_u\end{aligned}$$

Hence the connection is inadequate for combined dead load plus seismic load. The ratio of edge distance to bolt diameter is

$$\begin{aligned}d_e/d_b &= 8.5/0.75\\ &= 11.33\\ &> 10\end{aligned}$$

Hence the concrete capacity in shear is given by

$$\begin{aligned}\phi V_c &= 800\phi\lambda A_b(f'_c)\\ &= 800 \times 0.65 \times 1 \times 0.442 \times (4000)^{0.5}/1,000\\ &= 14.54 \text{ kips}\\ &< V_u\end{aligned}$$

Hence the connection is inadequate for the applied dead load.

5.2 SPECIAL MOMENT-RESISTING FRAMES

5.2.1 Design strength

The basic requirement of strength design is to ensure that the design strength of a member is not less than the required ultimate strength. For seismic loading, the required strength consists of the service level loads multiplied by the appropriate load factors specified in UBC Section 1921.2.7. The design strength of a member consists of the theoretical ultimate strength, or nominal strength, multiplied by the appropriate strength reduction factor ϕ. Thus

$$\phi \times (\text{nominal strength}) \geq U$$

UBC Section 1909.3.2 defines the reduction factor as

$$\begin{aligned}\phi &= 0.90 \text{ for flexure}\\ &= 0.85 \text{ for shear and torsion}\\ &= 0.75 \text{ for compression members with spiral reinforcement}\\ &= 0.70 \text{ for compression members with lateral ties}\end{aligned}$$

In seismic zones 3 and 4, UBC Section 1909.3.4 specifies a shear strength reduction factor ϕ of 0.6 for walls, topping slabs, and framing members with a nominal shear strength less than the shear corresponding to their nominal flexural strengths. For beam-to-column connections, the shear strength reduction factor ϕ is specified as 0.85.

The nominal strength of a member is determined in accordance with the principles defined in UBC Section 1910.2.7. These principles are employed in the calculator programs in Appendix III and in published design aids[1,2,3,4,5]. The nominal flexural capacity of a member with only tensile reinforcement is given by

$$M_n = A_s f_y d(1 - 0.59\rho f_y/f'_c)$$

where
A_s = area of tensile reinforcement
f_y = yield stress of the reinforcement
d = effective depth of section
ρ = reinforcement ratio
= A_s/bd
f'_c = compressive strength of the concrete
b = width of section

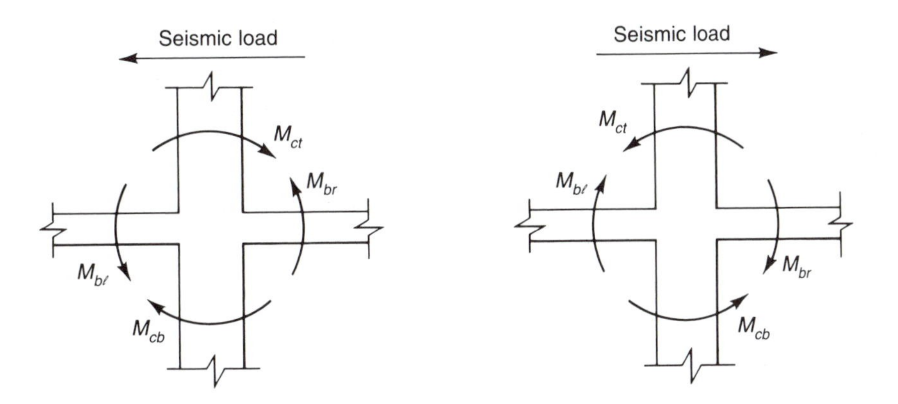

Figure 5–4 Strong-column/weak-beam concept

The formation of plastic hinges at both ends of the columns in a given story, due to seismic loads, may produce a sidesway mechanism which causes the story to collapse. To prevent this, the strong-column/weak-beam concept is required by UBC Section 1921.4.2. A column forming part of the lateral-force-resisting system and with factored axial force exceeding $A_g f'_c/10$ shall be designed to satisfy UBC Formula (21–1) which is

$$\Sigma M_e \geq 1.2\Sigma M_g$$

where
ΣM_e = sum of the design flexural strengths of columns at the center of a joint
ΣM_g = sum of the design flexural strengths of beams at the center of the joint and in the same plane as the columns
A_g = gross area of column section

As shown in Figure 5–4, the relationship applies to seismic loading from either direction. The sign convention adopted in the Figure is that bending moments at the ends of a member are shown acting from the joint to the member, i.e. the support reactions are considered. The arrow heads point towards the face of the member which is in tension.

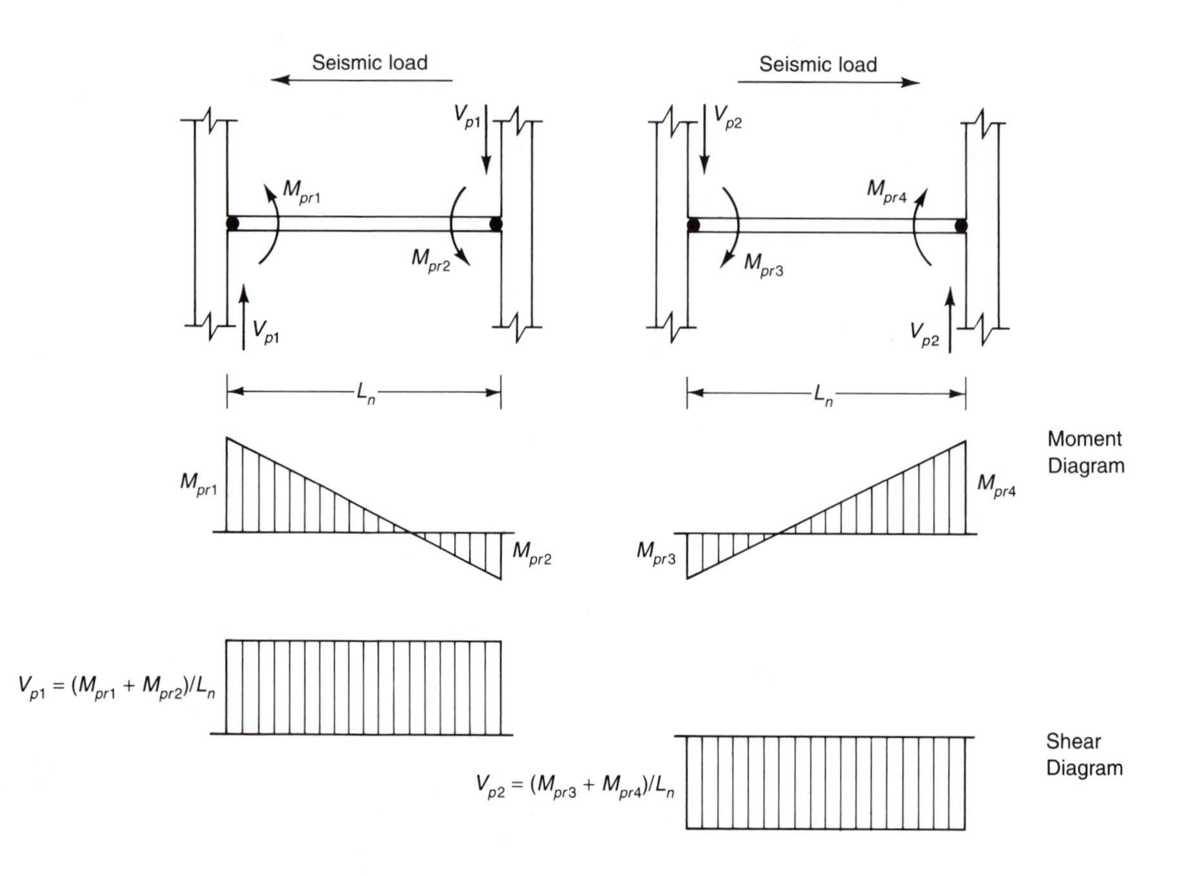

Figure 5-5 Beam shear due to probable flexural strength

To ensure ductile flexural failure of a member and prevent brittle shear failure, UBC Section 1921.3.4 requires the design shear force to be determined from the probable flexural strength that can be developed at the ends of the member plus the unfactored tributary gravity loads. The probable flexural strength is calculated[6,7] by assuming that strain hardening increases the effective tensile strength of the reinforcement by twenty-five percent and by using a strength reduction factor ϕ of 1.0, as specified in UBC Section 1921.0. The probable flexural strength is given by

$$M_{pr} = A_s(1.25f_y)d[1 - 0.59\rho(1.25f_y)/f'_c]$$
$$= A_sf_yd(1.25 - 0.92\rho f_y/f'_c)$$

As shown in Figure 5-5, moments of opposite sign act at the ends of a beam bent in double curvature and the sense of the moments reverses as the seismic loading reverses. Hence both the hogging and sagging probable flexural strengths must be calculated at both ends of the beam in order to determine the critical shear value. The design shear force at the left end of the beam for seismic load acting to the left is

$$V_e = (M_{pr1} + M_{pr2})/L_n + V_g$$

where L_n = beam clear span

V_g = shear due to unfactored tributary gravity loads at the left end of the beam.

The design shear force at the right end of the beam for seismic load acting to the right is

$V_e = (M_{pr3} + M_{pr4})/L_n + V_g$

where V_g = shear due to unfactored tributary gravity loads at the right end of the beam.

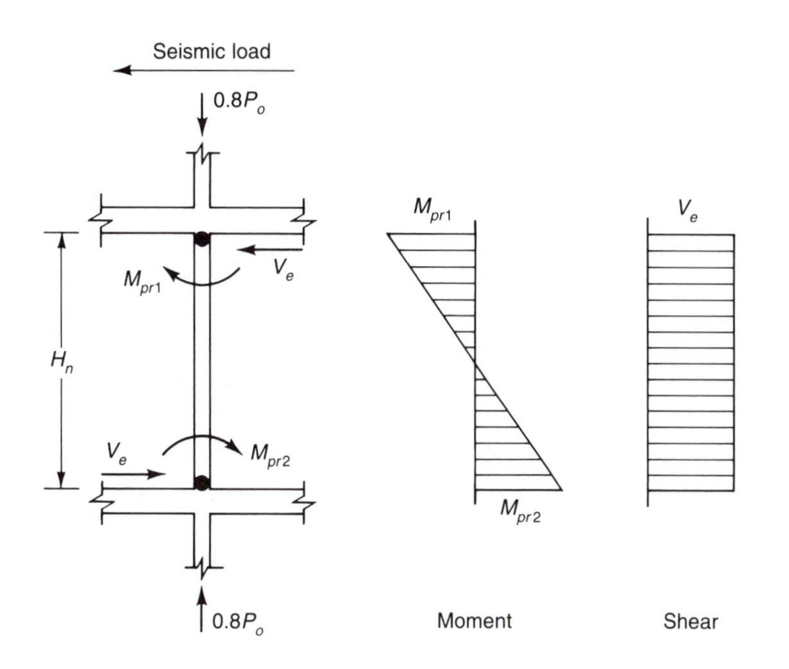

Figure 5–6 Column shear due to probable flexural strength

Similarly, in accordance with UBC Sections 1921.4.5 and 1921.4.4.7, the design shear force for columns shall be calculated using the probable moment strengths at the top and bottom of the column. As shown in Figure 5–6, the maximum probable moments are assumed to occur[7] under the maximum axial load of $0.8P_o$ which corresponds to the minimum accidental eccentricity. The design shear force at the top and bottom of the column is

$V_e = (M_{pr1} + M_{pr2})/H_n$

where H_n = column clear height

However, the column design shear need not exceed the value determined from the probable moment strengths of the beams framing into the top and bottom of the column. As shown in Figure 5–7, the design shear force for this condition is given by

$$V_e = (M_{pr1} + M_{pr2} + M_{pr3} + M_{pr4})/2H_n$$

The cyclical nonlinear effects produced by seismic loading necessitate additional shear requirements to ensure a ductile flexural failure. When the compressive force in a member is less than $A_g f'_c/20$ and the seismic induced shear is not less than half of the total design shear, the shear resistance of the concrete V_c shall be neglected. Shear reinforcement shall then be provided to resist the total design shear as required by UBC Section 1921.3.4.

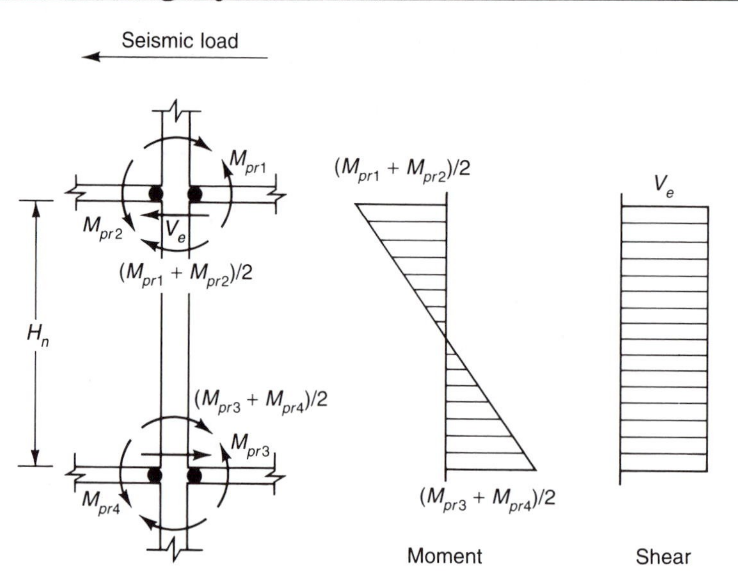

Figure 5–7 Column shear due to beam probable flexural strength

At a joint in a frame, the horizontal design shear force is determined as required by UBC Section 1921.5.1.1 and as shown in Figure 5–8. The shear force produced in the column by the probable moment strengths of the beams at the joint is

$$V = (M_{pr1} + M_{pr2})/H_c$$

where H_c = floor to floor height

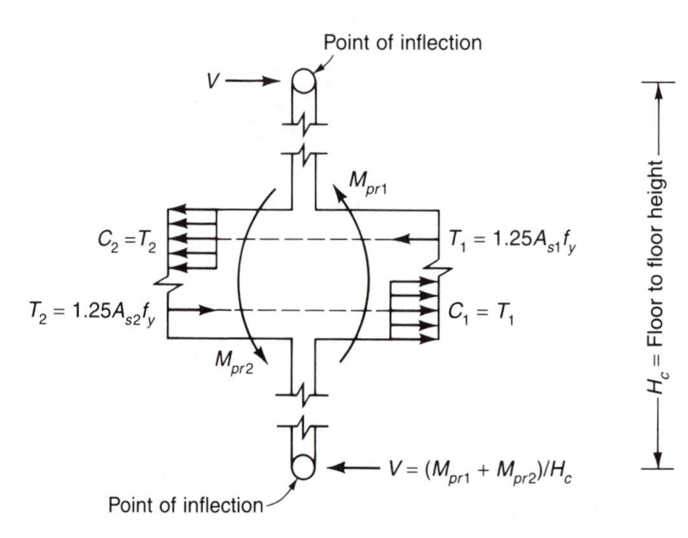

Figure 5–8 Forces acting at a joint

The probable tensile force in the tensile reinforcement in the right hand beam is

$$T_1 = 1.25A_{s1}f_y$$

where A_{s1} = area of tensile (top) reinforcement of right hand beam.

The probable compressive force in the concrete in the left hand beam is

$$C_2 = T_2 = 1.25A_{s2}f_y$$

where A_{s2} = area of tensile (bottom) reinforcement of left hand beam.

The net shear acting on the joint is given by

$$\begin{aligned} V_e &= T_1 + T_2 - V \\ &= 1.25f_y(A_{s1} + A_{s2}) - (M_{pr1} + M_{pr2})/H_c \end{aligned}$$

In accordance with UBC Section 1921.5.3 the nominal shear capacity of the joint depends on the concrete strength and the effective area of the joint and is given by

$$\begin{aligned} V_n &= 20A_j(f'_c)^{0.5} \text{ for joints confined on four faces} \\ &= 15A_j(f'_c)^{0.5} \text{ for joints confined on opposite faces or on three faces} \\ &= 12A_j(f'_c)^{0.5} \text{ for other conditions} \end{aligned}$$

where A_j = effective cross-sectional area within the joint.

As shown in Figure 5-9, the effective joint depth equals the overall depth of the column. Where a beam frames into a column of larger width, the effective joint width is given by

$$\begin{aligned} b_e &= b + h \\ &\leq b + 2x \end{aligned}$$

where
- b = beam width
- h = column depth
- x = smaller distance from edge of beam to edge of column

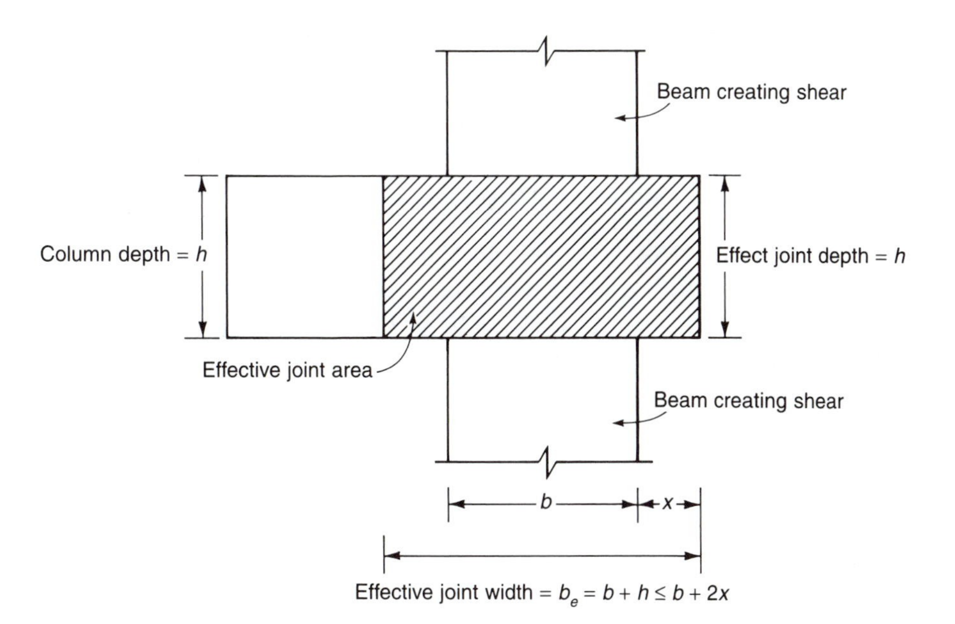

Figure 5-9 Effective area of a joint

5.2.2 Beam details

Flexural members are defined in UBC Section 1921.3.1 as elements having a clear span not less than four times the effective depth and with a factored axial compressive force not exceeding $A_g f'_c/10$. In order to provide a compact cross-section with good stability during nonlinear displacements, geometrical constraints are imposed in UBC Section 1921.3.1. These are:

$$b/h \geq 0.3$$
$$b \geq 10 \text{ inches}$$
$$\leq b_c + 0.75h \text{ on each side of the column}$$

where b = beam width
h = beam depth
b_c = column width

UBC Section 1921.3.2 stipulates limitations on the amount of longitudinal reinforcement to prevent steel congestion, ensure nonbrittle ductile behavior, and provide a minimum reinforcement capacity greater than the tensile strength of the concrete. These limitations are:

$$\rho_{min} \geq 200/f_y$$
$$\rho_{max} \leq 0.025$$

In addition, to allow for the possibility of moment reversals:

- A minimum of two reinforcing bars shall be provided at the top and bottom of the beam.
- At the ends of the member, positive moment capacity is required at least equal to fifty percent of the negative moment capacity.
- At any section, neither the positive nor the negative moment capacity shall be less than twenty five percent of the moment capacity at the ends of the member.

The location of reinforcement splices is not permitted in regions of plastic hinging as they are unreliable under inelastic cyclic loading conditions. Splices shall not be used:

- Within the joints.
- Within a distance of twice the beam depth from the face of the column.
- At locations of potential plastic hinging.

To prevent the spalling of concrete cover at splice locations, hoop reinforcement shall be provided over the lap length with a maximum spacing of d/4 or four inches.

UBC Section 1921.5.4, to account for cyclical loading and for the reinforcement stress exceeding the yield stress, specifies that the development length for a hooked bar in normal weight concrete shall be

$$l_{dh} = f_y d_b/65(f'_c)^{0.5}$$
$$\geq 8d_b$$
$$\geq 6 \text{ inches}$$

where d_b = bar diameter

The hook shall be located within the confined core of a column. For straight bars of sizes no. 3 through no. 11 the development length is given as

$$l_d = 2.5 l_{dh}$$

and when the depth of concrete beneath the bar exceeds twelve inches

$$l_d = 3.5 l_{dh}$$

For straight bars extending beyond the confined core of the column, the development length is given by

$$l_{dm} = 1.6 l_d - 0.6 l_{dc}$$

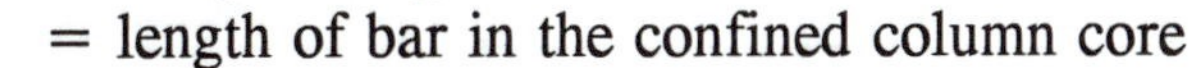

where l_{dc} = length of bar in the confined column core

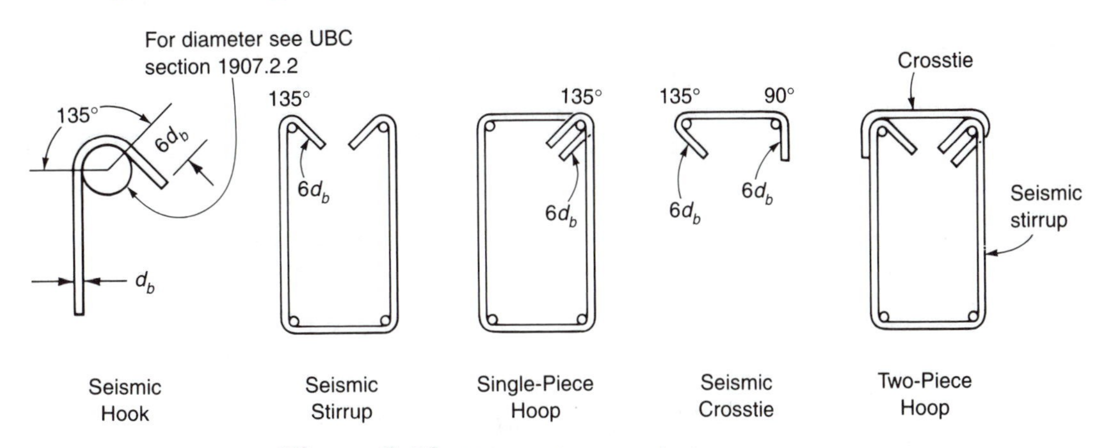

Figure 5–10 Seismic hoops and stirrups

Transverse reinforcement is required to provide shear resistance, to provide confinement to the concrete at locations of plastic hinging, and to control lateral buckling of longitudinal bars after the concrete cover has spalled. Closed hoops, as shown in Figure 5–10, are required to provide confinement and may also provide shear resistance. Seismic stirrups or links with 135 degree seismic hooks provide only shear resistance. Either single-piece or two-piece closed hoops may be provided. The two-piece hoop consists of a seismic stirrup and a seismic crosstie with one 135 degree seismic hook and one 90 degree hook. Adjacent crossties shall have the seismic hooks on opposite sides of the member. Hoops are required:

- Over a distance of 2d from face of the column.
- Over a distance of 2d on both sides of a section subjected to plastic hinging

The first hoop shall be located not more than two inches from the face of the column. The hoop spacing shall not exceed:

$$s_{max} \leq d/4$$
$$\leq 8d_b$$
$$\leq 24d_t$$
$$\leq 12 \text{ inches}$$

where d = beam effective depth
d_b = diameter of longitudinal bar
d_t = diameter of hoop bar

Where hoops are not required, stirrups with 135 degree seismic hooks shall be provided, throughout the length of the member, at a maximum spacing of d/2. Details of hoop and stirrup requirements are shown in Figure 5-11.

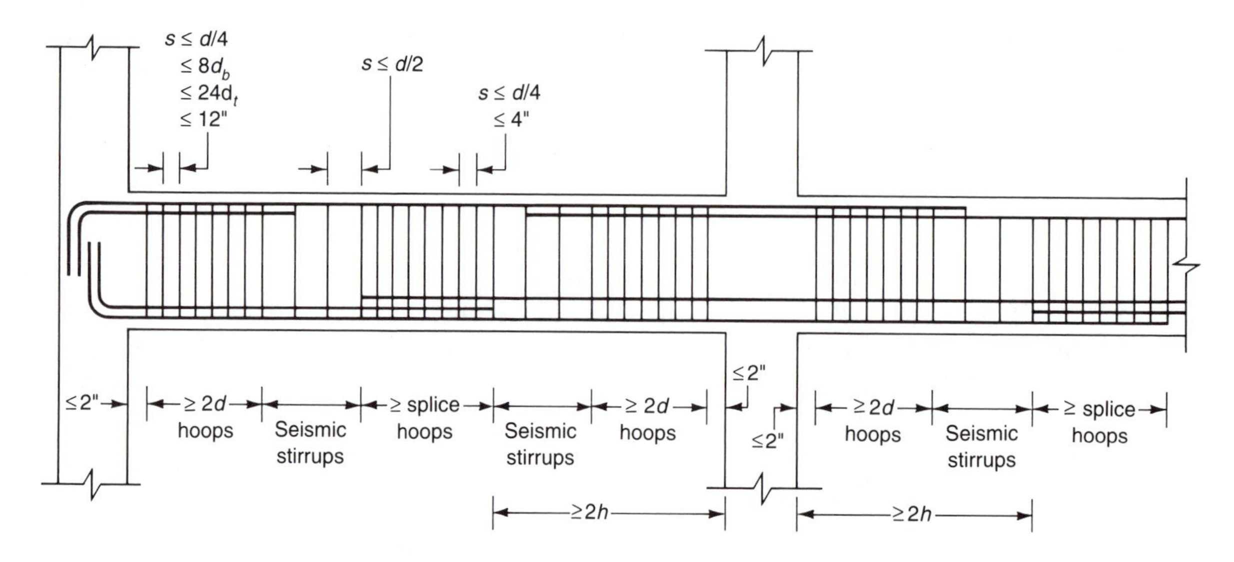

Figure 5-11 Hoop and stirrup location requirements

5.2.3 Column details

Columns are defined in UBC Section 1921.4 as members with a factored axial compressive force exceeding $A_g f'_c/10$. Geometrical constraints are imposed based on established design practice and these are:

$$h_{min} \geq 12"$$
$$h_{min}/h_{perp} \geq 0.4$$

where h_{min} = minimum cross-sectional dimension

h_{perp} = dimension perpendicular to minimum dimension

Longitudinal reinforcement limits are imposed by UBC Section 1921.4.3 in order to control creep, reduce steel congestion, and provide a flexural capacity in excess of the cracking moment. These limitations are

$$\rho_g \geq 0.1$$
$$\leq 0.6$$

where ρ_g = ratio of reinforcement area to cross-sectional area.

Spalling of the concrete cover typically occurs at the ends of columns which makes these areas undesirable for the location of lap splices. Lap splices, proportioned as tension lap splices, are restricted to the center half of the column where moment reversals are unlikely. It is recommended[8] that splices are proportioned as Class A tension lap splices with a minimum lap length of $30d_b$ and with confinement reinforcement at four inch spacing over the full length of the splice.

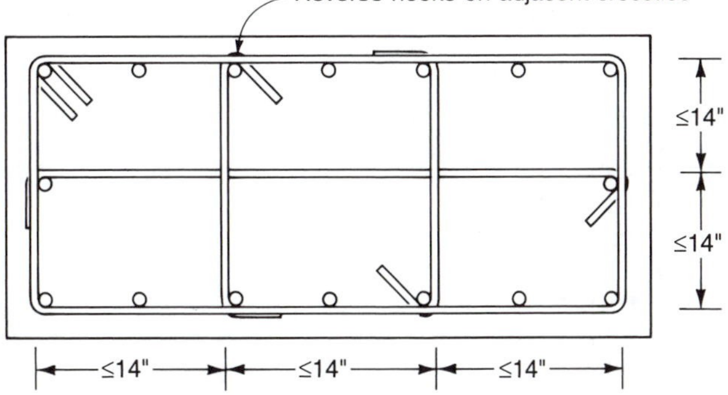

Figure 5–12 Column transverse reinforcement

Transverse reinforcement, consisting of closed hoops and crossties, shall be provided throughout the height of the column to provide shear resistance and confinement. The hoop spacing requirements in accordance with UBC Section 1921.4.4.6 are:

$$s_{max} \leq 6d_b$$
$$\leq 6 \text{ inches}$$

As shown in Figure 5–12, crossties or legs of overlapping hoops shall be spaced a maximum distance of fourteen inches on center and shall engage a longitudinal bar at each end. At the ends of the column, the area of the confinement reinforcement required is given by the greater value obtained from UBC Formulas (21–3) and (21–4) which are

$$A_{sh} = 0.35h_c(A_g/A_{ch} - 1)f'_c/f_y$$
$$A_{sh} = 0.09sh_cf'_c/f_y$$

where s = spacing of hoop reinforcement
A_g = gross area of column section
A_{ch} = cross–sectional area measured out–to–out of hoop reinforcement
h_c = dimension of core measured center-to-center of confining reinforcing

In accordance with UBC Section 1921.4.4.4, confinement reinforcement is required over a distance of l_o from each joint face given by

$$l_o \geq h$$
$$\geq H_n/6$$
$$\geq 18 \text{ inches}$$

where h = depth of column
H_n = column clear height

The spacing of the confinement reinforcement is limited by UBC Section 1921.4.4.2 to

$$s \leq h_{min}/4$$
$$\leq 4 \text{ inches}$$

where h_{min} = minimum column dimension

Details of column reinforcement are shown in Figure 5–13.

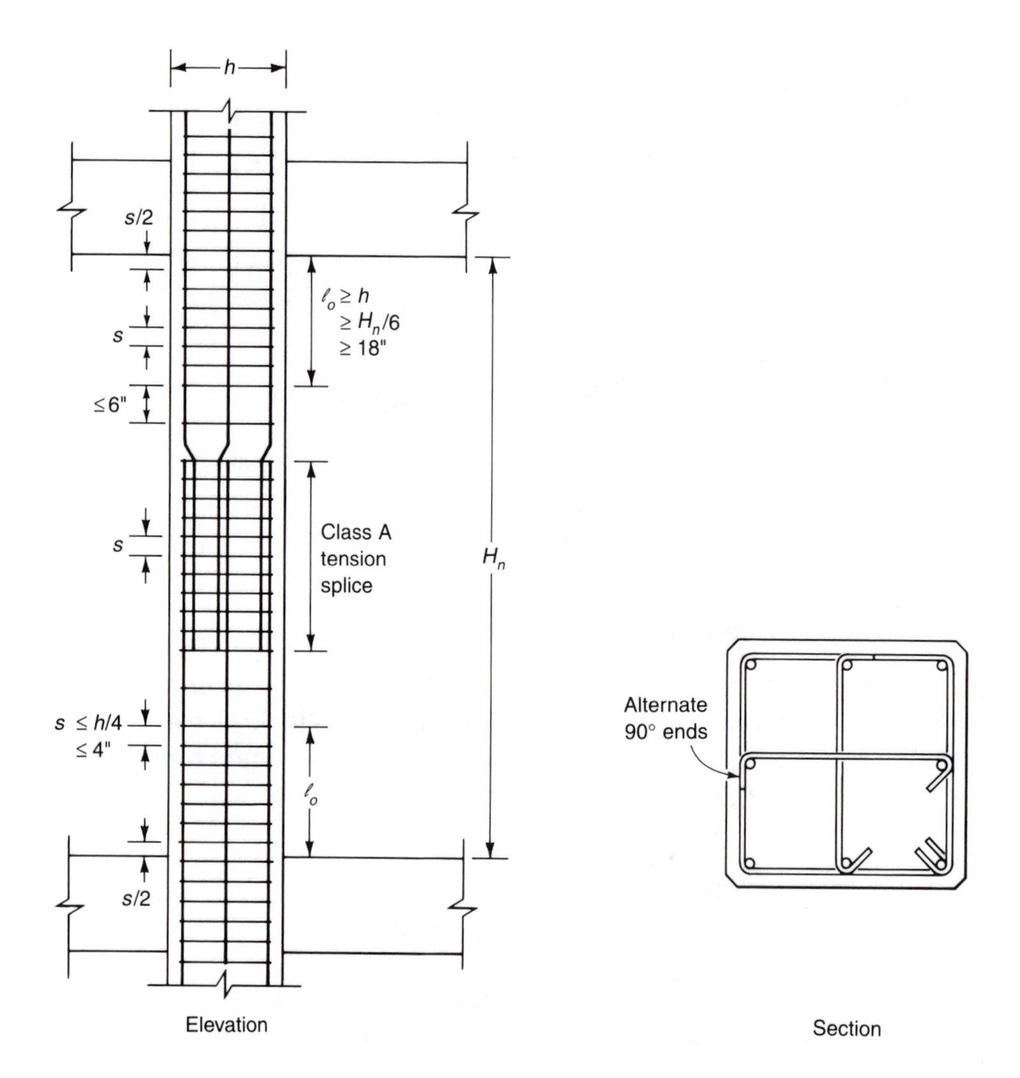

Figure 5–13 Column reinforcement details

Columns supporting discontinued walls are required by UBC Section 1921.4.4.5 to be provided with confinement reinforcement over their full height to provide the necessary ductility. The confinement reinforcement shall extend into the wall for the development length of the longitudinal bars. In addition, when the point of contraflexure is outside the center half of the column, the hinging region may extend over an extensive length of the column. For this situation, UBC Section 1921.4.4.1 requires confinement reinforcement to be provided over the full height of the column. Where the strong-column/weak-beam concept of the UBC Formula (21–1) is not satisfied at a joint, UBC Section 1921.4.2.3 requires confinement reinforcement to be provided over the full height of the column below the joint.

Example 5–2 (Structural Engineering Examination 1990, Section C, weight 6.0 points)

GIVEN: A three-story concrete special moment-resisting frame as shown in Figures 5–14 and 5–15. The summary of factored moments and factored axial load for column AB and factored moment for beam BC is tabulated below. Column AB has an ultimate moment capacity as shown in Figure 5–15.

Load Case	COLUMN AB			BEAM BC
	Axial Load	Moment End B	Moment End A	Moment End B
1.4(D + L + E)	288 Kips	258 k-ft.	266 k-ft.	-578 k-ft.

Note: Negative moment means producing tension on the top face of the beam.

CRITERIA: Seismic zone 4
Strength Design
Materials:

- Concrete f'_c = 4000 pounds per square inch
- Reinforcing steel f_y = 60,000 pounds per square inch
- Use number 4 bars for column tie bars
- Use number 9 bars for other bars

Assumptions: Column is classified as slender

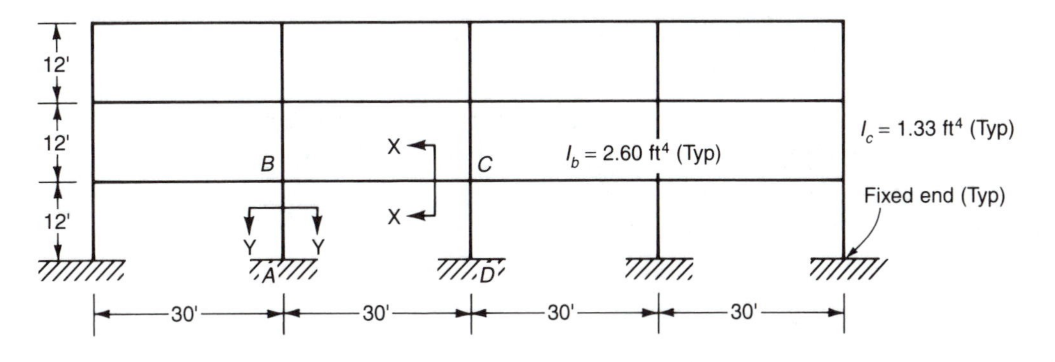

Figure 5–14 Elevation of frame for Example 5–2

REQUIRED:
1. Check the adequacy of the minimum flexural strength of column AB at joint B.
2. Determine the column transverse reinforcement requirements for shear.
3. Determine the column transverse reinforcement requirements for confinement.
4. Draw column AB reinforcing details showing vertical bars, tie bars, crossties, spacing, and lap splices.

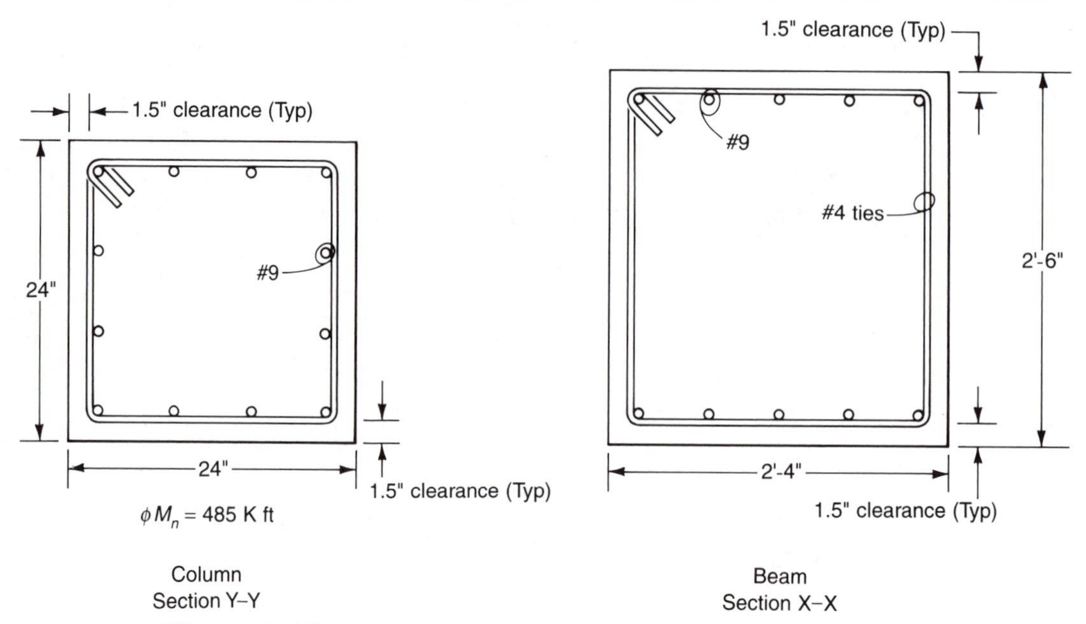

Figure 5–15 Cross-sections of members for Example 5–2

SOLUTION

1. COLUMN MOMENT CAPACITY

The factored axial compressive force in the column is given as

$$P_u = 288 \text{ kips}$$

This exceeds the value specified in UBC Sections 1921.4.1 and 1921.4.2 as

$$\begin{aligned} 0.1A_g f'_c &= 0.1 \times 24 \times 24 \times 4 \\ &= 230 \text{ kips} \end{aligned}$$

Then member AB qualifies as a column and the moment capacity of the columns relative to that of the beam must be determined in accordance with UBC Formula (21–1).

The design moment strength of the beam is obtained by using calculator program A.1.2 in Appendix III. The reinforcement ratio is

$$\begin{aligned} \rho &= A_s/bd \\ &= 5/(28 \times 27.44) \\ &= 0.0065 \\ &< 0.025 \\ &> 200/f_y \end{aligned}$$

Hence the requirements of UBC Section 1921.3.2 are satisfied.

The design moment strength, ignoring compression reinforcement, is

$$\begin{aligned} \phi M_n &= 0.9A_s f_y d(1 - 0.59\rho f_y/f'_c)/12 \\ &= 582 \text{ kip feet} \\ &= \Sigma M_g \end{aligned}$$

The design moment strength of the column is given as 485 kip feet. Then, the sum of the design moment strengths of the columns framing into the joint is

$$\Sigma M_c = 2 \times 485$$
$$= 970$$
$$> 1.2\Sigma M_g$$

Hence UBC Formula (21-1) is satisfied and hoop reinforcement is not required over the full height of the column.

2. COLUMN SHEAR

In accordance with UBC Sections 1921.4.5 and 1921.4.4.7 the design shear force for column AB may be calculated from the probable moment strength at the top and bottom of the column. The column probable moment strength is determined by assuming a strength reduction factor of zero and a tensile reinforcement stress of $1.25f_y$. The factored axial compressive force in the column is

$$P_u = 288 \text{ kips}$$
$$> 0.1A_gf'_c$$

The strength reduction factor for this condition of axial compression with flexure is given by UBC Section 1909.3.2 as

$$\phi = 0.7$$

The probable moment strength may be determined from the column design moment strength as

$$M_{pr} \approx \phi M_n \times 1.25/0.7$$
$$= 485 \times 1.25/0.7$$
$$= 866 \text{ kip feet}$$

The clear height of column AB is

$$H_n = 12 - 2.5$$
$$= 9.5 \text{ feet}$$

The design shear force is then

$$V_e = 2M_{pr}/H_n$$
$$= 2 \times 866/9.5$$
$$= 182 \text{ kips}$$

However, in accordance with UBC Section 1921.4.5.1, the maximum design shear force in the column need not exceed that determined from the probable flexural strength of girder BC which frames into joint B. The probable girder strength, assuming a strength reduction factor of unity and a tensile reinforcement stress of $1.25f_y$, may be derived[6] as

$$M_{pg} = A_sf_yd(1.25 - 0.92\rho f_y/f'_c)/12$$
$$= 5 \times 60 \times 27.44(1.25 - 0.92 \times 0.0065 \times 60/4)/12$$
$$= 796 \text{ kip feet}$$

This moment may be distributed equally to the upper and lower column at joint B and, for a sway displacement, the same value is carried over to end A giving

$$\begin{aligned} M_{AB} &= M_{BA} \\ &= M_{pg}/2 \\ &= 796/2 \\ &= 398 \text{ kip feet} \end{aligned}$$

Then the maximum probable design shear force acting on the column is

$$\begin{aligned} V_e &= (M_{AB} + M_{BA})/H_n \\ &= 2 \times 398/(12 - 2.5) \\ &= 83.79 \text{ kips} \\ &< 182 \text{ kips} \end{aligned}$$

Then $\quad V_e = 83.79$ kips governs

The compressive stress value given by

$$\begin{aligned} A_g f'_c/20 &= 24^2 \times 4/20 \\ &= 115 \text{ kips} \\ &< P_u \end{aligned}$$

Hence, since the total design shear is due to earthquake forces, in accordance with UBC Section 1921.4.5.2 the design shear strength provided by the concrete may be utilized and this is given by UBC Formula (11-3) as

$$\begin{aligned} \phi V_c &= 0.85 \times 2bd(f'_c)^{0.5} \\ &= 0.85 \times 2 \times 24 \times (24 - 1.5 - 0.5 - 0.56)(4000)^{0.5}/1{,}000 \\ &= 55.32 \text{ kips} \\ &< V_e \end{aligned}$$

The design shear strength required from shear reinforcement is given by UBC Formula (11-2) as

$$\begin{aligned} \phi V_s &= V_e - \phi V_c \\ &= 83.79 - 55.32 \\ &= 28.47 \text{ kips} \\ &< 4 \times \phi V_c \end{aligned}$$

Hence the proposed shear reinforcement, in accordance with UBC Section 1911.5.6.8, is satisfactory. The maximum hoop spacing, in accordance with UBC Section 1921.4.4.6, may not exceed the lesser value given by

$$\begin{aligned} s &= 6d_b \\ &= 6 \times 1.128 \\ &= 6.8 \text{ inches} \end{aligned}$$

or $\quad s = 6$ inches

Hence a spacing of six inches governs.

The minimum size of crosstie required for a number 9 longitudinal bar is specified, by UBC Section 1907.10.5, as a number 3 bar and at least one crosstie is required to satisfy the lateral support requirements of UBC Section 1921.4.4.3. The minimum area of shear reinforcement, which may be provided at a spacing of six inches, is given by UBC Formula (11-14) as

$$A_{v(min)} = 50b_w s/f_y$$
$$= 50 \times 24 \times 6/60{,}000$$
$$= 0.12 \text{ square inches}$$

The area of shear reinforcement which is required, at a spacing of six inches, to provide a shear strength of ϕV_s, is specified by Formula (11–17) as

$$A_v = \phi V_s s/\phi d f_y$$
$$= 28.47 \times 6/(0.85 \times 21.44 \times 60)$$
$$= 0.16 \text{ square inches}$$
$$> 0.12 \text{ square inches}$$

Providing a number 4 hoop and one number 4 crosstie gives an area of

$$A_v = 3 \times 0.2$$
$$= 0.6 \text{ square inches}$$
$$> 0.16 \text{ square inches}$$

3. CONFINEMENT REINFORCEMENT

To conform with UBC Section 1921.4.3 and SEAOC Section 3D.3, a Class A tension splice must be provided within the center half of the clear column height. The lap length required is specified by UBC Section 1912.15.1 as being equal to the tensile development length. The basic development length is given by UBC Section 1912.2.2 as

$$l_{db} = 0.04A_b f_y/(f'_c)^{0.5}$$
$$= 0.04 \times 1.0 \times 60{,}000/(4000)^{0.5}$$
$$= 38 \text{ inches}$$

Hoop reinforcement, at a maximum spacing of four inches, is provided over the splice length in accordance with SEAOC Section 3D.3. Hence, in accordance with UBC Section 1912.2.3.5 a modification factor of 0.75 is applicable. The modified development length is given by

$$l_d = 0.75l_{db}$$
$$= 0.75 \times 38$$
$$= 28.5 \text{ inches}$$

The minimum development length is specified by UBC Section 1912.2.3.6 as

$$l_{d(min)} = 0.03d_b f_y/(f'_c)^{0.5}$$
$$= 0.03 \times 1.128 \times 60{,}000/(4000)^{0.5}$$
$$= 32 \text{ inches}$$

The minimum lap length is recommended in SEAOC Commentary Section 3D.3 as

$$l_{(min)} = 30d_b$$
$$= 30 \times 1.128$$
$$= 34 \text{ inches}$$

Hence the required lap length is thirty four inches.

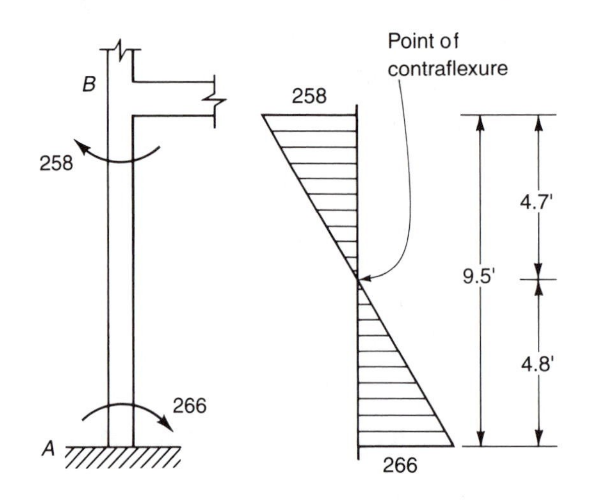

Figure 5–16 Factored moments in column AB

As shown in Figure 5–16, the point of contraflexure of the column, for the given loading, lies within the center half of the column clear height. Hence in accordance with UBC Section 1921.4.4.4 confinement reinforcement at a maximum spacing of four inches are required only for a distance from each joint face given by the greater of

$$l_o = h = 24 \text{ inches}$$

or $$l_o = H_n/6 = 9.5 \times 12/6 = 19 \text{ inches}$$

or $$l_o = 18 \text{ inches}$$

Hence the length of 24 inches governs.

Using number 4 hoop reinforcement bars, and providing one and a half inch clear cover to the bars, gives a core dimension, measured center-to-center of the bars, of

$$h_c = 24 - 3 - 0.5 = 20.5 \text{ inches}$$

The dimension measured out-to-out of the confining bars is, in both directions,

$$h'_c = h_c + 0.5 = 20.5 + 0.5 = 21 \text{ inches}$$

The area, calculated out-to-out of the confining bars, is

$$A_{ch} = (h'_c)^2 = 21^2 = 441 \text{ square inches}$$

The required area of confinement reinforcement is given by the greater value obtained from UBC Formulas (21–3) and (21–4)

Then $A_{sh} = 0.3sh_c(A_g/A_{ch} - 1)f'_c/f_y$
$= 0.3 \times 4 \times 20.5(576/441 - 1)4/60$
$= 0.50$ square inches

or $A_{sh} = 0.09sh_cf'_c/f_y$
$= 0.09 \times 4 \times 20.5 \times 4/60$
$= 0.49$ square inches

Hence three number 4 bars are required providing an area of confinement reinforcing of

$A_{sh} = 0.60$ square inches

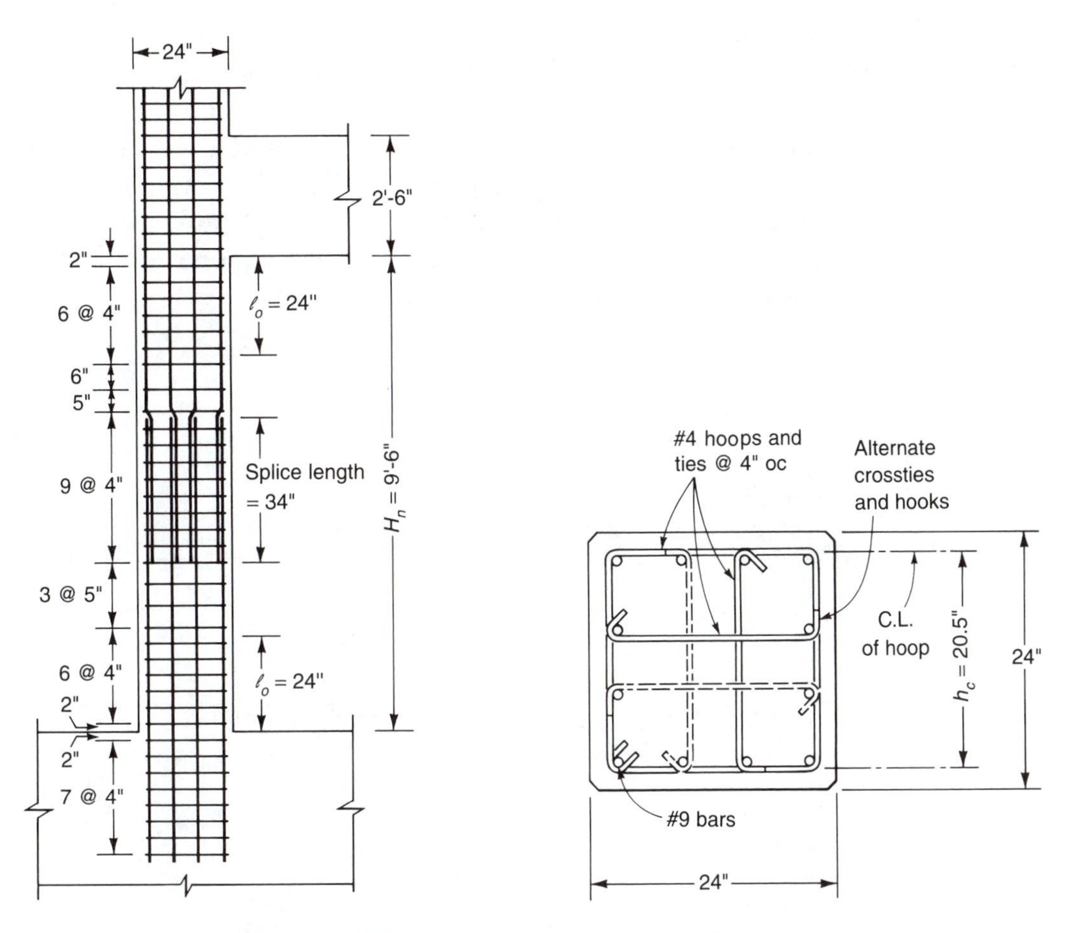

Figure 5–17 Column reinforcement for Example 5–2

The center–to–center spacing between the column longitudinal bars is

$s_l = (h_c - 1.128 - 0.5)/3$
$= 18.9/3$
$= 6.3$ inches

The maximum center–to–center spacing between the crossties and the hoop reinforcement is

$$
\begin{aligned}
s' &= 18.9 - s_l \\
&= 18.9 - 6.33 \\
&= 12.6 \text{ inches} \\
&< 14 \text{ inches}
\end{aligned}
$$

Hence the requirements of UBC Section 1921.4.4.3 are satisfied.
Details of the column reinforcement are shown in Figure 5–17.

5.2.4 Girder-to-column connection

Beam–column joints are subjected to high stress concentrations and require careful detailing to ensure confinement of the concrete. With the exception of the condition where beams frame into all four sides of the joint, hoop reinforcement A_{sh} shall be provided throughout the height of the joint as specified for the ends of a column in UBC Section 1921.4.4 at a maximum spacing of four inches. Where beams, with a width at least 75 percent of the column width, frame into all four sides of the joint hoop reinforcement with an area of $A_{sh}/2$ shall be provided at a maximum spacing of six inches. The hoop reinforcement provides concrete confinement and the shear resistance specified in UBC Section 1921.5.1.1

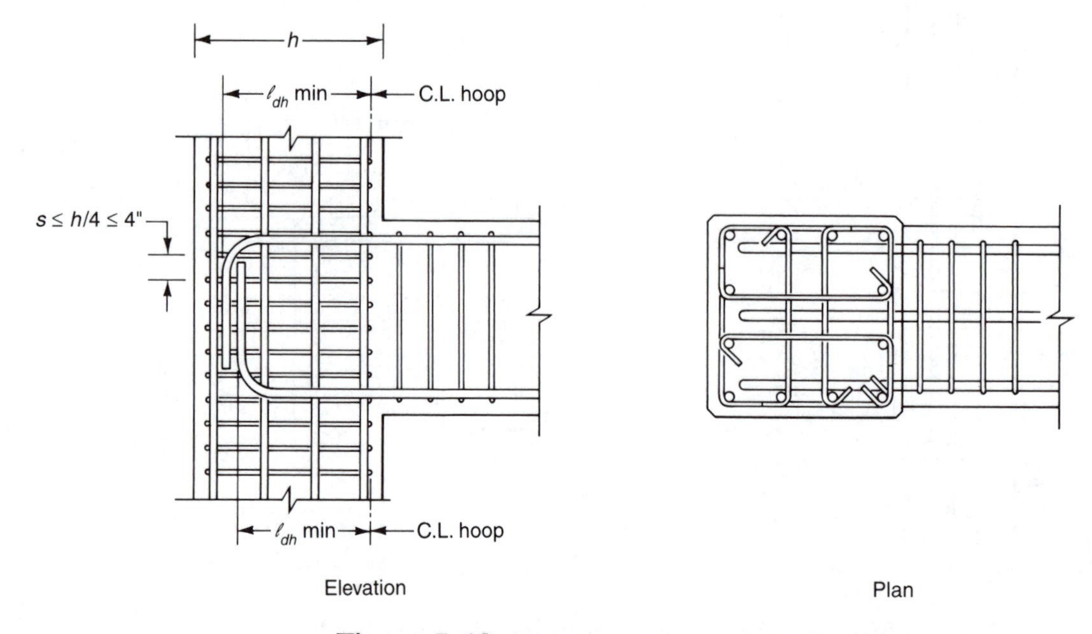

Figure 5–18 Girder to column connection

Beam reinforcement terminated in a column shall extend to the far face of the confined concrete core and be provided with the anchorage length specified in UBC 1921.5.4. Typical joint details are shown in Figure 5–18.

Example 5-3 (Structural Engineering Examination 1993, Section C, weight 5.0 points)

GIVEN: The beam-column joint indicated in Figure 5-19 and the 21" × 30" girders indicated in Figure 5-20 are part of the lateral force resisting system of a building located in seismic zone 4 with $R_W = 12$. The 21" × 28" beam is not part of the lateral load resisting system.

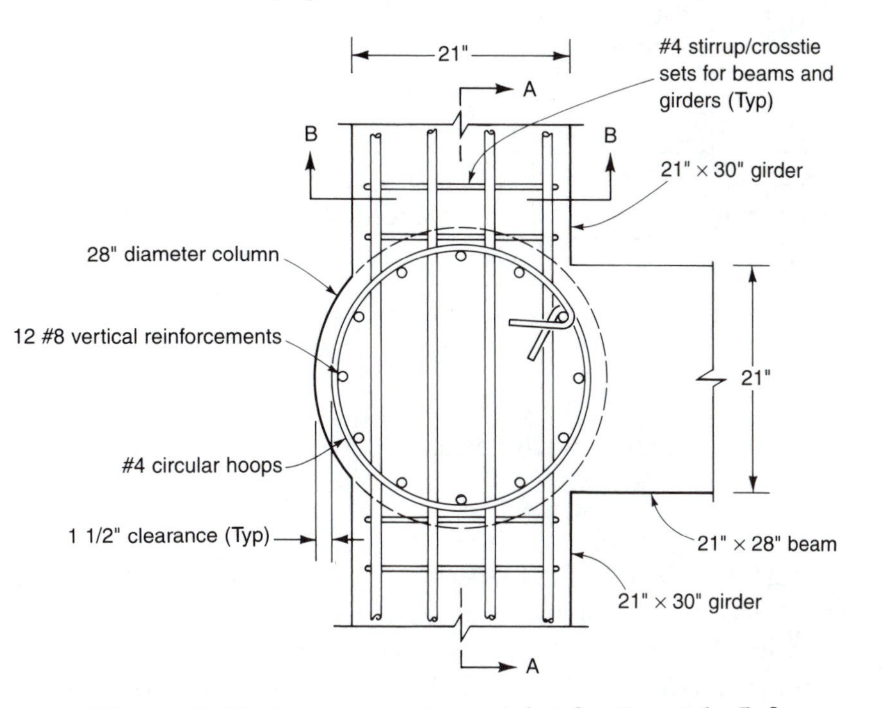

Figure 5-19 Beam-to-column joint for Example 5-3

CRITERIA: Materials:

- Concrete, normal weight, f'_c = 4,000 pounds per square inch
- Reinforcement, f_y = 60,000 pounds per square inch

Assumptions:

- Ignore the slab in all calculations.
- Ignore compression reinforcement for the girders in all calculations.
- The minimum factored axial compressive force in the column exceeds $A_g f'_c/10$.
- The maximum factored axial compressive force in the girders is less than $A_g f'_c/20$.

REQUIRED:

1. Determine the center-to-center spacing of the number 4 circular hoop reinforcement within the joint, indicated as "S1" in Section A.
2. Determine the minimum spacing of the number 4 stirrup and crosstie sets for the 21" × 30" girder, indicated as "S2" in Section A. Use a design shear force V_e equal to 98 kips.
3. Determine if "strong-column/weak-girder" requirements are satisfied at the joint. In making the calculation, take the sum of the design flexural strength moments of the columns framing into the joint to equal 920 kip feet.

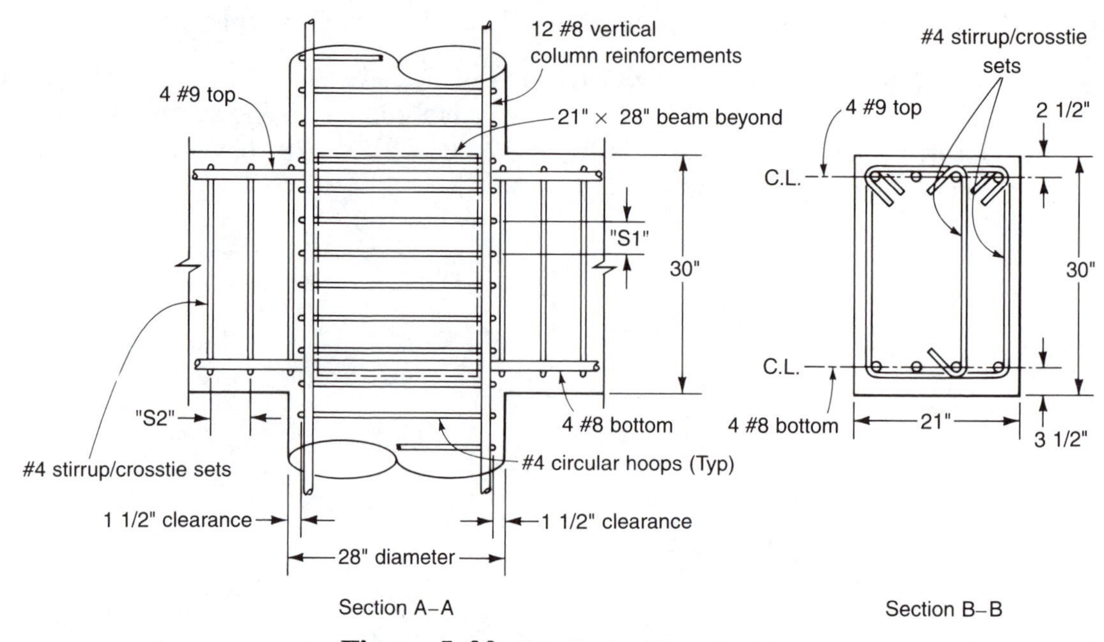

Figure 5–20 Details for Example 5–3

SOLUTION

1. BEAM–TO–COLUMN JOINT CONFINEMENT REINFORCEMENT

The factored axial force in the column exceeds $0.1A_gf'_c$ and confinement reinforcement is required as specified in UBC Section 1921.4.4. Since beams frame into only three sides of the joint, in accordance with UBC Section 1921.5.2, confinement reinforcement is required as specified in UBC Section 1921.4.4 at the ends of the columns. The maximum hoop spacing may not exceed the lesser value given by

$$s = h_{min}/4 = 28/4 = 7 \text{ inches}$$

or $$s = 4 \text{ inches}$$

Hence a spacing of four inches governs. The minimum required volumetric ratio of circular hoop reinforcement to volume of core is given by the greater value obtained from UBC Formulas (10–5) and (21–2). Formula (10–5) is

$$\rho_s = 0.45(A_g/A_c - 1)f'_c/f_y$$

where A_g = gross area of column

A_c = area of concrete core measured to outside of hoops

and $$A_g/A_c = (28/25)^2 = 1.25$$

Hence $$\rho_s = 0.45(1.25 - 1)4/60 = 0.0076 = A_v\pi(D_c - D_v)sA_c$$

where A_v = area of hoop reinforcement
D_c = diameter of core measured to outside of hoop = 25 inches
D_v = diameter of hoop reinforcement = 0.5 inches
s = pitch of hoop reinforcement

and $A_v/A_c = (0.5/25)^2$
$= 0.0004$

Hence $s = 0.004\pi(25 - 0.5)/0.0076$
$= 4.05$ inches

Formula (21-2) is

$\rho_s = 0.12f'_c/f_y$
$= 0.12 \times 4/60$
$= 0.0080$

and $s = 4.05 \times 0.0076/0.0080$
$= 3.85$ inches

A 3.5 inch spacing of the hoop reinforcement is adequate for confinement of the joint.

2. BEAM CONFINEMENT REINFORCEMENT

In accordance with UBC Section 1921.3.3.2, closed hoops are required at the ends of the beam for confinement and the hoop spacing may not exceed the lesser value given by

$s = d/4$
$= (30 - 3.5)/4$
$= 6.625$ inches

or $s = 8d_b$
$= 8 \times 1.0$
$= 8$ inches

or $s = 24d_t$
$= 24 \times 0.5$
$= 12$ inches

or $s = 12$ inches

Hence the required spacing for the confinement reinforcement is 6.625 inches.

Since the axial force in the beam is less than $0.1A_gf'_c$ Section 1921.4.4.1 of the UBC is not applicable and the required area of transverse reinforcement is based on the design shear force acting on the beam. The design shear force may be determined, as shown in Figure 5-5, from the probable flexural strengths which can be developed at the ends of the beam. The design shear force is given in the problem statement as

$V_e = 98$ kips.

Since the factored axial force in the beam is less than $0.05A_gf$ and assuming that seismic induced shear force exceeds one-half of the total design shear, the shear resistance of the concrete V_c is neglected, in accordance with UBC Section 1921.4.5.2. Hence the hoop reinforcement must be designed to resist the total design shear force. The design shear strength required from the hoop reinforcement is given by the UBC Formula (11-2) as

$$\phi V_s = V_e - \phi V_c = 98 - 0 = 98 \text{ kips}$$

The design shear strength provided by transverse reinforcement is limited, by UBC Section 1911.5.6.8, to a maximum value of

$$\phi V'_s = 8\phi bd(f'_c)^{0.5} = 8 \times 0.85 \times 21 \times 26.5(4000)^{0.5}/1000 = 239 \text{ kips} > 98 \text{ kips}$$

Hence the requirement is satisfied.

The cross-sectional area provided by one number 4 hoop and one number 4 crosstie is

$$A_v = 3 \times 0.2 = 0.6 \text{ square inches}$$

The required pitch of the transverse reinforcement for shear resistance is given by UBC Formula (11-17) as

$$s' = \phi df_y A_v/\phi V_s = 0.85 \times 26.5 \times 60 \times 0.6/98 = 8.27 \text{ inches} > 6.625 \text{ inches}$$

The maximum permissible pitch is given by UBC Formula (11-14) as

$$s_{max} = f_y A_v/50b = 60{,}000 \times 0.6/50 \times 21 = 34 \text{ inches} > 6.625 \text{ inches}$$

The required spacing of the beam transverse reinforcement is controlled by confinement requirements and is given by

$$s = 6.625 \text{ inches}$$

3. STRONG-COLUMN/WEAK-BEAM CONCEPT

For seismic load acting to the left, the top reinforcement in the right hand beam is in tension and the bottom reinforcement in the left-hand beam is in tension. The tensile reinforcement ratio of the right hand beam is

$$\rho = A_s/bd = (4 \times 1.0)/21(30 - 3.5) = 4/21 \times 26.5 = 0.0072$$

The design flexural strength of the right hand beam, ignoring compression reinforcement, is obtained from calculator program A.1.2 in Appendix III as

$$\phi M_n = 0.9A_sF_yd(1 - 0.59\rho f_y/f'_c)/12$$
$$= 446.7 \text{ kip feet}$$

The tensile reinforcement ratio of the left hand beam is

$$\rho = A_s/bd$$
$$= 4 \times 0.79/21(30 - 2.5)$$
$$= 3.16/21 \times 27.5$$
$$= 0.00547$$

The design flexural strength of the left hand beam is

$$\phi M_n = 372.1 \text{ kip feet}$$

The sum of the design flexural strengths of the beams framing into the joint is

$$\Sigma M_g = 446.7 + 372.1$$
$$= 818.8 \text{ kip feet}$$

The sum of the design flexural strengths of the columns framing into the joint is given in the problem statement as

$$\Sigma M_c = 920 \text{ kip feet}$$
$$< 1.2\Sigma M_g$$

Hence, the strong-column/weak beam requirements of UBC Formula (21-1) are not satisfied at the joint.

5.3 SHEAR WALLS AND WALL PIERS

5.3.1 Shear capacity of shear walls

The nominal shear strength of a shear wall may be determined as specified in UBC Section 1921.6.4.2 and given in the UBC Formula (21-6) as

$$V_n = A_{cv}[2(f'_c)^{0.5} + \rho_n f_y]$$

where A_{cv} = area of web over the horizontal length considered
= $12b_w$ per foot of wall
b_w = web width
ρ_n = reinforcement ratio of horizontal shear reinforcement

In accordance with UBC Section 1921.6.2.1, when the design shear force V_u exceeds $A_{cv}(f'_c)^{0.5}$ the minimum reinforcement ratios for the horizontal and vertical reinforcement shall be

$$\rho_n = A_{sn}/A_{cn} \geq 0.0025$$
$$\rho_v = A_{sv}/A_{cv} \geq 0.0025$$

where A_{sn} = area of horizontal reinforcement over the vertical length considered
A_{cn} = area of web over the vertical length considered
A_{sv} = area of vertical reinforcement over the horizontal length considered

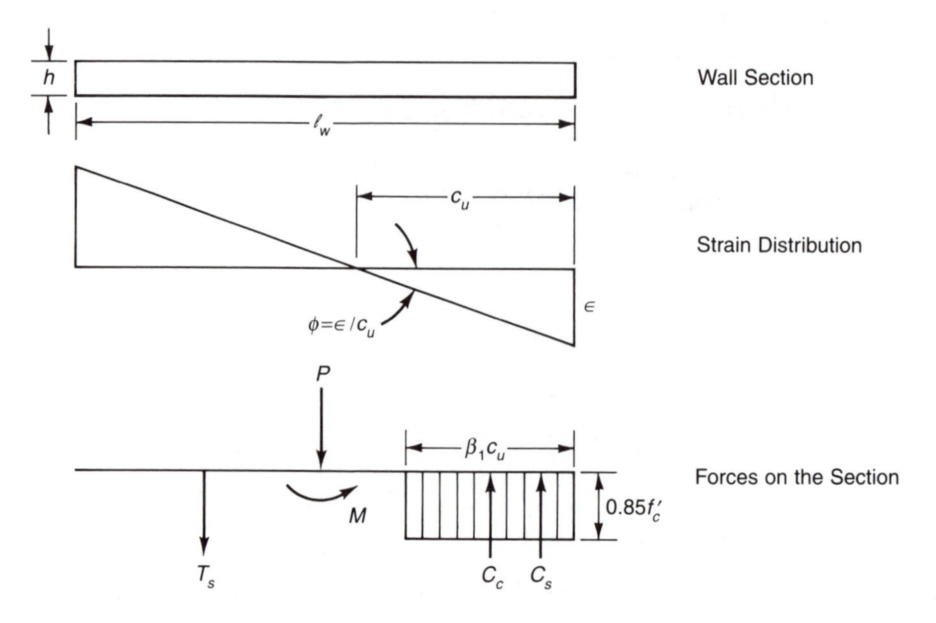

Figure 5–21 Assumptions used in shear wall design

In addition, the spacing of shear reinforcement shall not exceed eighteen inches each way and horizontal reinforcement shall terminate at the edge of the wall with a standard hook or a U-stirrup. In order to control cracking and inhibit fragmentation of the wall due to cyclical loading in the inelastic range, UBC Section 1921.6.2.2 specifies two curtains of reinforcement when the design shear force is given by

$$V_u > 2A_{cv}(f'_c)^{0.5}$$

When the ratio of wall height h_w to base length l_w is less than 2 the nominal shear strength of the wall may be determined from UBC Formula (21-7) which is

$$V_n = A_{cv}[\alpha_c(f'_c)^{0.5} + \rho_n f_y]$$

where α_c $= 2.0$ for $h_w/l_w = 2.0$
$= 3.0$ for $h_w/l_w = 1.5$

5.3.2 Boundary members

For shear walls subjected to combined flexural and axial load[9] UBC Section 1921.6.5.1 requires the wall to be designed as a tied column with a strength reduction factor ϕ of 0.6. The strain distribution across the section is assumed linear with a maximum concrete compressive strain of 0.003. The assumptions used[10] are shown in Figure 5–21.

The effective width of flanged sections contributing to the compressive strength of the section is specified in UBC Section 1921.6.5.2 as half the distance between adjacent walls but not more than ten percent of the wall height as shown in Figure 5–22.

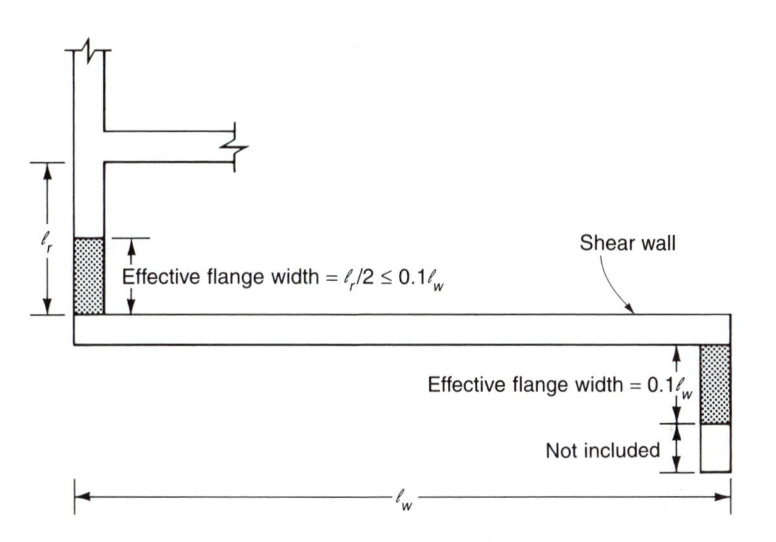

Figure 5–22 Effective flange widths

UBC Section 1921.6.5.3 imposes a limit on the design axial force above which the wall is no longer considered effective in resisting lateral loads. This limit is given by

$P_u = 0.35P_o$

where P_o = nominal axial load strength at zero eccentricity

$= 0.85f'_c(A_g - A_{st}) + f_yA_{st}$

A_g = gross area of section

A_{st} = area of vertical reinforcement

It is assumed that the nominal axial load at balanced strain conditions is given by

$P_b = 0.35P_o$

This is the point on the interaction diagram for a column at which the maximum strain in the concrete reaches 0.003 simultaneously with yielding of the tensile reinforcement. Increasing the factored axial load beyond this value results in a sudden, brittle, concrete compression mode of failure. For this condition, P-delta effects must be investigated and the wall shown to be adequate for vertical load carrying when displaced $0.375R_W$ times the displacement resulting from the prescribed seismic forces.

Special treatment of boundary zone areas are not required when concrete strains do not exceed 0.003. In accordance with UBC Section 1921.6.5.4 this condition is realized when

$P_u \le 0.10A_gf'_c$ ... for symmetrical walls

when $P_u \le 0.05A_gf'_c$ ... for unsymmetrical walls

and $M_u/V_ul_w \le 1.0$

or $V_u \le 3l_wh(f'_c)^{0.5}$

When boundary zone treatment is required, empirical detailing principles may be adopted which eliminate the necessity for further strain analysis of the wall. The required boundary zone dimensions are shown in Figure 5–23 and are:

- A minimum length of $0.15l_w$ for $P_u \leq 0.15P_o$
- A length of $0.25l_w$ for $P_u = 0.35P_o$ reducing linearly to
- A length of $0.15l_w$ for $P_u = 0.15P_o$

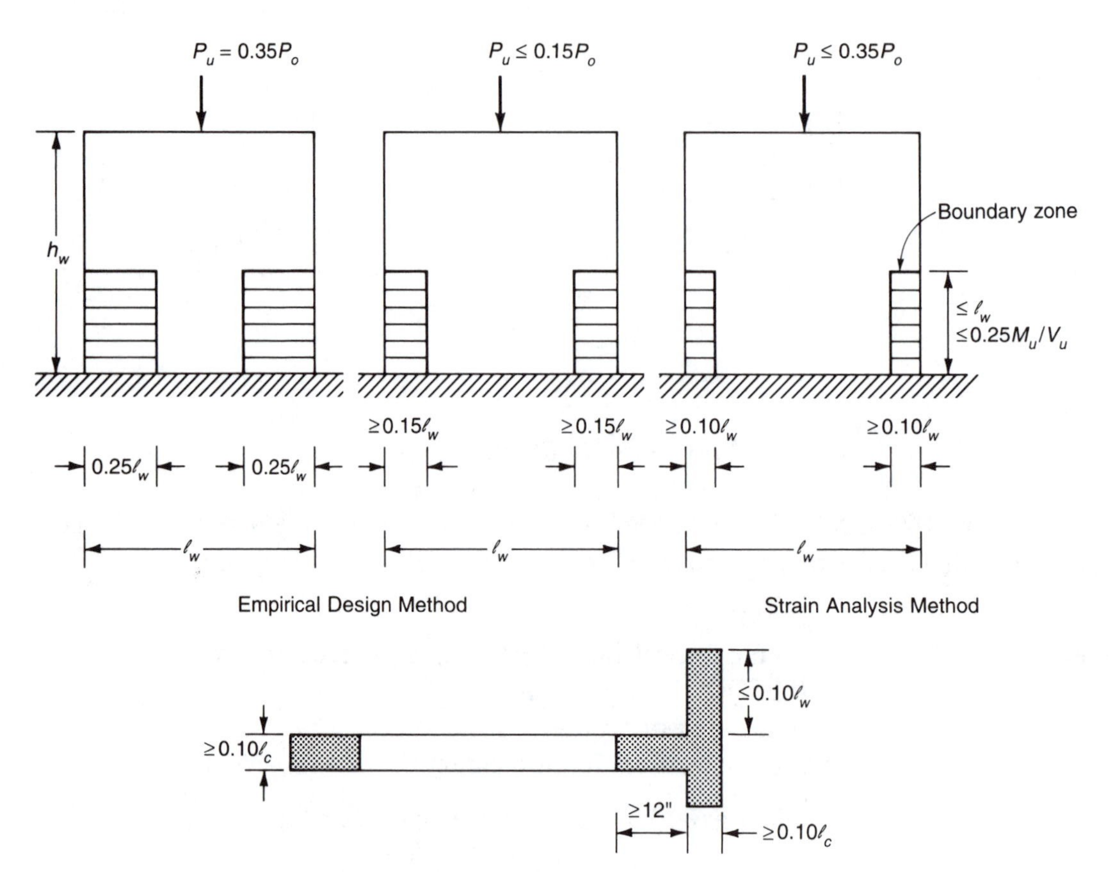

Figure 5–23 Boundary zone dimensions

In accordance with UBC Section 1921.6.5.6, and as shown in Figure 5–24, confinement reinforcement to prevent inelastic buckling of the vertical reinforcement consists of closed hoops and crossties such that:

- Minimum area provided $A_{sh} = 0.09sh_cf'_c/f_y$
- Maximum vertical spacing $s = 6d_b$
  ≤ 6 inches
- Horizontal spacing between the legs of hoops or crossties is $h_t \leq 12$ inches.
- Alternate vertical bars must be contained by the corner of a hoop or by a crosstie.
- The aspect ratio of the hoop ≤ 3

In addition, horizontal reinforcement shall not be lap spliced within the boundary zone and vertical reinforcement must be provided to satisfy fully all compression and tension requirements. The area of the vertical reinforcement in the boundary zone to control excessive elongation is limited to

$A_{sz} \geq 0.005A_{cz}$
$\geq$ two number 5 bars at each edge

where A_{cz} = area of boundary zone

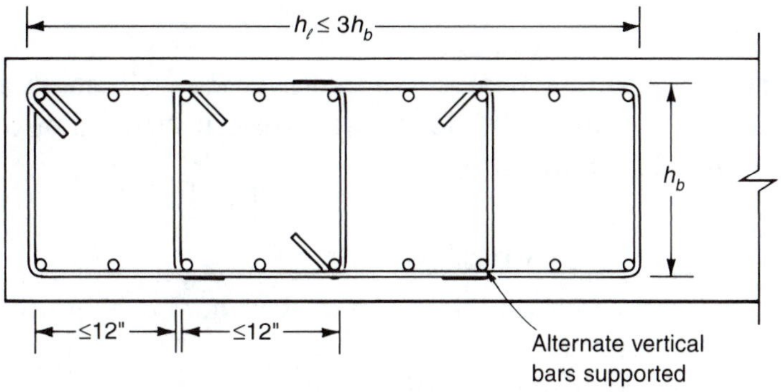

Figure 5–24 Confinement reinforcement in shear walls

To prevent spalling of concrete cover during cyclical deformations in the inelastic range, lap splices of vertical reinforcement shall have confinement reinforcement at four inches maximum spacing.

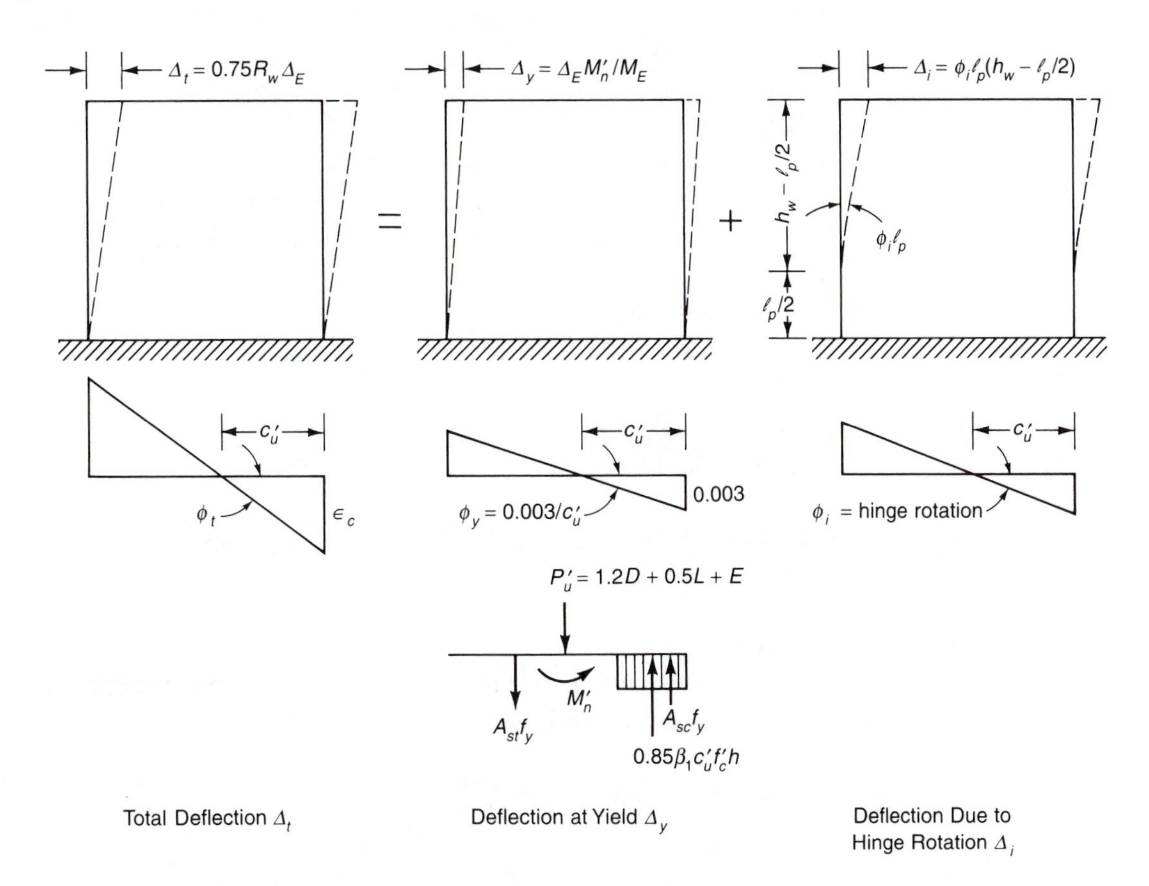

Figure 5–25 Total curvature demand in a shear wall

The strain analysis technique for cantilever shear walls is detailed in UBC Section 1921.6.5.5 and illustrated in Figure 5–25. The maximum permissible extreme fiber compressive strain on the section is given by UBC Formula (21–8) as

$$\varepsilon_{max} = 0.0015R_W \leq 0.015$$

Boundary zone confinement reinforcement is required over the wall length in which compressive strains exceed 0.003. To determine this length requires an estimation of the total curvature demand by UBC Formula (21–9) which is

$$\phi_t = \phi_i + \phi_y = \Delta_i/(h_w - l_p/2)l_p + \phi_y$$

where
- Δ_i = inelastic deflection at top of wall due to hinge rotation $= \Delta_t - \Delta_y$
- Δ_t = total deflection at top of wall $= 0.75R_W\Delta_E$
- R_W = structure response modification factor given in UBC Table 16–N
- Δ_E = elastic deflection at top of wall due to the prescribed seismic force $ZI_pC_pW_p/R_W$ acting on the gross wall section
- Δ_y = yield deflection at top of wall
  = linear elastic deflection at top of wall corresponding to the extreme fiber compressive strain of 0.003
  = deflection at top of wall due to the yield moment M'_n
  $= \Delta_E M'_n/M_E$
- M_E = moment due to the prescribed seismic forces of $ZICW/R_W$
- M'_n = nominal flexural strength of the section with an applied axial force of P'_n which may be obtained from the interaction diagram or from the strain diagram of the section
- P'_u = factored axial load $= 1.2D + 0.5L + E$
- ϕ_i = hinge rotation
  $= \Delta_i/(h_w - l_p/2)l_p$
  $= 2\Delta_i/(h_w - l_w/4)l_w$
- h_w = wall height
- l_p = height of plastic hinge $= 0.5l_w$
- l_w = wall length
- ϕ_y = yield rotation
  = rotation corresponding to the extreme fiber compressive strain of 0.003
  = rotation due to the yield moment M'_n
  $= 0.003/c'_u$
- c'_u = neutral axis depth due to P'_u
  $= (P'_u + A_{st}f_y + 0.85A_{sc}f'_c - A_{sc}f_y)/0.85\beta_1 f'_c h$
- A_{st} = area of tension reinforcement
- A_{sc} = area of compression reinforcement

β_1 = compression zone factor
h = wall thickness
ε_c = compressive strain at the extreme fiber
= $c'_u \phi_t$
$\leq 0.0015R_W$
≤ 0.015
l_z = length over which confinement is required
= zone in which compressive strain exceeds 0.003
= $c'_u - 0.003/\phi_t$
$\geq 0.10l_w$
$\geq$ 12 inches plus effective flange width

In addition, the height of the boundary zone shall extend above the section at which the compressive strain equals 0.003, a distance equal to the development length of the largest vertical bar. However, the boundary zone reinforcement need not extend above the base a distance greater than the larger of

$h_z = l_w$
or $h_z = M_u/4V_u$
where M_u and V_u are the factored forces due to 1.4 (D + L + E) or 0.9D - 1.4E.

All portions of the boundary zones shall have a thickness of

$h \geq l_c/10$
where l_c = vertical distance between supports

Confinement reinforcement is as specified for the empirical design method.

Example 5-4 (Structural Engineering Examination 1992, Section C, weight 7.0 points)

GIVEN: A new three-story building has 12 inch thick concrete perimeter shear walls with an independent vertical load carrying frame. A typical perimeter shear wall is depicted in Figure 5-26

CRITERIA: Seismic Zone 4
Materials:
- Concrete, normal weight 150 pounds per cubic foot, f'_c = 4,000 pounds per square inch
- Reinforcement, f_y = 60,000 pounds per square inch

Assumptions:
- The lateral loads shown in Figure 5-26 are unfactored seismic forces.
- Vertical gravity loads tributary to the perimeter shear walls are carried by the independent vertical load carrying frames.
- Include self weight of shear wall in all calculations.
- Assume the unbraced length of all compression members is adequate.

REQUIRED:

1. Using two curtains of number 4 reinforcing bars, determine the maximum spacing of horizontal and vertical reinforcing in the bottom story of the shear wall.
2. Determine the boundary zone detailing requirements for the bottom story shear wall.
3. Using number 8 reinforcing bars, determine the required number of vertical bars in the bottom story for flexural resistance.
4. Determine the embedment depth of number 8 dowels with standard hooks into the footings for the vertical reinforcement. Provide all calculations.
5. Determine the required lap splice length for the vertical number 8 reinforcement and number 8 dowels embedded into the footing. Provide all calculations.

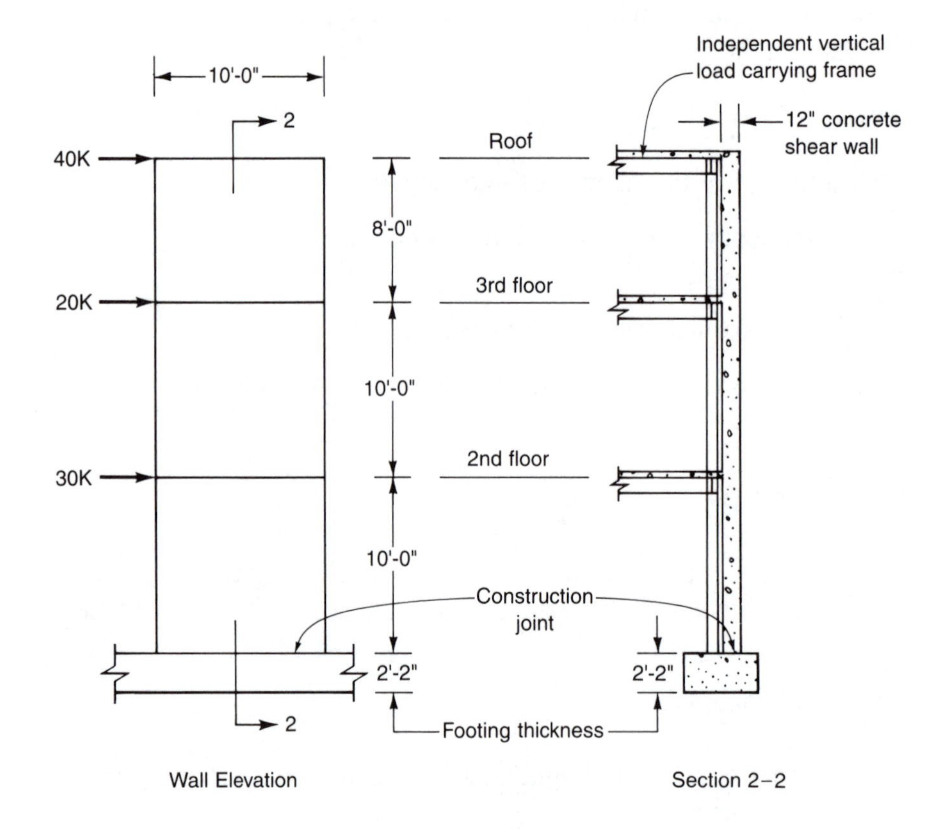

Figure 5–26 Shear wall details for Example 5–4

SOLUTION

DESIGN FORCES

In the bottom story, the factored shear force is given by

$$V_u = 1.4(40 + 20 + 30)$$
$$= 126 \text{ kips}$$

The factored moment is

$$M_u = 1.4(40 \times 28 + 20 \times 20 + 30 \times 10)$$
$$= 2548 \text{ kip feet}$$

The factored axial load is

$$P_u = 1.4 \times 0.150 \times 1.0 \times 10 \times 28$$
$$= 58.8 \text{ kips}$$

or $$P_u = 0.9 \times 0.150 \times 1.0 \times 10 \times 28$$
$$= 37.8 \text{ kips}$$

1. SHEAR REINFORCEMENT REQUIRED

The design shear force limit given in UBC Section 1921.6.2.1 is given by

$$A_{cv}(f'_c)^{0.5} = 120 \times 12(4000)^{0.5}/1000$$
$$= 91 \text{ kips}$$
$$< V_u$$

Hence the required minimum reinforcement ratios along the longitudinal and transverse axes are

$$\rho_v = \rho_n = 0.0025$$

Using number 4 bars in each face at a spacing of thirteen inches provides a reinforcement ratio of

$$\rho = 2 \times 0.2/12 \times 13$$
$$= 0.00256$$
$$> 0.0025$$

The proposed spacing is less than the maximum permissible value of eighteen inches and is satisfactory. The height to width ratio of the wall is

$$h_w/l_w = 28/10$$
$$= 2.8$$
$$> 2.0$$

Hence UBC Section 1921.6.4.2 applies and the design shear force is given by

$$\phi V_n = 0.6A_{cv}[2(f'_c)^{0.5} + \rho_n f_y]$$
$$= 0.6 \times 1440[2(4000)^{0.5} + 0.00256 \times 60{,}000]/1000$$
$$= 242 \text{ kips}$$
$$> V_u$$

Hence number 4 bars in each face at a spacing of thirteen inches is satisfactory.

2. BOUNDARY ZONE REQUIREMENTS

The nominal axial load strength at zero eccentricity is

$$P_o = 0.85f'_c(A_g - A_{st}) + f_yA_{st}$$
$$= 0.85 \times 4 \times 1440(1 - 0.00256) + 60 \times 1440 \times 0.00256$$
$$= 5105 \text{ kips}$$

and $$P_u < 0.35\, P_o$$

Hence, in accordance with UBC Section 1921.6.5.3 the wall may be considered effective in resisting lateral loads and may be designed as a shear wall.

An exemption from the provision of boundary zone confinement reinforcement is given by UBC Section 1921.6.5.4 provided that

$$
\begin{aligned}
P_u &\leq 0.10A_g f'_c \\
&= 0.10 \times 1440 \times 4 \\
&= 576 \\
&> P_u
\end{aligned}
$$

and

$$
\begin{aligned}
V_u &\leq 3l_w h(f'_c)^{0.5} \\
&= 3 \times 120 \times 12(4000)^{0.5} \\
&= 273 \\
&> V_u
\end{aligned}
$$

Both conditions are satisfied and boundary zone detail requirements need not be provided.

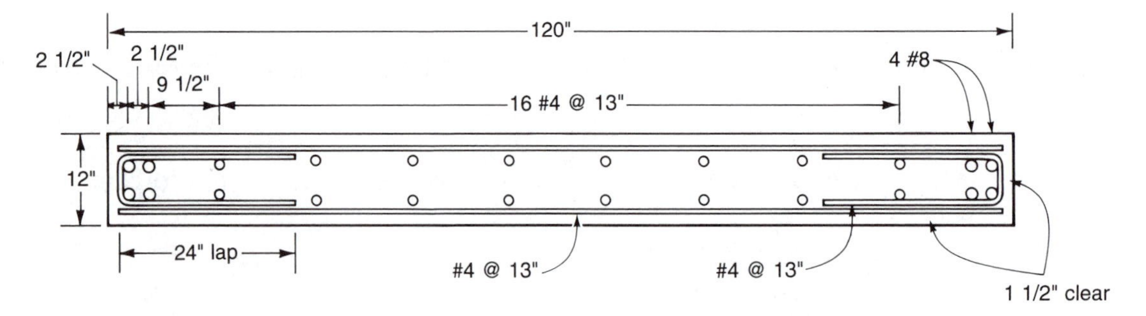

Figure 5–27 Shear wall reinforcement for Example 5–4

3. FLEXURAL REINFORCEMENT

To resist the factored moment, provide four number 8 reinforcing bars at each end and eight number 4 bars at thirteen inch spacing in each face as shown in Figure 5–27. To determine the flexural capacity of the wall, an interaction diagram must be derived. The total steel area is

$$
\begin{aligned}
A_{st} &= 8 \times 0.79 + 16 \times 0.20 \\
&= 9.52 \text{ square inches}
\end{aligned}
$$

AXIAL LOAD ONLY

The nominal axial load strength at zero eccentricity is

$$
\begin{aligned}
P_o &= 0.85f'_c(A_g - A_{st}) + f_y A_{st} \\
&= 0.85 \times 4(1440 - 9.52) + 60 \times 9.52 \\
&= 5435 \text{ kips}
\end{aligned}
$$

The design axial load strength at zero eccentricity is

$$
\begin{aligned}
\phi P_o &= 0.7 \times 5435 \\
&= 3804 \text{ kips}
\end{aligned}
$$

and this is plotted on the interaction diagram in Figure 5–28.

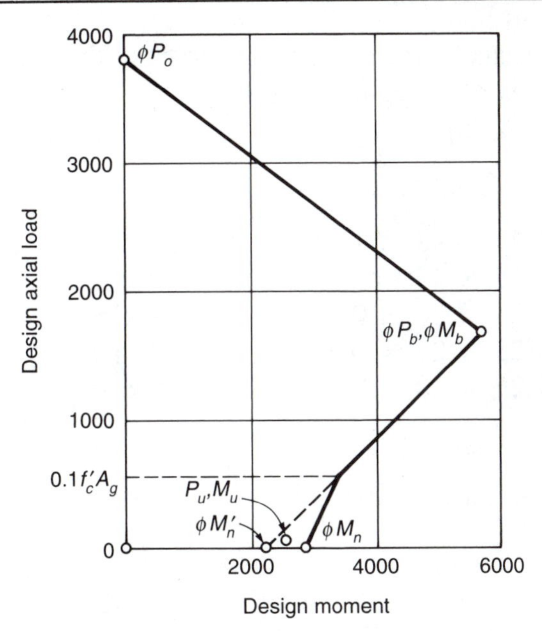

Figure 5–28 Interaction diagram for Example 5–4

BALANCED DESIGN

For the balanced strain condition under combined flexure and axial load, the maximum strain in the concrete and in the tension reinforcement must simultaneously reach the values specified in UBC Section 1910.3.2 as

$$\varepsilon_c = 0.003$$

$$\varepsilon_s = f_y/E_s = 60/29{,}000 = 0.00207$$

The depth to the neutral axis is given by

$$c_b = d\varepsilon_c/(\varepsilon_c + \varepsilon_s) = 117.5 \times 0.003/(0.003 + 0.00207) = 69.5 \text{ inches}$$

In accordance with UBC Section 1910.2.7.1 the depth of the equivalent rectangular concrete stress block is

$$a = c_b\beta_1$$

where β_1 = compression zone factor
= 0.85 as defined in UBC Section 1910.2.7.1

then $a = 69.5 \times 0.85 = 59$ inches

The strain distribution across the section and the forces developed are shown in Figure 5–29 and the force produced in a reinforcing bar is given by

$$F = eA_sE_s \times 0.003/c_b = eA_s \times 29{,}000 \times 0.003/69.5 = 1.25eA_s$$

where e = distance of a reinforcing bar from the neutral axis
A_s = area of the reinforcing bar
and F_{max} = $60A_s$ kips

The tensile forces in the reinforcement are

$$T_1 = 1.25 \times 48 \times 1.58 = 95 \text{ kips}$$
$$T_2 = 1.25 \times 45.5 \times 1.58 = 90 \text{ kips}$$
$$T_3 = 1.25 \times 36 \times 0.4 = 18 \text{ kips}$$
$$T_4 = 1.25 \times 23 \times 0.4 = 12 \text{ kips}$$
$$T_5 = 1.25 \times 10 \times 0.4 = 5 \text{ kips}$$

The sum of the tensile forces is

$$\Sigma T = 220 \text{ kips}$$

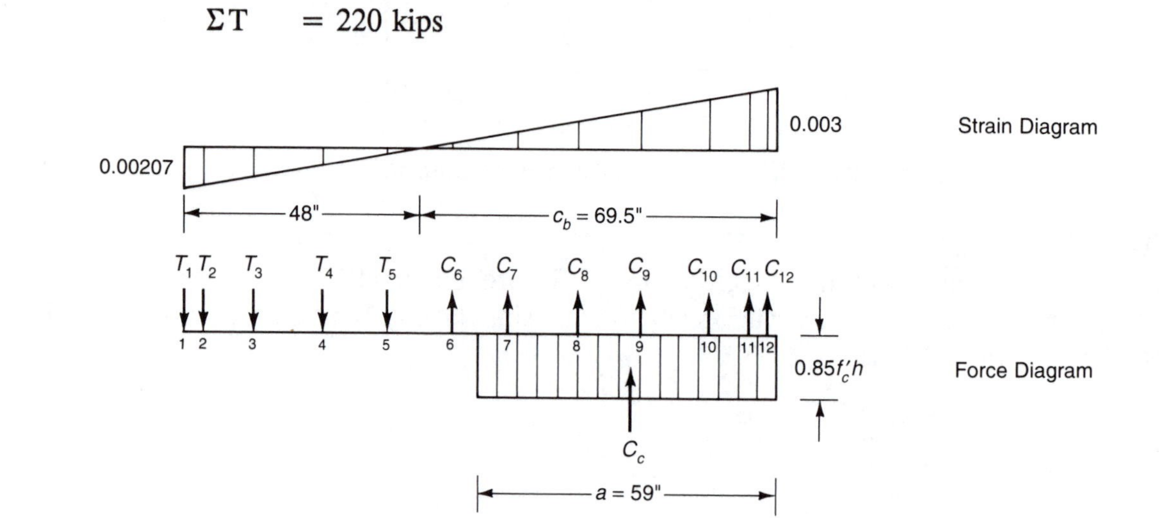

Figure 5–29 Balanced strain conditions for Example 5–4

The compressive forces in the reinforcement are

$$C_6 = 1.25 \times 3 \times 0.4 = 2 \text{ kips}$$
$$C_7 = 1.25 \times 16 \times 0.4 = 8 \text{ kips}$$
$$C_8 = 1.25 \times 29 \times 0.4 = 15 \text{ kips}$$
$$C_9 = 1.25 \times 42 \times 0.4 = 21 \text{ kips}$$
$$C_{10} = 60 \times 0.4 = 24 \text{ kips}$$
$$C_{11} = 60 \times 1.58 = 95 \text{ kips}$$
$$C_{12} = 60 \times 1.58 = 95 \text{ kips}$$

The sum of the compressive forces in the reinforcement is

$$\Sigma C = 260 \text{ kips}$$

The force in the concrete stress block is

$$C_c = 0.85f'_c(ah - A'_s)$$
$$= 0.85 \times 4(59 \times - 8 \times 0.2 - 4 \times 0.79)$$
$$= 2391 \text{ kips}$$

The nominal axial load capacity at the balanced strain condition is

$$
\begin{aligned}
P_b &= C_c + \Sigma C - \Sigma T \\
&= 2391 + 260 - 220 \\
&= 2431 \text{ kips}
\end{aligned}
$$

The design axial load capacity at the balanced strain condition is

$$
\begin{aligned}
\phi P_b &= 0.7 \times 2431 \\
&= 1702 \text{ kips}
\end{aligned}
$$

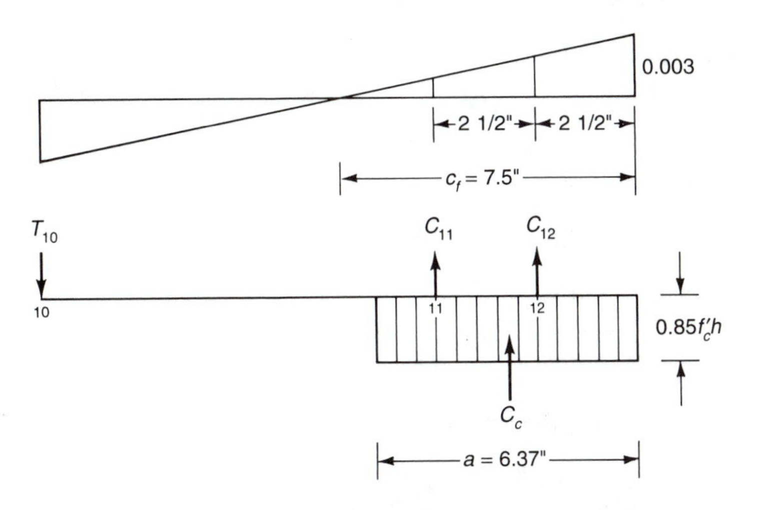

Figure 5–30 Flexure only strain conditions for Example 5–4

The nominal moment capacity at the balanced strain condition is obtained by summing moments about the mid–depth of the section and is given by

$$
\begin{aligned}
M_b &= 30.5C_c + 57.5T_1 + 55T_2 + 45.5T_3 + 32.5T_4 + 19.5T_5 - 6.5C_6 \\
&\quad + 6.5C_7 + 19.5C_8 + 32.5C_9 + 45.5C_{10} + 55C_{11} + 57.5C_{12} \\
&= 97438 \text{ kip inches} \\
&= 8120 \text{ kip feet}
\end{aligned}
$$

The design moment capacity at the balanced strain condition is

$$
\begin{aligned}
\phi M_b &= 0.7 \times 8120 \\
&= 5684 \text{ kip feet}
\end{aligned}
$$

and the point corresponding to the balanced strain condition is plotted on the interaction diagram.

FLEXURE ONLY

Assume the depth to the neutral axis is

$$c_f = 7.5 \text{ inches}$$

Then the depth of the equivalent rectangular concrete stress block is

$$
\begin{aligned}
a &= 7.5 \times 0.85 \\
&= 6.37 \text{ inches}
\end{aligned}
$$

The strain distribution across the section and the forces developed are shown in Figure 5-30. All bars from 1 to 10 are in tension and are stressed to the yield stress. Bars 11 and 12 are in compression and the compressive forces are

$$
\begin{aligned}
C_{12} &= eA_sE_s \times 0.003/c_f \\
&= 1.58 \times 29{,}000 \times 0.003 \times (7.5 - 2.5)/7.5 \\
&= 18.32 \times 5 \\
&= 91 \text{ kips} \\
C_{11} &= 18.32 \times 2.5 \\
&= 46 \text{ kips}
\end{aligned}
$$

The force in the concrete stress block is

$$
\begin{aligned}
C_c &= 0.85f'_c(ah - A'_s] \\
&= 0.85 \times 4(6.37 \times 12 - 4 \times 0.79) \\
&= 249 \text{ kips}
\end{aligned}
$$

The total compressive force on the section is

$$
\begin{aligned}
\Sigma C &= C_{12} + C_{11} + C_c \\
&= 91 + 46 + 249 \\
&= 386 \text{ kips}
\end{aligned}
$$

The sum of the tensile forces on the section is

$$
\begin{aligned}
\Sigma T &= 60(4 \times 0.79 + 16 \times 0.2) \\
&= 382 \text{ kips} \\
&\approx \Sigma C
\end{aligned}
$$

Hence the assumed depth to the neutral axis is satisfactory. The nominal moment capacity of the section, without axial load, is obtained by summing moments about the mid-depth of the section and is given by

$$
\begin{aligned}
M_n &= C_c(60 - 6.37/2) + 95(55 + 57.5) + 48(45.5 + 32.5 + 19.5 + 6.5) \\
&\quad + 46 \times 55 + 91 \times 57.5 \\
&= 37587 \text{ kip inches} \\
&= 3132 \text{ kip feet}
\end{aligned}
$$

In accordance with UBC Section 1909.3.2 the design moment capacity without axial load is

$$
\begin{aligned}
\phi M_n &= 0.9M_n \\
&= 2819 \text{ kips}
\end{aligned}
$$

At an axial load of $0.1f'_cA_g$ the strength reduction factor ϕ reduces to 0.7 and using this value gives a design moment capacity of

$$
\begin{aligned}
\phi M'_n &= 0.7M \\
&= 2193 \text{ kips}
\end{aligned}
$$

The values of ϕM_n, $\phi M'_n$ and $0.1f'_cA_g$ are plotted on the interaction diagram. The design forces P_u and M_u are also plotted on the interaction diagram and lie within the valid region for an under-reinforced section. Hence the proposed reinforcement is satisfactory.

4. EMBEDMENT LENGTH

The basic development length for a hooked bar in tension is specified in UBC Section 1912.5 as

$$l_{hb} = 1200d_b/(f'_c)^{0.5}$$
$$= 19 \text{ inches}$$

The modification factor for increased cover is 0.7 giving a development length of

$$l_{dh} = 0.7 \times 19$$
$$= 13.3 \text{ inches}$$

or

$$\geq 8d_b$$
$$\geq 6 \text{ inches}$$

Hence the required embedment depth is 19 inches.

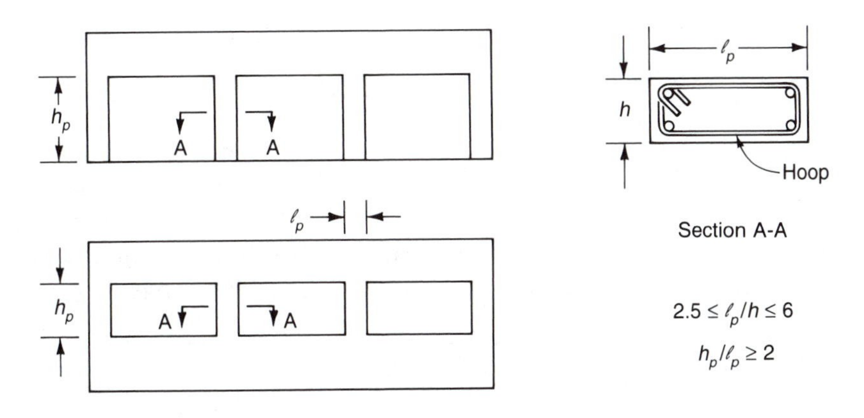

Figure 5–31 Wall pier proportions

5. SPLICE LENGTH

The basic development length for a straight bar in tension is specified in UBC Section 1912.2 as

$$l_{db} = 0.04A_bf_y/(f'_c)^{0.5}$$
$$= 30 \text{ inches}$$

No modification factors are applicable and the development length is

$$l_d = 30 \text{ inches}$$

or

$$\geq 12 \text{ inches}$$

or

$$\geq 0.03d_bf_y/(f'_c)^{0.5}$$
$$\geq 28.5 \text{ inches}$$

Then

$$l_d = 30 \text{ inches controls}$$

For a Class B splice, UBC Section 1912.15 specifies a lap length of

$$1.3l_d = 1.2 \times 30$$
$$= 39 \text{ inches}$$

5.3.3 Wall piers

As shown in Figure 5-31, wall piers are defined in UBC Section 1921.1 as being a wall segment with a horizontal length to thickness ratio of

$$l_p/h \geq 2.5$$

and

$$\leq 6$$

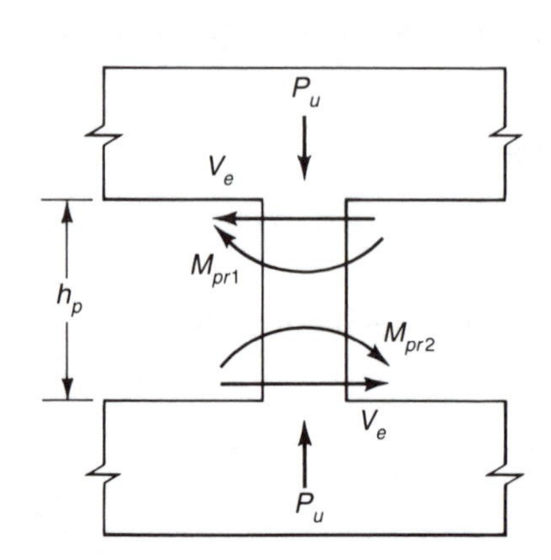

Figure 5-32 Wall pier shear

For ratios smaller than 2.5 the segment is classified as a column and for ratios greater than 6 the segment is classified as a shear wall. In addition, the ratio of clear height to horizontal length is defined as

$$h_p/l_p \geq 2$$

As specified in UBC Section 1921.6.12, transverse reinforcement shall be provided to resist the design shear force V_e determined by considering the probable flexural strengths at the top and bottom of the pier using a tensile reinforcement stress of $1.25f_y$ and a strength-reduction factor ϕ of 1.0. As shown in Figure 5-32 the design shear force is given by

$$V_e = (M_{pr1} + M_{pr2})/h_p$$

The probable flexural strength is readily determined by ignoring axial load and neglecting compression reinforcement[11].

When the compressive force P_u in a pier is less than $A_g f'_c/20$ and the seismic induced shear is not less than half of the total design shear, the shear resistance of the concrete V_c shall be neglected. When the axial compressive force, including seismic effects is not less than $A_g f'_c/20$ transverse reinforcement shall consist of closed hoops at a maximum spacing of six inches. As shown in Figure 5-33, the hoops shall extend beyond the ends of the pier a distance equal to the development length of the longitudinal reinforcement in the pier. When the axial force is less than $A_g f'_c/20$, the transverse reinforcement may consist of straight bars terminating at the edge of the pier with a standard hook. In addition, when a wall pier is braced by means of a shear wall with a stiffness at least six times greater than the stiffness of the pier, special transverse reinforcement is not required.

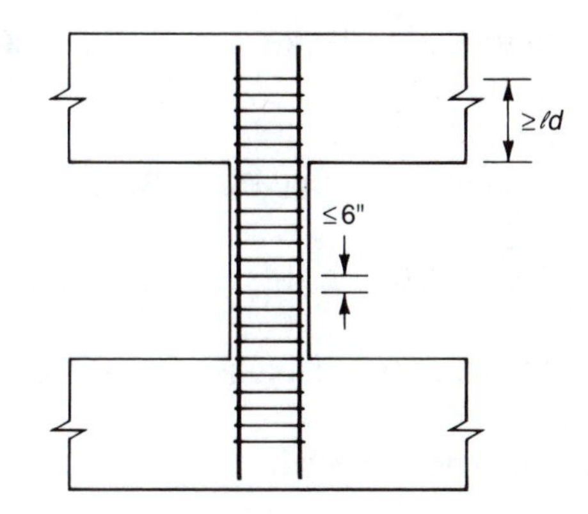

Figure 5–33 Wall pier transverse reinforcement

5.4 WALLS WITH OUT-OF-PLANE LOADING

5.4.1 General requirements

The design of slender reinforced concrete walls, where tension controls the design, is covered in UBC Section 1914.8. To qualify for this design method, the following limitations are imposed:

- The axial stress at the location of the maximum moment, due to the applied service loads, is given by

$$f_a \leq 0.04f'_c$$

- The reinforcement ratio is given by

$$\rho \leq 0.60\rho_b$$

where ρ_b = reinforcement ratio producing balanced strain conditions as defined in UBC Section 1908.4.3
$= 87{,}000 \times 0.85\beta_1 f'_c/f_y(87{,}000 + f_y)$

β_1 = compression zone factor is defined in UBC Section 1910.2.7

- The design moment strength of the section is given by

$$\phi M_n > M_{cr}$$

where M_n = nominal moment strength determined by adding the axial load to the force in the reinforcement[12]
$= A_{se}f_y(d - a/2)$

$A_{se} = (P_u + A_sf_y)/f_y$

P_u = factored axial load

A_s = tensile steel area

a = depth of the equivalent rectangular concrete stress block defined in UBC Section 1910.2.7
$= (P_u + A_sf_y)/0.85f'_cb$
$= A_{se}f_y/0.85f'_cb$

	ϕ	= strength–reduction factor defined in UBC Section 1909.3
		$= 0.9 - 2.0P_u/f'_cA_g$
	A_g	= gross area of section
	M_{cr}	= cracking moment defined in UBC Section 1914.0
		$= 5I_g(f'_c)^{0.5}/y_t$
	I_g	= moment of inertia of gross concrete section
		$= bt^3/12$
	y_t	= distance from centroidal axis of gross section to extreme fibre in tension
		$= t/2$
	t	= overall thickness of wall

5.4.2 Design strength

The factored design moment M_u at midheight of the wall must include the effects of the factored axial loads and eccentricities, the factored lateral load, and the P–delta effect. As shown in Figure 5–34 the ultimate moment is given by

	M_u	$= w_ul_c^2/8 + P_{u1}e/2 + (P_{u1} + P_{u2}/2)\Delta_n$
where	P_{u1}	= factored applied axial load
	P_{u2}	= factored self weight of the wall
	w_u	= factored lateral load
	e	= eccentricity of applied axial load
and	Δ_n	= wall displacement at midheight
		$= 5M_nl_c^2/48E_cI_{cr}$
where	I_{cr}	= moment of inertia of cracked section transformed to concrete
		$= nA_{se}(d - c)^2 + bc^3/3$
	n	= modular ratio of elasticity
		$= E_s/E_c$
	c	= depth to neutral axis
		$= a/\beta_1$
	b	= width of wall
	l_c	= vertical distance between supports
and	M_u	$\leq \phi M_n$

The maximum permissible deflection at midheight Δ_s due to service vertical and lateral loads is given by

$$\Delta_s \leq l_c/150$$

When the applied service moment M_s exceeds the cracking moment M_{cr} the service deflection is given by

$$\Delta_s = \Delta_{cr} + (\Delta_n - \Delta_{cr})(M_s - M_{cr})/(M_n - M_{cr})$$

where $\Delta_{cr} = 5M_{cr}l_c^2/48E_cI_g$

When the applied service moment M_s is less than the cracking moment M_{cr} the service deflection is given by

$$\Delta_s = 5M_sl_c^2/48E_cI_g$$

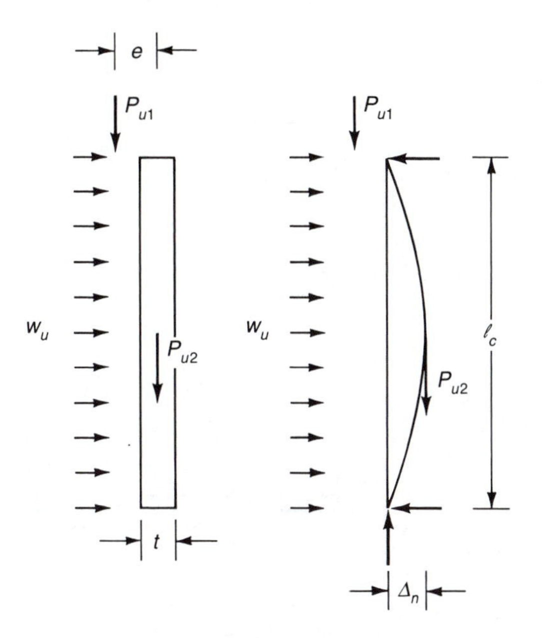

Figure 5-34 Analysis of slender wall

Example 5-5 (Structural Engineering Examination 1990, Section C, weight 7.0 points)

GIVEN: Precast concrete wall as shown in Figure 5-35.

CRITERIA: Seismic zone 4, I_p=1.0
Slender wall design.
Materials:

- Concrete f'_c = 3000 pounds per square inch
- Reinforcing steel f_y = 60,000 pounds per square inch
- Normal weight concrete E = 3.16 × 10^6 pounds per square inch

Assumptions:

- Wind does not govern
- Ignore effect of parapet
- Wall is pinned at top and bottom
- Vertical reinforcing is at center line of the wall

REQUIRED:

1. Determine the ultimate moment at midheight. Include the P-delta moment and assume A_{se} = 0.24 square inches per foot for calculating Δ_n. The value of "c" used in calculating I_{cr} shall be 0.84 inches.
2. Determine the adequacy of vertical reinforcement consisting of number 4 bars at 10 inches on center and show that the first three items of limitations in the UBC Section 1914.8.2 for slender wall design have been complied with.

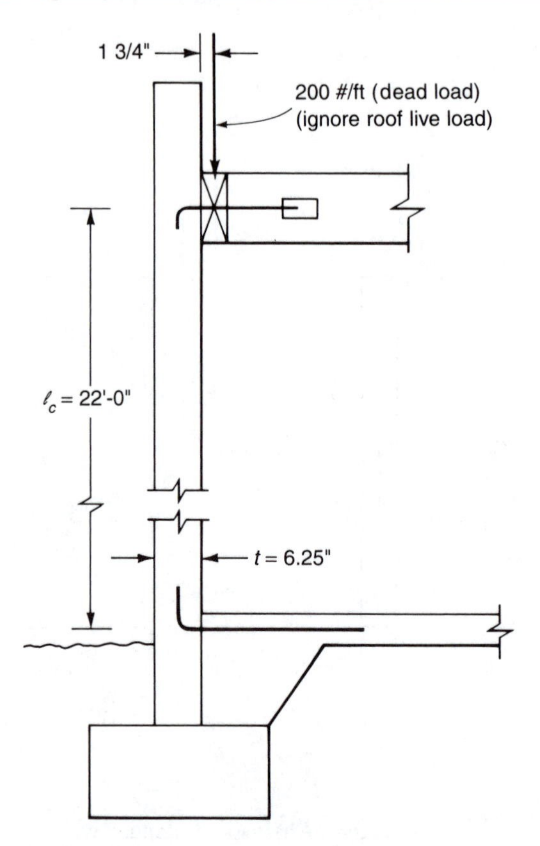

Figure 5-35 Wall details for example 5-5

SOLUTION

FACTORED LOADS

Since the concrete panel is located in seismic zone number 4, in accordance with UBC Section 1921.2.7, the required strength is

$$U = 1.4(D + E)$$

where D = dead load

and E = seismic load

Considering a one foot width of panel, the factored roof dead load is given by

$$P_{u1} = 1.4 \times 200 = 280 \text{ pounds}$$

The eccentricity of the roof load about the panel center line is

$$e = 1.75 + 6.25/2 = 4.875 \text{ inches}$$

The factored self weight of the panel is

$$q_u = 1.4 \times 150 \times 6.25/12 = 109.375 \text{ pounds per foot}$$

The factored self weight of the panel, neglecting the parapet, is

$$P_{u2} = q_u l_c$$
$$= 109.375 \times 22$$
$$= 2406 \text{ pounds}$$

The total factored axial loads at midheight of the wall, neglecting the parapet, is

$$P_u = P_{u1} + P_{u2}/2$$
$$= 280 + 1203$$
$$= 1483 \text{ pounds}$$

The factored seismic load on the panel is given by UBC Section 1630.2, Formula (30–1) as

$$w_u = ZI_pC_pq_u$$

where Z = 0.4 as given
I_p = 1.0 as given
C_p = 0.75 from UBC Table 16–O, item I.1.b
q_u = 109.375 as calculated

Then
$$w_u = 0.4 \times 1 \times 0.75 \times 109.375$$
$$= 32.81 \text{ pounds per foot}$$

MOMENT STRENGTH

The depth of the equivalent rectangular stress block is given by UBC Section 1910.27 as

$$a = A_{se}f_y/0.85f'_cb$$

where A_{se} = 0.24, the given effective reinforcement area
b = 12 inches

Then
$$a = 0.24 \times 60/0.85 \times 4 \times 12$$
$$= 0.47 \text{ inches}$$

The nominal moment strength is given by[12]

$$M_n = A_{se}f_y(d - a/2)$$
$$= 0.24 \times 60(6.25/2 - 0.47/2)$$
$$= 41.61 \text{ kip inches}$$

In accordance with UBC Section 1909.3.1, the strength reduction factor for combined axial load and flexure is given by

$$\phi = 0.9 - 2P_u/f'_cA_g$$
$$= 0.9 - 2 \times 1483/(3000 \times 12 \times 6.25)$$
$$= 0.887$$

Hence, the design moment strength is given by

$$\phi M_n = 0.887 \times 41.61$$
$$= 36.91 \text{ kip inches}$$

1. ULTIMATE MOMENT

The modular ratio is

$$n = E_s/E_c$$
$$= 29{,}000/3160$$
$$= 9.18$$

From UBC Section 1914.8.4, the moment of inertia of the cracked section is

$$I_{cr} = nA_{se}(d - c)^2 + bc^3/3$$
$$= 9.18 \times 0.24(6.25/2 - 0.84)^2 + 12 \times 0.84^3/3$$
$$= 13.87 \text{ inches}^4$$

The deflection occurring at the midheight of the wall under the effects of the nominal moment is

$$\Delta_n = 5M_n l_c^2/48E_c I_{cr}$$
$$= 5 \times 41.61 \times 22^2 \times 144/(48 \times 3.16 \times 13.87)$$
$$= 6.89 \text{ inches}$$

Then, the applied ultimate moment at the midheight of the panel is given by[12]

$$M_u = w_u l_c^2/8 + P_{u1}e/2 + (P_{u1} + P_{u2}/2)\Delta_n$$
$$= 32.81 \times 22^2 \times 12/8000 + 0.280 \times 4.875/2 + 1.483 \times 6.89$$
$$= 34.72 \text{ kip inches}$$

2. SECTION CAPACITY

For number 4 vertical bars at 10 inch spacing the reinforcement area per foot width is

$$A_s = 0.2 \times 12/10$$
$$= 0.24 \text{ square inches}$$

The effective reinforcement area is given by UBC Section 1914.8.4 as

$$A_{se} = (P_u + A_s f_y)/f_y$$
$$= (1.483 + 0.24 \times 60)/60$$
$$= 0.265 \text{ square inches}$$

The depth of the equivalent rectangular stress block is

$$a = A_{se} f_y/0.85 f'_c b$$
$$= 0.265 \times 60/(0.85 \times 3 \times 12)$$
$$= 0.52 \text{ inches}$$

The design moment strength is, then,

$$\phi M_n = \phi A_{se} f_y(d - a/2)$$
$$= 0.887 \times 0.265 \times 60(6.25/2 - 0.52/2)$$
$$= 40.41 \text{ kip inches}$$
$$> M_u$$

Hence the reinforcement provided is adequate.

2A. SERVICE LOAD STRESS

The total service axial load at midheight of the wall is

$$P = P_u/1.4 = 1{,}483/1.4 = 1059 \text{ pounds}$$

The service axial load stress at midheight of the wall is

$$f_a = P/A_g = 1059/(12 \times 6.25) = 14.12 \text{ pounds per square inch} < 0.04f'_c$$

and the required limitation is satisfied.

2B. REINFORCEMENT RATIO

The reinforcement ratio producing balanced strain conditions is given by UBC Section 1908.4.3, Formula (8–1) as

$$\rho_b = 87 \times 0.85\beta_1 f'_c/(87 + f_y)f_y = 87 \times 0.85 \times 0.85 \times 3/60(87 + 60) = 0.0214$$

The actual reinforcement ratio provided is

$$\rho = A_s/bd = 0.24/(12 \times 3.125) = 0.0064 < 0.6\rho_b$$

and the required limitation is satisfied.

2C. CRACKING MOMENT

The moment of inertia of the gross concrete section about its centroidal axis, neglecting reinforcement is

$$I_g = bt^3/12 = 12 \times 6.25^3/12 = 244 \text{ inches}^4$$

The section modulus of the gross section is given by

$$S_g = 2I_g/t = 2 \times 244/6.25 = 78 \text{ inches}^3$$

The cracking moment is given by UBC Section 1914.0 as

$$M_{cr} = 5S_g(f'_c)^{0.5} = 5 \times 78 \times (3000)^{0.5}/1000 = 21.36 \text{ kip inches} < \phi M_n$$

and the required limitation is satisfied.

References

1. American Concrete Institute. *Design handbook in accordance with the strength design method.* Detroit, MI, 1985.

2. Ghosh, S. K. and Domel, A. W. *Design of concrete buildings for earthquake and wind forces.* Portland Cement Association, Skokie, IL, 1992.

3. Portland Cement Association. *Strength design of reinforced concrete columns.* Skokie, IL, 1978.

4. Concrete Reinforcing Steel Institute. *CRSI handbook.* Chicago, IL, 1984.

5. Portland Cement Association. *Notes on ACI 318-89: Building code requirements for reinforced concrete.* Skokie, IL, 1990.

6. Cole, E. E. et al. *Seismic design examples in seismic zones 4 and 2A.* Concrete Reinforcing Steel Institute. Schaumburg, IL, 1993.

7. Lai, S. L. Detailing provisions for concrete structures. *Proceedings of the 1988 seminar of the Structural Engineers Association of Southern California.* Los Angeles, CA, November 1988.

8. Structural Engineers Association of California. *Recommended lateral force requirements and commentary.* Sacramento, CA, 1990.

9. Zsutty, T and Merovich, A. New SEAOC design provisions for shear walls. *Proceedings of the Structural Engineers Association of Southern California seminar on changes to the building code.* Los Angeles, CA, October 1993.

10. Strand, D. R. Proposed concrete shearwall boundary member design. *Proceedings of the Los Angeles Tall Buildings Structural Design Council Seminar.* Los Angeles, CA, May 1993.

11. Lai, J. S. Special provisions for design of wall piers. *Proceedings of the 59th annual convention of the Structural Engineers Association of California.* Sacramento, CA, 1990.

12. American Concrete Institute and Structural Engineers Association of Southern California. *Report of the task committee on slender walls.* Los Angeles, CA, 1982.

6

Seismic design of wood structures

6.1 CONSTRUCTION REQUIREMENTS

6.1.1 Diaphragms

Diaphragm rotation

Lateral loads may be distributed to the shear walls by diaphragm rotation, as shown in Figure 6–1, when one end of a structure is without shear walls. By equating moments about the corner O the shear forces in the diaphragm may be obtained.

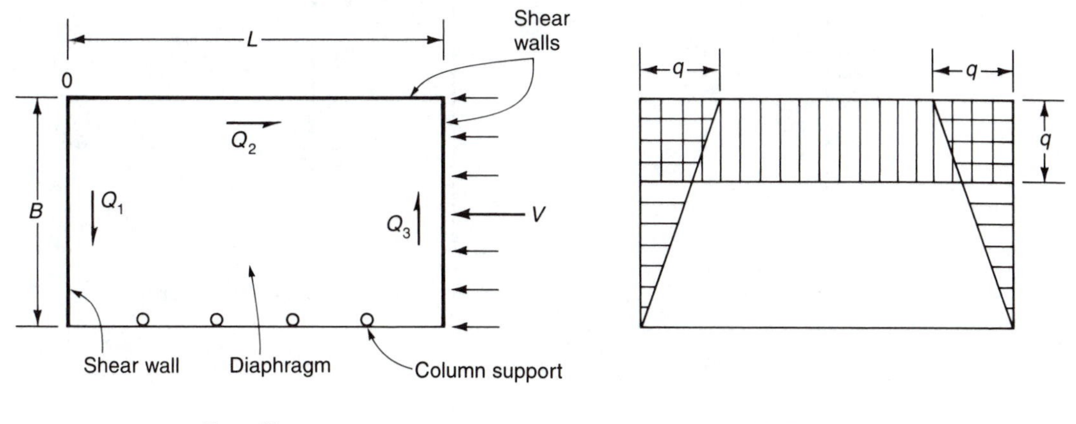

Figure 6–1 Diaphragm rotation

$$Q_3 = VB/2L$$
$$= Q_1$$

and $$Q_2 = V$$

The unit shear is given by

$$q = V/L$$

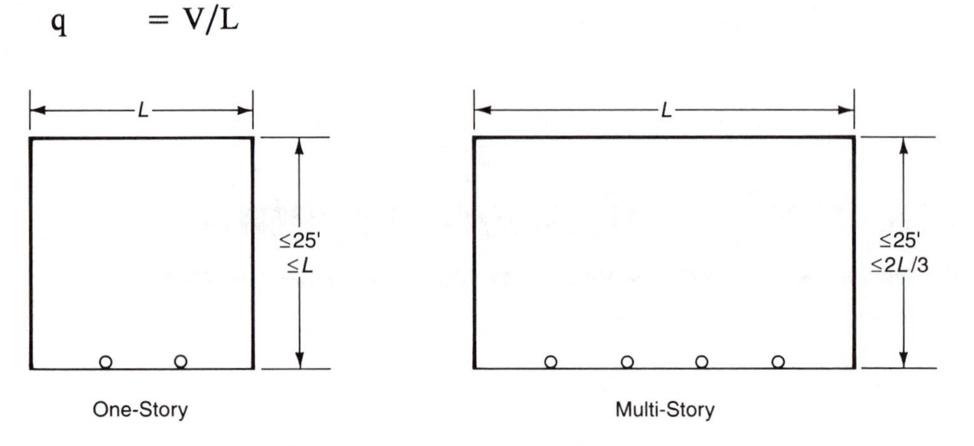

Figure 6–2 Code limitations

Rotation, together with translation may cause excessive deflections and to control this UBC Section 2314.1 imposes limitations on the diaphragm dimensions which are illustrated in Figure 6–2. Rotation is also produced, as shown in Figure 6–3, when the diaphragm cantilevers over a shear wall. The diaphragm shear forces are obtained as

$$Q_1 = Ve/L$$

and $$Q_2 = V(1 + e/L)$$

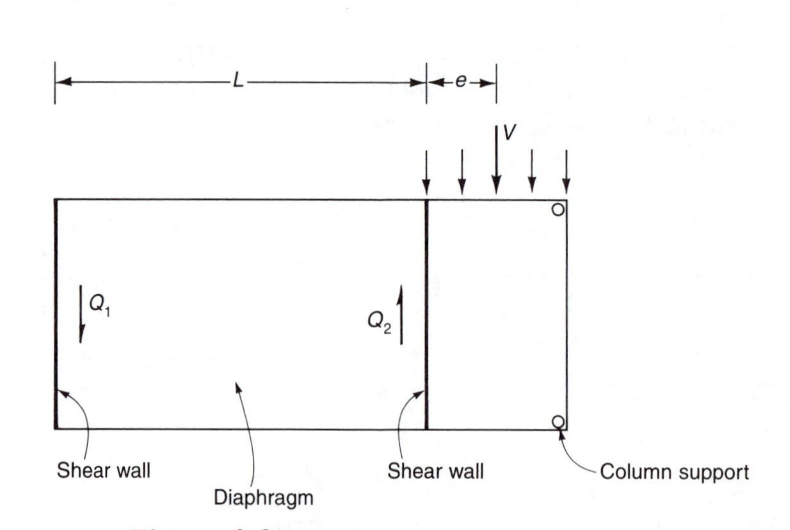

Figure 6–3 Cantilever diaphragm

Extensive damage occurred, during the 1994 Northridge earthquake in California, to buildings employing these types of construction. As a result, diaphragm rotation to distribute lateral loads in wood frame construction may no longer be employed[1,2] in seismic zones 3 and 4.

Diaphragm nailing
The allowable unit shear values for a plywood diaphragm, given in UBC Table 23-I-J-1, are based on the use of common nails. Current practice is to use sinker nails which are easier to drive, more adaptable to nailing guns, and less likely to split lumber than common nails. For the equivalent pennyweight classification, sinker nails have a smaller shank diameter than common nails. This gives sinker nails a lateral load capacity of fifteen percent less than common nails and requires a corresponding reduction[3] in the unit shear values of UBC Table 23-I-J-1.

Lateral support of concrete and masonry walls
The seismic response of a structure with a large wood diaphragm and walls of masonry or concrete is governed by the characteristics of the diaphragm not the wall system. In order to improve the seismic performance of this type of structure UBC Section 1631.2.9 specifies that design forces shall be determined using a structure response modification factor of

$$R_w \quad \leq 6$$

In addition, the acceleration response over the central half of a diaphragm supporting concrete or masonry walls is greater than at the ends. To compensate for this, UBC Table 16-O requires the horizontal force factor C_p to be increased by 50 percent over the central half of the diaphragm.

High-load horizontal plywood diaphragms
The maximum allowable shear tabulated in UBC Table 23-I-J-1 is 820 pounds per foot. This requires 19/32-inch C-D grade plywood, 10 pennyweight common nails at two inch spacing, and three inch nominal framing members. With large concrete tilt-up structures, this value is frequently exceeded. To provide increased shear values requires the use of multiple lines of closely-spaced and staggered nails and increased size of framing members. Test results for these configurations have been published[4,5,6] and a shear value of 1810 pounds per foot may be obtained by using ⅝-inch Structural I plywood, 10 pennyweight common nails at 2½-inch staggered spacing in three lines, and four inch nominal framing members.

6.1.2 Collector elements
The function of a collector is to ensure that all parts of the structure act as a part of the overall structure without localized separation or loss of support. In accordance with UBC Section

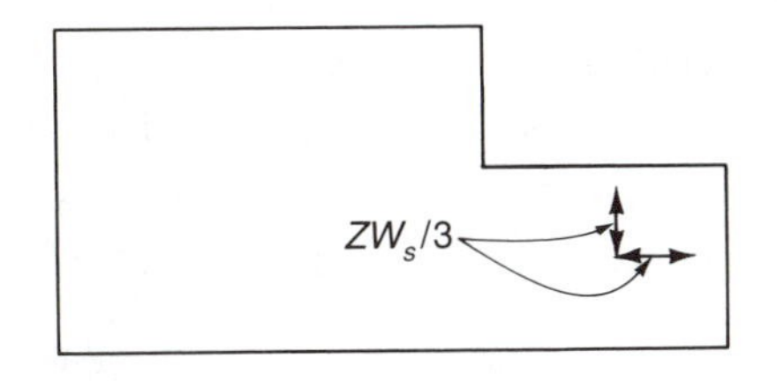

Figure 6–4 Force on portion of a structure

1631.2.5, and as shown in Figure 6–4, the smaller section of a building shall be tied to the larger section with members designed to resist a force of

	V	$= ZW_s/3$
where	Z	= seismic zone factor
	W_s	= weight of the smaller section

In addition, connecting members shall be designed to resist a horizontal force parallel to the member of

	F	$= ZW_m/5$
where	W_m	= tributary dead plus live load

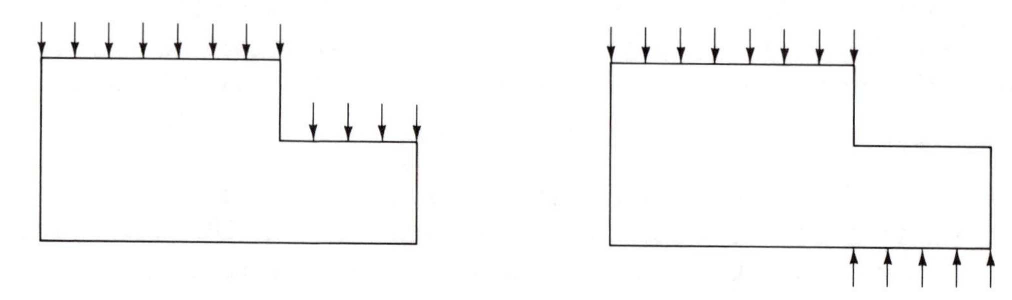

Figure 6–5 Design of a structure with an irregularity

Chords and collectors, for structures in seismic zones 3 and 4 with a plan irregularity type 2 in UBC Table 16–M, shall be designed as specified in UBC Section 1631.2.9 for independent movement of the projecting wings of the structure. As shown in Figure 6–5 this may consist of either motion of the projecting wings in the same direction or motion of the projecting wings in the opposite direction. In addition, for structures with a plan irregularity type 1, 2, 3, or 4 UBC Section 1631.2.9 allows no increase in stress in designing the connections of diaphragms and collectors.

Example 6–1 (Structural Engineering Examination 1993, Section D, weight 6.0 points)

<u>GIVEN</u>: A tilt–up concrete building with plywood roof shown in Figure 6–6

<u>CRITERIA</u>: Materials:

Roof:

- Built–up roofing with gravel
- Panelized plywood roof
- 15/32–inch plywood, Structural I, Blocked
- Plywood nailing at critical locations consists of 10d common nails spaced at 2 inches at all edges and at 12 inches at intermediate framing. Subpurlins are 3× members.

Walls:

- Normal weight concrete = 150 pounds per cubic foot
- Wall thickness = 6 inches

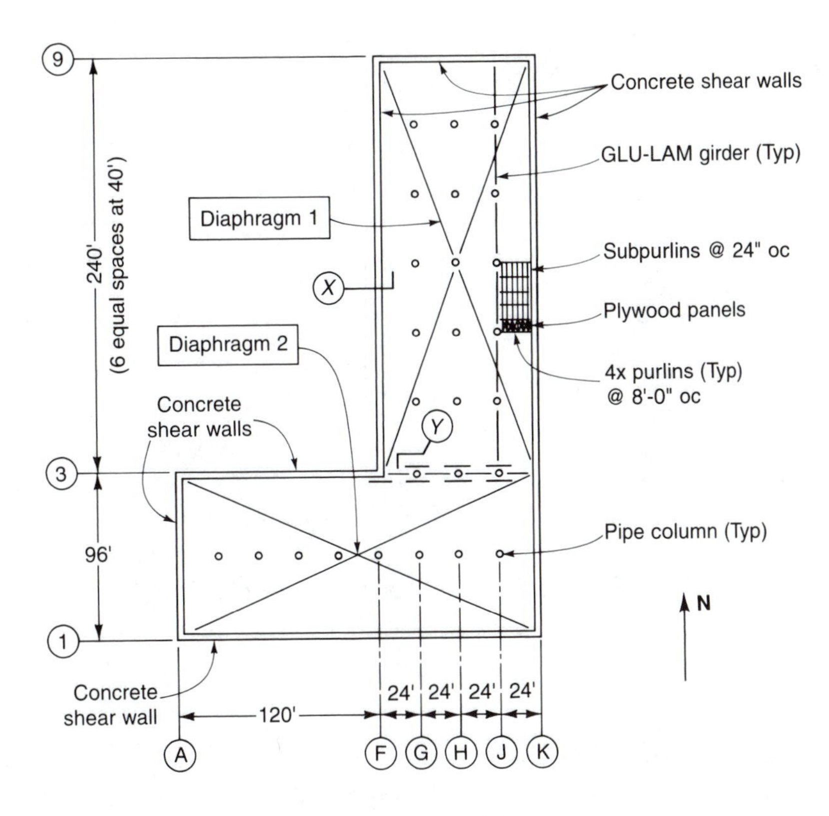

Figure 6–6 Building plan for Example 6–1

Assumptions:

- Seismic zone 4, I = 1.0
- Total roof dead load = 16 pounds per square foot
- Neglect wind loads

REQUIRED:

1. For "Diaphragm 1" shown in Figure 6–6, calculate for east–west seismic forces:
    a. Maximum diaphragm shear force. Is the diaphragm adequate?
    b. Maximum chord force.
    c. Maximum out-of-plane force for design of roof-to-wall anchors. See Figure 6–7.
    d. Provide a sketch showing an arrangement of subdiaphragms and continuity ties which meets UBC requirements using the minimum number of ties. No construction details are necessary.
2. Calculate collector forces at grid F–3 for east–west seismic forces.
3. Calculate collector forces at grid G–3 for east–west seismic forces. Determine the adequacy of two rows of ⅞-inch bolts as indicated in Figure 6–8 on each side of the glued-laminated girder. Take into account UBC Section 1631 and use modification factors for metal side plates and number of bolts in a row.

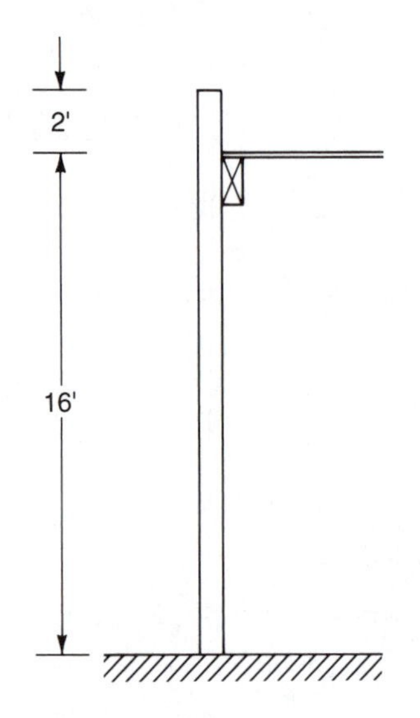

Figure 6–7 Section X–X for Example 6–1

SOLUTION

1a. DIAPHRAGM 1 SHEAR FORCE

The relevant dead load tributary to the roof diaphragms in the east–west direction is due to the east and west walls and the roof dead load and is given by

$$\text{Roof} = 16 \times 96 = 1536 \text{ pounds per linear foot}$$
$$\text{Walls} = 2 \times 150 \times 18^2/(2 \times 2 \times 16) = 1519 \text{ pounds per linear foot}$$

The total dead load tributary to the roof diaphragm is

$$W_1 = (1536 + 1519)240/1000$$
$$= 733 \text{ kips}$$

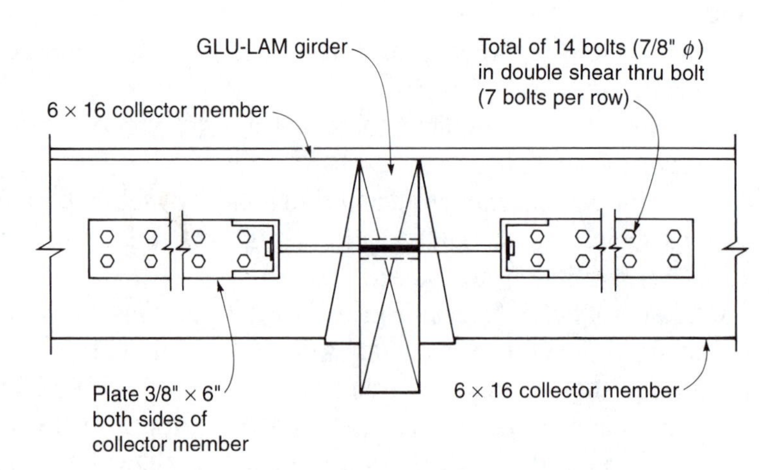

Figure 6–8 Section Y–Y for Example 6–1

For a single story structure, the seismic load acting at roof level is given by UBC Formula (28-1) as

$V = (ZIC/R_W)W_1$

where
Z = 0.4 for zone 4 from UBC Table 16-I
I = 1.0 as given
C = 2.75 from UBC Section 1628.2.1
R_W = 6 from UBC Table 16-N item 1.2.b
= maximum value allowed by UBC Section 1631.2.9 for a flexible diaphragm supporting a concrete wall

The total seismic load acting on the diaphragm is

$$V_1 = (0.4 \times 1.0 \times 2.75/6)733$$
$$= 0.183 \times 733$$
$$= 134 \text{ kips}$$

The shear force along the diaphragm boundaries at grid lines 3 and 9 is

$$Q_1 = V_1/2$$
$$= 67 \text{ kips}$$

The unit shear along the diaphragm boundaries is

$$q_1 = Q_1/B_1$$
$$= 67 \times 1000/96$$
$$= 700 \text{ pounds per linear foot}$$

The allowable unit shear is obtained from UBC Table 23-I-J-1 for 10 pennyweight common nails spaced at two inches at all edges with a case 5 plywood Structural I layout and three inch nominal subpurlins as

$$q_a = 820 \text{ pounds per linear foot}$$
$$> q_1$$

Hence the diaphragm is adequate.

1b. CHORD FORCE

The bending moment at the midpoint of east and west boundaries due to the east-west seismic force is

$$M = V_1L_1/8$$
$$= 134 \times 240/8$$
$$= 4020 \text{ kip feet}$$

The corresponding chord force is

$$F_c = M/B_1$$
$$= 4020/96$$
$$= 41.9 \text{ kips}$$

1c. WALL ANCHOR FORCE

The seismic loading on the concrete wall at roof diaphragm level is given by UBC Formula (30–1) as

$F_p = ZI_pC_pW_p$

where
$Z = 0.4$ for zone 4 from UBC Table 16–I
$I_p = 1.0$ as given
$C_p = 1.5 \times 0.75$ from UBC Table 16–O item 1.1.b and footnote 3
$W_p = 1519/2 = 759.5$ pounds per linear foot

The anchorage force at roof diaphragm level is

$$F_p = 0.4 \times 1.0 \times 1.5 \times 0.75 \times 759.5$$
$$= 342 \text{ pounds per linear foot}$$
$$> 200$$

Hence the minimum anchorage force of 200 pounds per linear foot of wall, stipulated in UBC Section 1611 does not govern and the force on each wall anchor for a spacing of eight feet is

$$P = 8F_p$$
$$= 2736 \text{ pounds}$$

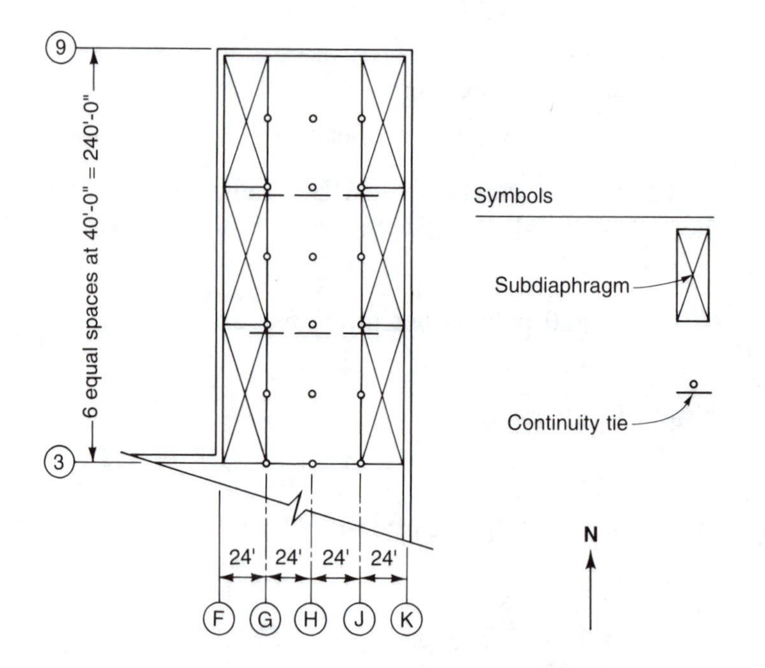

Figure 6–9 Subdiaphragms for Example 6–1

1d. SUBDIAPHRAGM REQUIREMENTS

Since the wall anchor spacing exceeds four feet, in accordance with UBC Section 1611, the wall must be designed to span between anchors. The wall anchor force is transferred to the four inch nominal purlins which span twenty-four feet from the wall to the glued-laminated girder and act as the subdiaphragm ties. As shown in Figure 6–9, the subdiaphragms are bounded by the glued-laminated girders on grid lines G and J and by the four inch nominal purlins on grid lines

5 and 7. The glued–laminated girders on grid lines G and J form the inner chord of the subdiaphragm. The subdiaphragm aspect ratio is

$$b/d = 80/24$$
$$= 3.33$$
$$< 4$$

Hence the requirements of UBC Table 23–I–I are satisfied. The subdiaphragms span 80 feet between the continuous, full depth crossties which consist of the four inch nominal purlins at 80 feet on center at grid lines 5 and 7. The design force in the crossties is

$$P_t = 80F_p$$
$$= 27{,}360 \text{ pounds}$$

Details of the subdiaphragms are shown on Figure 6–9.

2. COLLECTOR FORCES

DIAPHRAGM 2 SHEAR FORCE

The dead load tributary to diaphragm 2 is given by

$$\text{Roof} = 16 \times 216 = 3456 \text{ pounds per linear foot}$$
$$\text{Walls} = 1519 \text{ pounds per linear foot}$$

The total dead load tributary to diaphragm 2 is

$$W_2 = (3456 + 1519)96/1000$$
$$= 478 \text{ kips}$$

The seismic load acting at roof level is

$$V_2 = 0.183 \times 478$$
$$= 87.4 \text{ kips}$$

The shear force along the diaphragm boundary at grid line 3 is

$$Q_2 = V_2/2$$
$$= 43.7 \text{ kips}$$

The unit shear in diaphragm 2 along grid line 3 is

$$q_2 = Q_2/B_2$$
$$= 43.7 \times 1000/216$$
$$= 202 \text{ pounds per linear foot}$$

COLLECTOR FORCE AT F3

The total unit shear along grid line 3 between grid lines F and K as shown in Figure 6–10 is

$$q_t = q_1 + q_2$$
$$= 700 + 202$$
$$= 902 \text{ pounds per linear foot}$$

The collector force at grid line F is

$$F_F = 96q_t/1000 = 86.6 \text{ kips}$$

3. BOLTED CONNECTION

The collector force at grid line G on line 3 is

$$F_G = 72q_t/1000 = 64.9 \text{ kips}$$

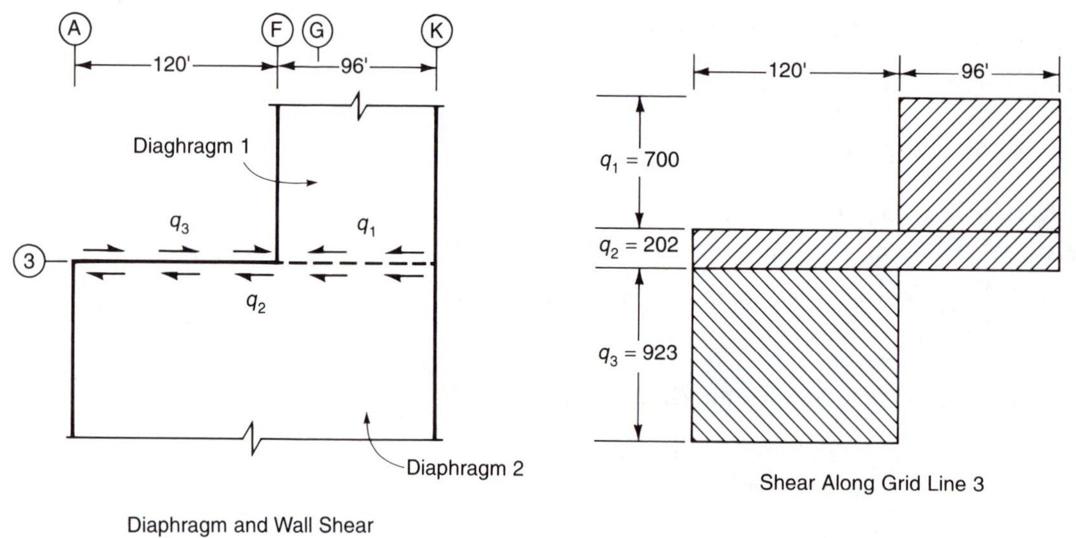

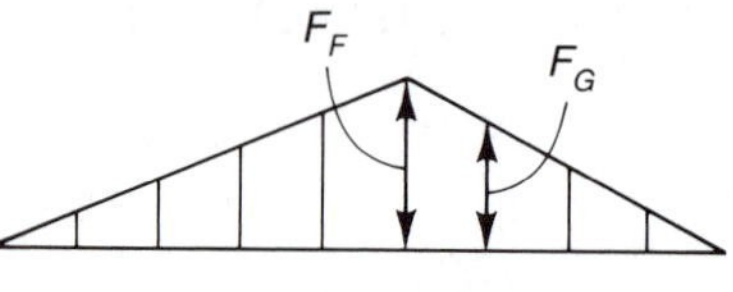

Figure 6–10 Collector forces for Example 6-1

The nominal single ⅞-inch bolt design value for a 5½-inch thick Douglas fir-larch main member, in double shear with steel side plates is given in UBC Table 23-III-O as

$$Z_{\parallel} = 4260 \text{ pounds}$$

Assuming the bolt edge distance, end distance, and spacing shown in Figure 6-8, are sufficient to develop the full design value, the group action factor may be determined from UBC Table 23-III-E. The cross-sectional area of the main member is

$$A_m = 5.5 \times 15.5 = 85.25 \text{ square inches}$$

The sum of the cross-sectional areas of the side members is

$$A_s = 2 \times 6 \times 0.375$$
$$= 4.5 \text{ square inches}$$

The ratio of the two areas is

$$A_m/A_s = 85.25/4.5$$
$$= 18.9$$

From UBC Table 23–III–E the group action factor for seven bolts in a row is given as

$$C_g = 0.94$$

Since the structure has a plan irregularity type 2 in UBC Table 16–M no increase in stress is allowed for seismic loads. Hence the allowable design value for the bolt group is

$$F = 14 \times C_g \times Z_{\parallel}$$
$$= 14 \times 0.94 \times 4260/1000$$
$$= 56.06 \text{ kips}$$
$$< F_g$$

Hence the connection is inadequate.

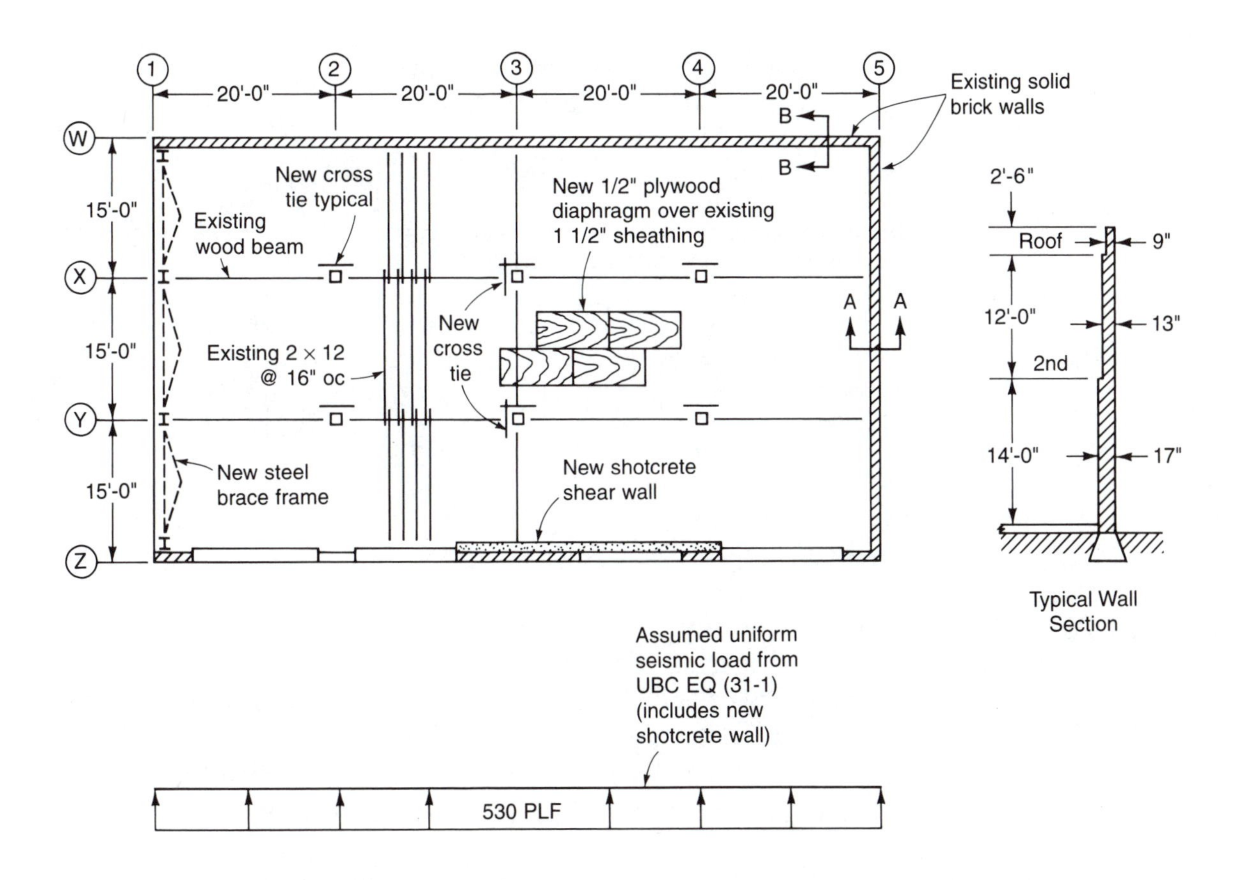

Figure 6–11 Second floor framing plan for Example 6–2

Example 6–2 (Structural Engineering Examination 1991, Section D, weight 8.0 points)

GIVEN: An existing two story building with unreinforced brick walls and wood floor and roof is to be strengthened to improve its seismic resistance. The second floor framing plan is shown in Figure 6–11.

CRITERIA: Seismic zone 4

Standard occupancy

Materials:

- Use the following allowable stresses for all new and existing wood:
  F_b = 1250 pounds per square inch, F_v = 95 pounds per square inch,
  $F_{c\perp}$ = 625 pounds per square inch,
- Bolts are ASTM A307.
- Brick wall density is 120 pounds per cubic foot.

Assumptions:

- The new steel braced frame on line "1" and the new shotcrete shear wall on line "Z" are adequate.
- The existing brick walls on lines "W" and "5" are adequate for shear and are adequate for out-of-plane bending if properly tied at the floor and roof.

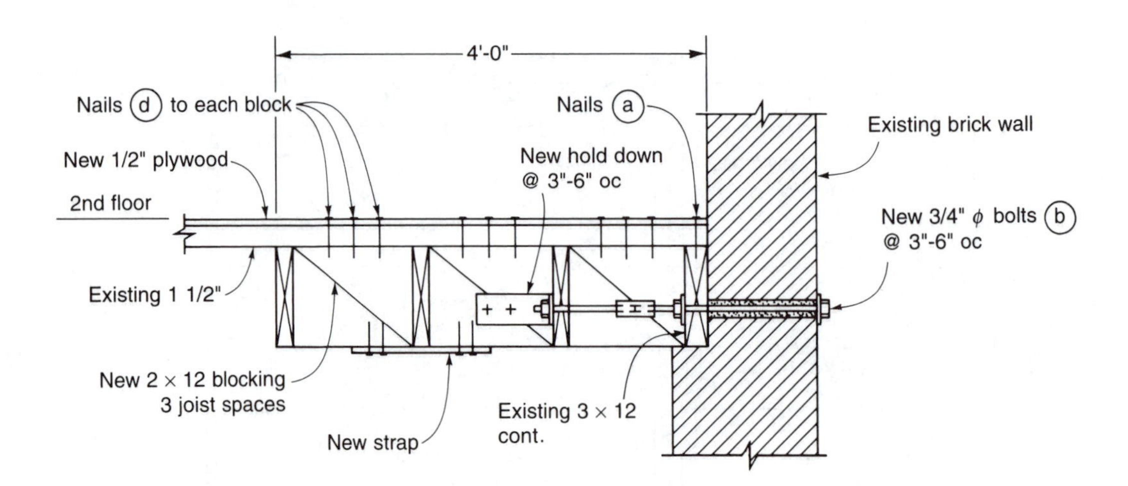

Figure 6–12 Detail A for Example 6–2

REQUIRED: Refer to Figure 6–12 for questions a through e.

a. Find the required spacing of nails 'a'. Use 16d common nails.
b. Check the adequacy of bolts 'b' to transfer diaphragm shear to the brick wall. (Check wood only, bolts are adequate in brick).
c. Find the tension in bolts 'b' due to seismic out-of-plane wall loads.
   Assume the tension in bolts 'b' is 1700 pounds per bolt to answer parts d and e.
d. Find the number of nails 'd' required through the plywood to each block. Use 16d common nails. The strap shown is adequate.

e. The blocking acts as a continuous crosstie for a subdiaphragm 4 feet wide. Find the design shear in the subdiaphragm, assuming adequate continuous crossties are provided across the building on lines "X" and "Y".
   Refer to Figure 6-13 for questions f through h. Assume the seismic out-of-plane wall load is 500 pounds per linear foot.
f. Check block 'f' for shear and bending due to out-of-plane wall loads. To allow for construction tolerance, assume the bolt may be as much as 3 inches from the center of the joist space as shown in detail B. Neglect reduction in the cross section due to the bolt hole and neglect hanger length.
g. Find the required size of square plate washer 'g'. Assume the ¼-inch thickness is adequate. Account for the reduced bearing area due to a ⅞-inch diameter hole drilled in the block.
h. Find the subdiaphragm design shear assuming a single continuous crosstie is provided along line "3" from line "W" to line "Z". The typical joist laps at lines "X" and "Y" have no tension capacity.

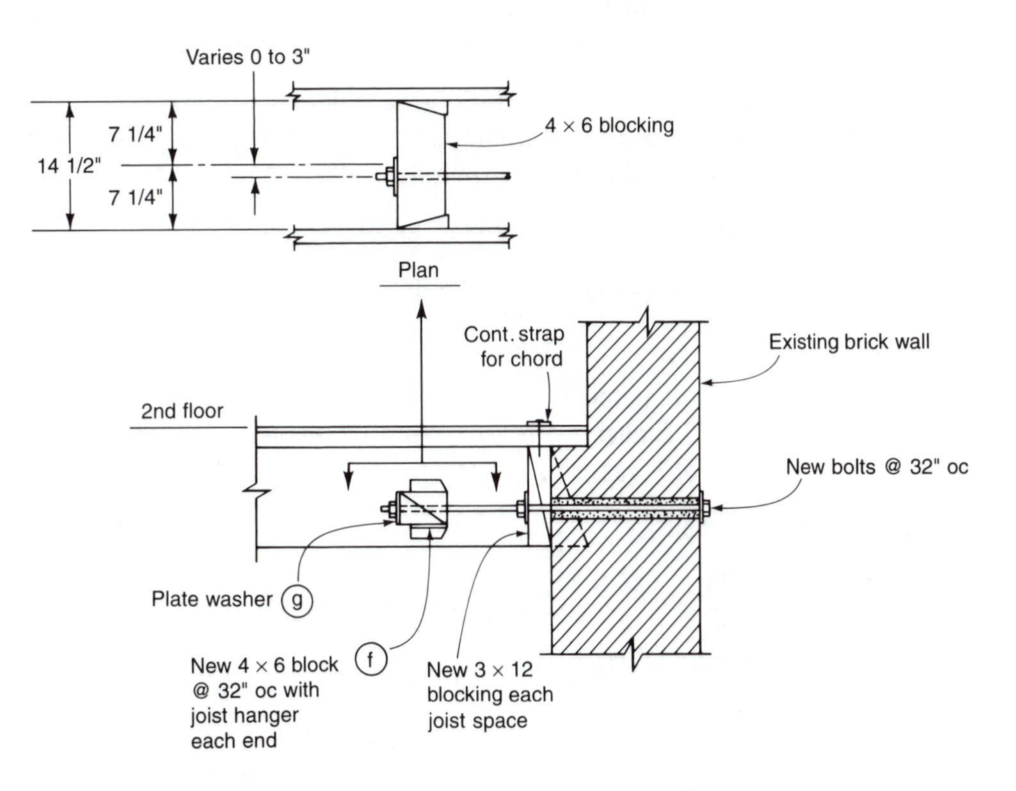

Figure 6-13 Detail B for Example 6-2

SOLUTION

a. DIAPHRAGM NAILING

The seismic load tributary to the second floor diaphragm is given by

$$V = 530 \times 80$$
$$= 42{,}400 \text{ pounds}$$

The unit shear in the diaphragm along grid line 5, assuming a flexible diaphragm, is

$$q = V/(2L) = 42{,}400/(2 \times 45) = 471 \text{ pounds per linear foot}$$

The allowable lateral load on a 16d common nail, with a penetration of 1.75 inches is obtained from UBC Table 23-I-G as

$$Z = 108 \text{ pounds}$$

This value must be adjusted by the following factors[7,8,9].

C_D = load duration factor
= 1.33 for seismic loading from UBC Section 2304.3.4

C_{di} = diaphragm factor
= 1.30 in accordance with NDS Commentary[10] Section 12.3.6 when a load duration factor of 1.33 is used for seismic loading

1.0 for 3 inch nominal framing[7,9]

1.0 for a single line of nails with spacing exceeding two inches[9]

The actual penetration of the nail into the framing is obtained from the nail length, which is given in UBC Table 23-I-G as 3.5 inches, and the thickness of the existing sheathing and the new plywood diaphragm. Thus, the actual penetration is

$$p = 3.5 - 1.5 - 0.5 = 1.5 \text{ inches}$$

and, in accordance with UBC Section 2340.3.4 the penetration depth factor is given by Formula (40-5) as

$$C_d = p/12D$$

where D = nail diameter
= 0.162 inches from UBC Table 23-III-II

Then $C_d = 1.5/(12 \times 0.162) = 0.77$

Hence, the adjusted lateral load is

$$Z' = 108 \times C_D \times C_{di} \times C_d = 108 \times 1.33 \times 1.3 \times 0.77 = 144 \text{ pounds}$$

The required spacing is then

$$s = Z' \times 12/q = 144 \times 12/471 = 3.7 \text{ inches}$$

The strength of the ½-inch Structural I plywood diaphragm is fully developed by a 10d nail[8]. The allowable shear on ½-inch Structural I plywood using three inch nominal framing members and 10d nails at 4 inch centers is obtained from UBC Table 23-I-J-1, for case 1 loading, as

q_a = 480 pounds per linear foot
> q

Hence the 3.7 inch spacing of the 16d nails is satisfactory.

b. LEDGER BOLT

The lateral seismic force acting on the bolt spaced at 3.5 feet centers is

$$F = 3.5q = 3.5 \times 471 = 1649 \text{ pounds}$$

The allowable, parallel-to-grain load on a ¾-inch diameter anchor bolt in the two and a half inch thick Douglas fir ledger is, in accordance with UBC Section 2311.2, obtained as half the double shear value tabulated in UBC Table 23-I-F for a five inch thick member. Hence the allowable seismic load is

$$Z'_{\parallel} = 1.33 \times 2845/2 = 1892 \text{ pounds} > F$$

Hence the bolts are satisfactory.

c. WALL ANCHORAGE

The anchorage force is due to that portion of the wall dead load which is tributary to the second floor diaphragm. The tributary dead load is

$$W_p = 170 \times 7 + 130 \times 6 = 1970 \text{ pounds per linear foot}$$

The pull-out force between the wall and the diaphragm is given by UBC Formula (30-1) as

$$F_p = ZI_pC_pW_p$$

where Z = 0.4 for zone 4 from UBC Table 16-I
I_p = 1.0 from UBC Table 16-K for a standard occupancy structure
C_p = 0.75 from UBC Table 16-O item I.1.b.
W_p = 1970 as calculated

Then $F_p = 0.4 \times 1 \times 0.75 \times 1970$
= 591 pounds per linear foot

and this value governs as it exceeds the minimum value of 200 pounds per linear foot specified in UBC Section 1611.

The force on each anchor bolt, for a spacing of three feet six inches, is given by

$$P = 3.5F_p = 3.5 \times 591 = 2069 \text{ pounds}$$

d. SUBDIAPHRAGM NAILING

For a pull-out force of 1700 pounds, the force to be transferred to each block is given by

$$P = 1700/3$$
$$= 567 \text{ pounds.}$$

The allowable lateral load on a 16d common nail with a penetration of 1.5 inches, and allowing for the increase of one-third for seismic loading, is given by

$$Z' = 108 \times 1.33 \times 0.77$$
$$= 111 \text{ pounds}$$

Hence, the number of nails required in each block is given by

$$n = P/Z'$$
$$= 567/111$$
$$= 5 \text{ nails per block}$$

e. SUBDIAPHRAGM SHEAR

For an anchorage force of 1700 pounds and an anchor spacing of three feet six inches, the pull-out force between the wall and the diaphragm is

$$F_p = 1700/3.5$$
$$= 486 \text{ pounds per linear foot}$$

For a subdiaphragm of dimensions four feet by fifteen feet, the unit design shear is given by

$$q = 15F_p/(2 \times 4)$$
$$= 15 \times 486/(2 \times 4)$$
$$= 911 \text{ pounds per linear foot}$$

f. ANCHOR BLOCK

The pull-out force between the wall and the diaphragm is given as

$$F_p = 500 \text{ pounds per linear foot}$$

The force on each anchor bolt, for a spacing of thirty-two inches, is given by

$$P = F_p \times 32/12$$
$$= 500 \times 32/12$$
$$= 1333 \text{ pounds}$$

The maximum bending moment on the anchor block is

$$M = Pl/4$$
$$= 1333 \times 14.5/4$$
$$= 4833 \text{ pound inches}$$

The corresponding maximum bending stress in the block is given by

$$f_b = M/S = 4833/17.6 = 275 \text{ pounds per square inch} < 1.33F_b$$

The maximum shear force on the block occurs with the bolt located three inches from the center of the block to give a shear force of

$$V = P \times 10.25/14.5 = 1333 \times 10.25/14.5 = 942 \text{ pounds}$$

The shear stress in the block is given by

$$f_v = 1.5V/A = 1.5 \times 942/19.25 = 73 \text{ pounds per square inch} < 1.33\,F_v$$

g. BEARING PLATE

In accordance with UBC Section 2306.6 no increase is allowable in bearing stress for seismic forces and, assuming a plate size of 1.5 inches, a size factor of 1.25 is applicable. Hence, the net bearing area required is given by

$$A_n = P/1.25F_{c\perp} = 1333/(1.25 \times 625) = 1.706 \text{ square inches}$$

The area of the ⅞- inch diameter hole in the bearing plate is

$$A_h = 0.601 \text{ square inches}$$

The gross area of bearing plate required is given by

$$A = A_n + A_h = 1.706 + 0.601 = 2.307 \text{ square inches}$$

Hence the required bearing plate size is

$$a = (A)^{0.5} = (2.307)^{0.5} = 1.5 \text{ inches}$$

h. SUBDIAPHRAGM SHEAR

For a subdiaphragm of dimensions fifteen feet by forty feet, the unit design shear is given by

$$q = 40F_p/(2 \times 15) = 40 \times 500/(2 \times 15) = 667 \text{ pounds per linear foot}$$

Example 6–3 (Ledger anchor bolts)
For the Douglas fir–larch ledger shown in Figure 6–14, determine the required anchor bolt spacing. The following details are applicable:

Roof live load = 20 pounds per square foot
Roof dead load = 11 pounds per square foot
Masonry strength f'_m = 1500 pounds per square inch, solid grouted with special inspection
Seismic lateral force acting on the ledger is
q_s = 206 pounds per linear foot

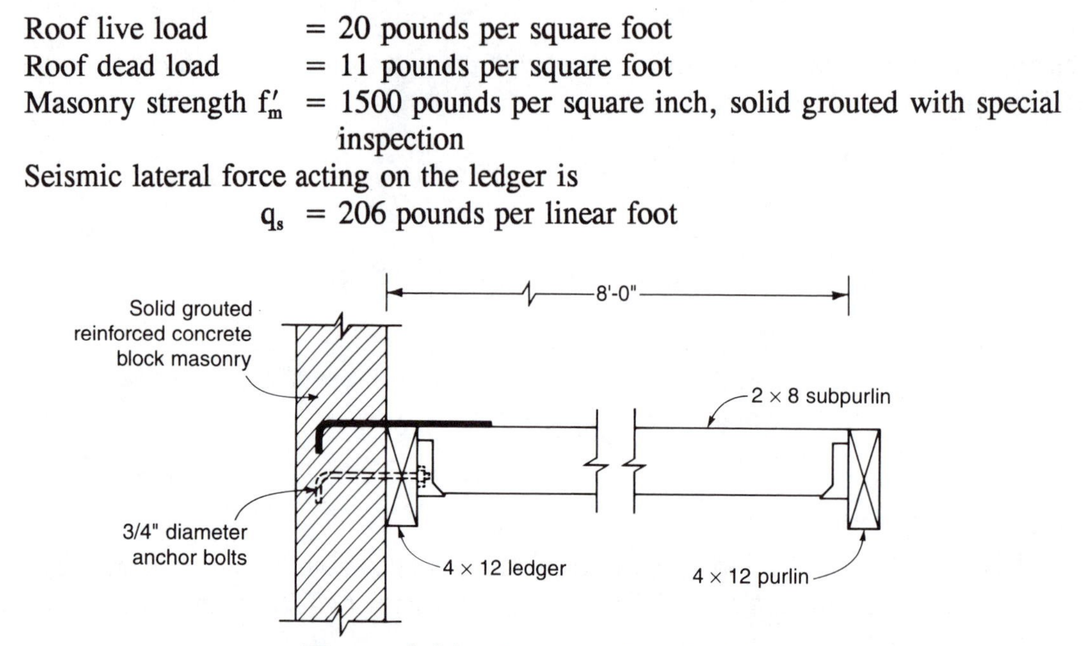

Figure 6–14 Details for Example 6–3

Solution
The loading acting on the ledger is due to the vertical dead load and live load on the roof diaphragm and the horizontal load due to the seismic forces. In accordance with UBC Section 1603.6, the two loading combinations, dead plus seismic and dead plus live, must be investigated. For both the ledger and the masonry wall, a stress increase of one–third is allowed for the first loading combinations, in accordance with UBC Section 1603.5. For the ledger, a twenty–five percent increase in stresses is allowed for the roof live load, in accordance with UBC Section 2304.3.4.

COMBINATION DEAD PLUS SEISMIC LOAD
The dead load acting on the ledger is

$w = 11 \times 8/2$
$= 44$ pounds per linear foot

The seismic load acting on the ledger is given as

$q_s = 206$ pounds per linear foot

The resultant load is given by

$r = (w^2 + q_s^2)^{0.5}$
$= (44^2 + 206^2)^{0.5}$
$= 211$ pounds per linear foot, and this acts at an angle to the grain of

$$\theta = \tan^{-1}(w/q) = \tan^{-1}(44/206) = 12°$$

The allowable loads on a ¾-inch diameter anchor bolt in the three and a half inch thick Douglas fir ledger are obtained from UBC Table 23-I-F and UBC Section 2311.2 as half the double shear value for a seven inch wide member and are

Allowable load parallel to grain p	= 2860/2	= 1430 pounds
Allowable load perpendicular to grain q	= 1835/2	= 918 pounds

From UBC Section 2336.2.1, the allowable load when inclined at an angle to the grain of $\theta°$, is obtained from Hankinson's Formula as

$$F_n = pq/(p\sin^2\theta + q\cos^2\theta) = 1430 \times 918/[1430 \times \sin^2(12°) + 918 \times \cos^2(12°)] = 1396 \text{ pounds}$$

Taking into account the allowable one-third stress increase, the required bolt spacing, determined by ledger stress is given by

$$s = 1.33F_n/r = 1.33 \times 1396/211 = 8.80 \text{ feet}$$

The allowable shear capacity of a ¾-inch diameter anchor bolt in a masonry wall with a specified compressive strength of 1500 pounds per square inch is obtained from UBC Table 21-F as

$$B_v = 1780 \text{ pounds}$$

Taking into account the allowable one-third stress increase, the required bolt spacing, determined by masonry stress is given by

$$s = 1.33B_v/r = 1.33 \times 1780/211 = 11.22 \text{ feet}$$

COMBINATION DEAD PLUS LIVE LOAD

The dead plus live load acting on the ledger is

$$w = (11 + 20) \times 8/2 = 124 \text{ pounds per linear foot}$$

The required bolt spacing, determined by masonry stresses is given by

$$s = B_v/w = 1780/124 = 14.35 \text{ feet}$$

Taking into account the allowable twenty-five percent stress increase the required bolt spacing, determined by ledger stresses, is given by

$$s = 1.25q/w$$
$$= 1.25 \times 918/124$$
$$= 9.25 \text{ feet}$$

Hence the governing bolt spacing is

$$s = 8.80 \text{ feet}$$

6.1.3 Shear walls

Aspect ratio of shear walls

The height–width ratio of a plywood shear wall is limited by UBC Table 23–I–I to a maximum value of 3½:1. Walls with this aspect ratio were found to behave unsatisfactorily during the 1994 Northridge earthquake in California. As a result, the height–width ratio of a plywood shear wall is now restricted[1,2], in seismic zones 3 and 4, to a maximum value of 2:1.
The height–width ratio for Portland cement plaster (stucco) and gypsum wallboard (drywall) shear walls is limited by UBC Section 2513 to a maximum value of 2:1. In seismic zones 3 and 4, this is now restricted[1,2] to a maximum value of 1:1.

Allowable shear values

As a result of the 1994 Northridge earthquake, allowable shear values have also been reduced[1,2] in seismic zones 3 and 4. The shear values for plywood shear walls given in UBC Table 23–I–K–1 must be reduced by 25 percent. In addition, three inch nominal boundary and panel edge members are required for plywood shear walls with a shear value exceeding 300 pounds per foot.
Similarly, the maximum allowable shear value for Portland cement plaster is limited to 90 pounds per square foot and for gypsum wallboard to 30 pounds per square foot. In addition, the use of shear walls of Portland cement plaster or gypsum wallboard is limited to single story construction or the upper story of a multi–story structure.

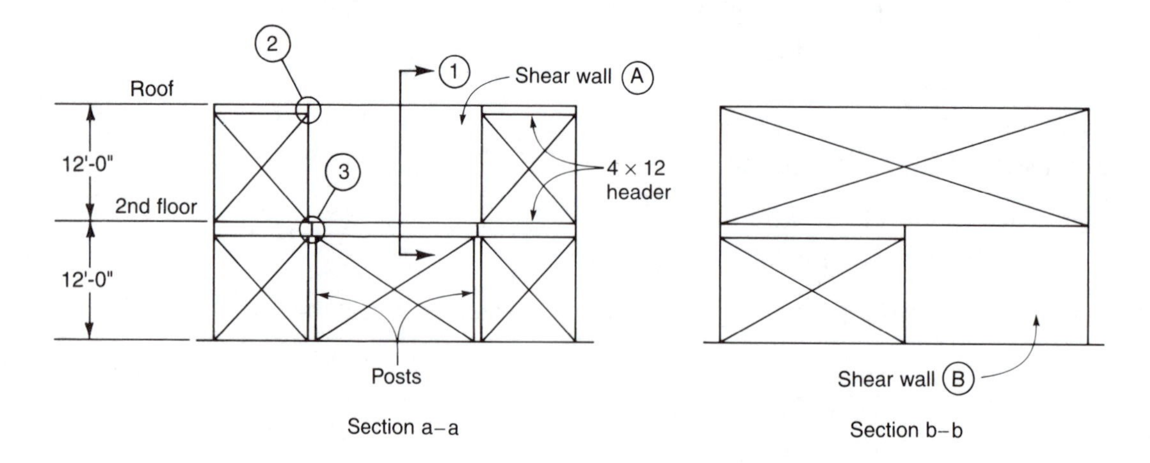

Figure 6–15 Sectional elevations for Example 6–4

Example 6–4 (Structural Engineering Examination 1987, Section D, weight 7.0 points)

GIVEN: A two story wood framed building with plywood shear walls shown in Figures 15 and 16.
The total building lateral forces have been determined for each level and are as shown in Figure 17.

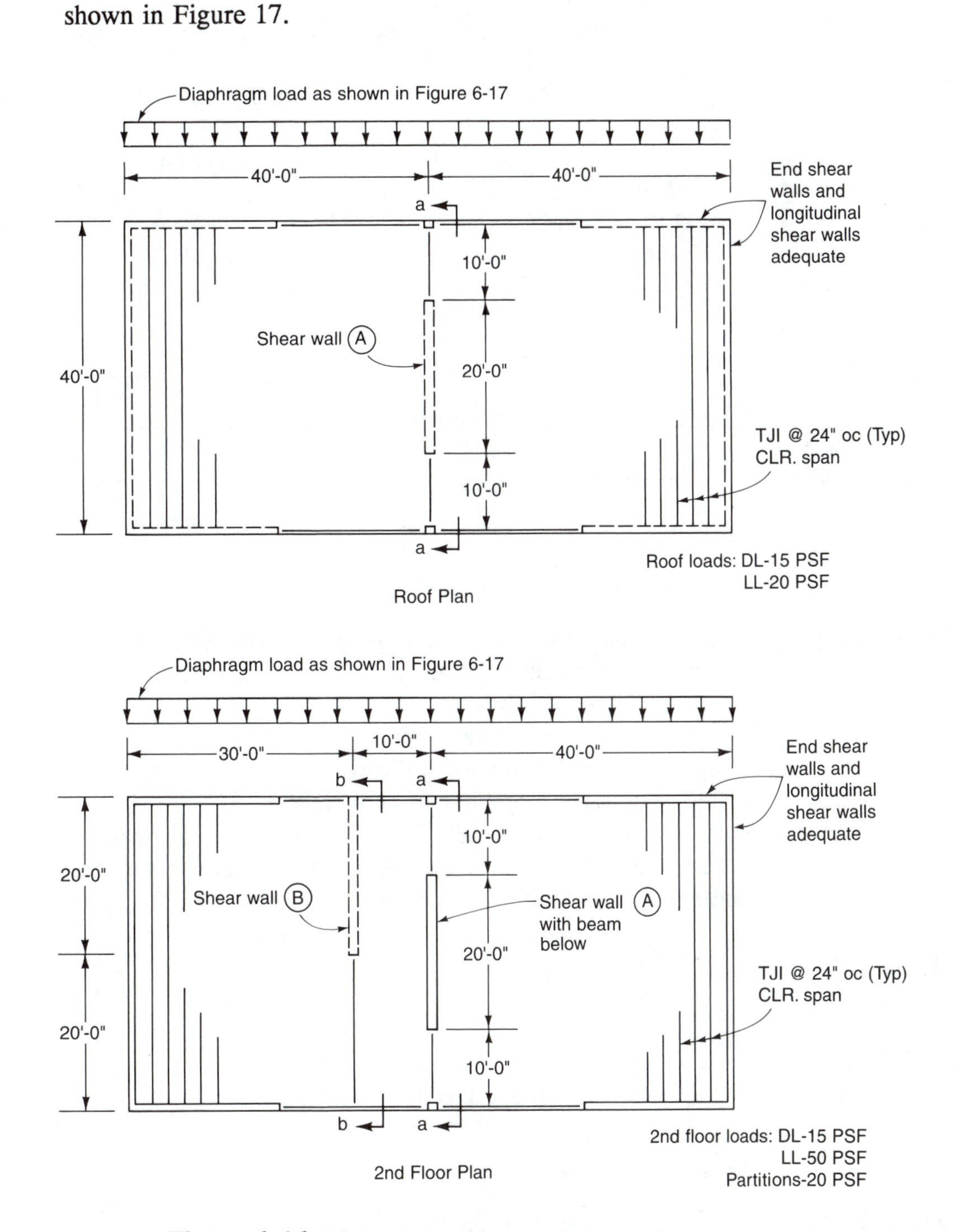

Figure 6–16 Roof and floor plans for Example 6–4

CRITERIA: Materials:

- Wood framing: Douglas fir–larch No 1 grade
- Plywood: Walls – ½-inch C–D, 32/16 panel index, Exposure 1
  Floor and roof – ⅝-inch C–D, 32/16 panel index, Exposure 1
- Machine bolts: ASTM A307
- Nails: Common as required except use 10d at roof and floor plywood and 8d at shear wall plywood
- Framing connectors: may be off the shelf with ICBO approval

Assumptions:

- Adequate shear walls in the longitudinal direction.
- End transverse shear walls are adequate.
- Roof and floor framing adequate for DL + LL.
- Foundations adequate for DL, LL, and lateral loads.

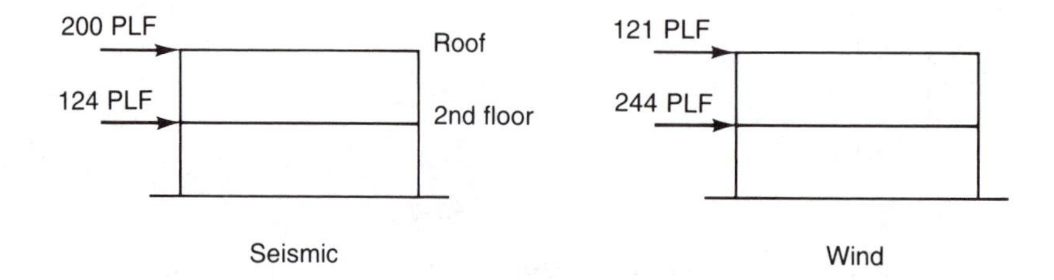

Figure 6–17 Total building lateral forces

REQUIRED: a. Determine lateral forces resisted by shear walls A and B.

b. For the transverse direction of the 2nd floor diaphragm:
   1. Sketch shear diagram.
   2. Sketch nailing diagram.

c. Design shear wall A for the critical lateral load determined above. Include nailing requirements and boundary members.

d. Sketch Section 1 and Details 2 and 3 required to adequately transfer lateral loads into and out of shear wall A. If framing connectors are utilized, show configuration and required minimum load capacity.

SOLUTION

a. SHEAR WALL FORCES

SHEAR WALL A

Due to wind load, the applied force in shear wall A is

$$V_A = 121 \times 40$$
$$= 4840 \text{ pounds}$$

Due to seismic load, the applied force in shear wall A is

$$V_A = 200 \times 40$$
$$= 8000 \text{ pounds, which governs.}$$

SHEAR WALL B

Due to seismic load, the applied force in shear wall B is

$$V_B = 124 \times 40 + 8000 \times 40/50$$
$$= 11{,}360 \text{ pounds}$$

Due to wind load, the applied force in shear wall B is

$$V_B = 244 \times 40 + 4840 \times 40/50$$
$$= 13{,}632 \text{ pounds, which governs.}$$

b. SECOND FLOOR DIAPHRAGM

SHEAR DIAGRAM

Shear wall B effectively subdivides the second floor diaphragm into two simply supported segments. These are shown in Figure 6–18 as span 12 and span 23. Wind loading governs and the wind loads acting on the diaphragm are indicated and produce the support reactions

$$R_{12} = 244 \times 15 = 3660 \text{ pounds}$$
$$R_{21} = 3660 \text{ pounds}$$
$$R_{23} = 244 \times 25 + 4840 \times 40/50 = 9972 \text{ pounds}$$
$$R_{32} = 244 \times 25 + 4840 \times 10/50 = 7068 \text{ pounds}$$

The resultant shear force diagram is shown in Figure 6–18

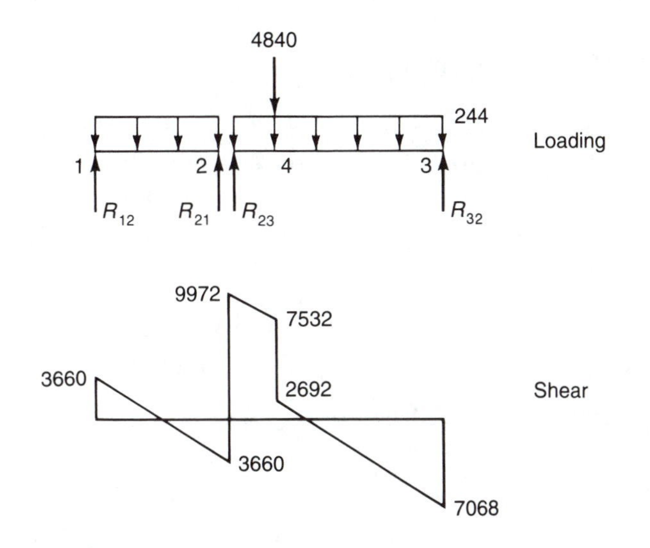

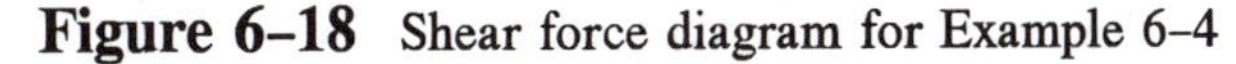

Figure 6–18 Shear force diagram for Example 6–4

NAILING REQUIREMENTS

The diaphragm unit shears are given by

$$q_{12} = R_{12}/40 = 3660/40 = 92 \text{ pounds per linear foot}$$
$$q_{23} = R_{23}/40 = 9972/40 = 249 \text{ pounds per linear foot}$$
$$q_{32} = R_{32}/40 = 7068/40 = 177 \text{ pounds per linear foot}$$
$$q_{42} = 7532/40 = 188 \text{ pounds per linear foot}$$
$$q_{43} = 2692/40 = 67 \text{ pounds per linear foot}$$

The required nail spacing is obtained from Table 23–I–J–1 with a case 1 plywood layout applicable, all edges blocked, and with 3½–inch framing. Using ⅝–inch grade C–D plywood and 10d nails with 1⅝–inch penetration, the nail spacing required is:

at diaphragm boundaries	= 6 inches
at all other edges	= 6 inches
at intermediate framing	= 12 inches

and this provides a shear capacity of 360 pounds per linear foot which exceeds the required capacity of 249 pounds per linear foot.

c. SHEAR WALL A

NAILING REQUIREMENTS

The aspect ratio of shear wall A is

$$\begin{aligned} l/h &= 20/12 \\ &= 1.7 \\ &< 3.5 \end{aligned}$$

This conforms to the requirements of UBC Table 23–I–I, for plywood panels nailed at all edges, and the unit shear in wall A for seismic loading is given by

$$\begin{aligned} q &= V_A/l \\ &= 8000/20 \\ &= 400 \text{ pounds per linear feet} \end{aligned}$$

The nail spacing required to provide a shear capacity of 400 pounds per linear foot using ½–inch grade C–D plywood on one side only and 8d nails with 1½– inch minimum penetration, may be obtained from UBC Table 23–I–K–1. With two inch nominal Douglas fir–larch vertical studs at sixteen inches on center and all panel edges backed with two inch nominal blocking, the required nail spacing is:

at all panel edges	= 3 inches
at intermediate framing members	= 12 inches

and this provides a shear capacity of 490 pounds per linear foot.

FLOOR ANCHORAGE

Provide a 2 × 4 sill plate with ⅝–inch diameter lag screws, five inches long, to anchor the shear wall to the four inch nominal Douglas fir–larch supporting joist. The combined thickness of the floor diaphragm and the sill plate is

$$\begin{aligned} t &= 1.5 + 0.625 \\ &= 2.125 \text{ inches} \end{aligned}$$

The total penetration of the lag screw into the main member, after allowing for an ⅛–inch thick washer is given by

	p	= T - E + S - t - 0.125
where	T	= thread length given in UBC Table 23-III-UU
	E	= length of tapered tip given in UBC Table 23-III-UU
	S	= unthreaded shank length given in UBC Table 23-III-UU
Then	p	= 2.594 + 2 - 2.125 - 0.125
		= 2.344

The penetration depth factor is given in UBC Section 2337.3.3 and Formula (37-5) as

	C_d	= p/8D
where	D	= diameter of lag screw
		= 0.625
Hence	C_d	= 2.344/(8 × 0.625)
		= 0.469

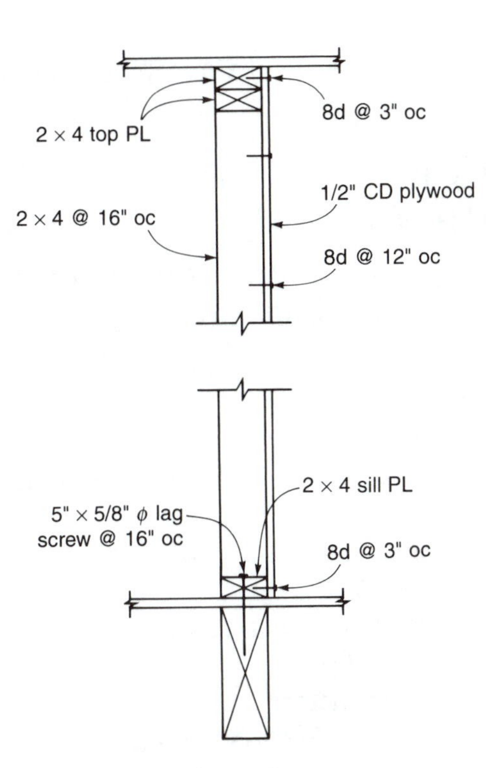

Figure 6–19 Shear wall details

The load duration factor for seismic load is obtained from UBC Section 2304.3.4 as

$$C_D = 1.33$$

The nominal design lateral load for a ⅝-inch diameter lag screw in a 1½-inch thick side member of Douglas fir–larch and a main member of Douglas fir–larch is given by UBC Table 23-III-T as

$$Z_{\parallel} = 920 \text{ pounds}$$

The adjusted design lateral load is

$$
\begin{aligned}
Z'_{\parallel} &= Z_{\parallel} \times C_D \times C_d \\
&= 920 \times 1.33 \times 0.469 \\
&= 574 \text{ pounds}
\end{aligned}
$$

The required lag screw spacing is then

$$
\begin{aligned}
s &= Z'_{\parallel} \times 12/q \\
&= 574 \times 12/400 \\
&= 17.2 \text{ inches}
\end{aligned}
$$

Hence a spacing of sixteen inches is satisfactory, as the edge distance of 1.75 inches provided exceeds the requirements of UBC Section 2337.4 which is 1.5D. The floor anchorage details are shown in Figure 6–19.

END STUDS

Neglecting vertical loads acting on the wall, the maximum compression or tension on the end studs is

$$
\begin{aligned}
P &= qh \\
&= 400 \times 12 \\
&= 4800 \text{ pounds}
\end{aligned}
$$

Provide 4 × 6 Douglas fir–larch visually graded sawn lumber end posts, braced in the strong direction, the relevant properties of which are

d = 5.5 inches

b = 3.5 inches

A = 19.25 square inches

F_t = 675 pounds per square inch for a Douglas fir–larch post from UBC Table 23–I–A–1

F_c = 1450 pounds per square inch for a Douglas fir–larch post from UBC Table 23–I–A–1

$E = E'$ = 1.7×10^6 pounds per square inch for a Douglas fir–larch post from UBC Table 23–I–A–1

C_D = load duration factor
= 1.33 from UBC Section 2304.3.4

C_F = size factor for a compression member
= 1.10 from UBC Table 23–I–A–1 for a 4 × 6 post

c = column eccentricity factor
= 0.80 from UBC Section 2307.3 for sawn lumber

K_{cE} = Euler buckling coefficient
= 0.30 from UBC Section 2307.3 for visually graded lumber

The compressive stress in the end strut is

$$
\begin{aligned}
f_c &= P/A \\
&= 4800/19.25 \\
&= 249 \text{ pounds per square inch}
\end{aligned}
$$

The effective unbraced length about the weak axis is

$$l_e = K_e l$$
$$= 1 \times 12$$
$$= 12 \text{ feet}$$

The corresponding slenderness ratio is

$$l_e/b = 12 \times 12/3.5$$
$$= 41.4$$
$$< 50$$

Using the NDS design method[11], the relevant design factors are obtained from calculator program A.3.3 in Appendix III. The tabulated compression value multiplied by the applicable adjustment factors is

$$F_c^* = F_c C_D C_F$$
$$= 1450 \times 1.33 \times 1.10$$
$$= 2121 \text{ pounds per square inch}$$

The critical buckling design value for the member is

$$F_{cE} = K_{cE} E'/(l_e/b)^2$$
$$= 0.3 \times 1{,}700{,}000/41.4^2$$
$$= 298 \text{ pounds per square inch}$$

The stress ratio is

$$F = F_{cE}/F_c^*$$
$$= 298/2121$$
$$= 0.140$$

The column stability factor is

$$C_p = (1 + F)/2c - \{[(1 + F)/2c]^2 - F/c\}^{0.5}$$
$$= (1 + 0.140)/(2 \times 0.80) - \{[(1 + 0.140)/(2 \times 0.80)]^2 - 0.140/0.80\}^{0.5}$$
$$= 0.136$$

Hence the allowable compressive stress is

$$F_c' = F_c C_D C_p C_F$$
$$= 1450 \times 1.33 \times 0.136 \times 1.10$$
$$= 289 \text{ pounds per square inch}$$
$$> f_c$$

By inspection, the tensile stress in the end post is also less than the allowable value. Hence, the 4 × 6 end studs are adequate.

HOLD-DOWN ANCHORS

A Simpson type HD2 hold-down with two ⅝-inch diameter bolts must be installed on each side of the supporting column and on each side of the shear wall end stud. The capacity in the four inch nominal member is

$$T = 2 \times 3300$$
$$= 6600 \text{ pounds}$$
$$> P$$

The minimum end distance, in accordance with UBC Table 23–III–H, is

$$l_n = 7D$$
$$= 7 \times 0.625$$
$$= 4.375 \text{ inches}$$

and a ⅝-inch diameter tie rod is required between hold-downs. The hold-down details are shown in Figure 6–20.

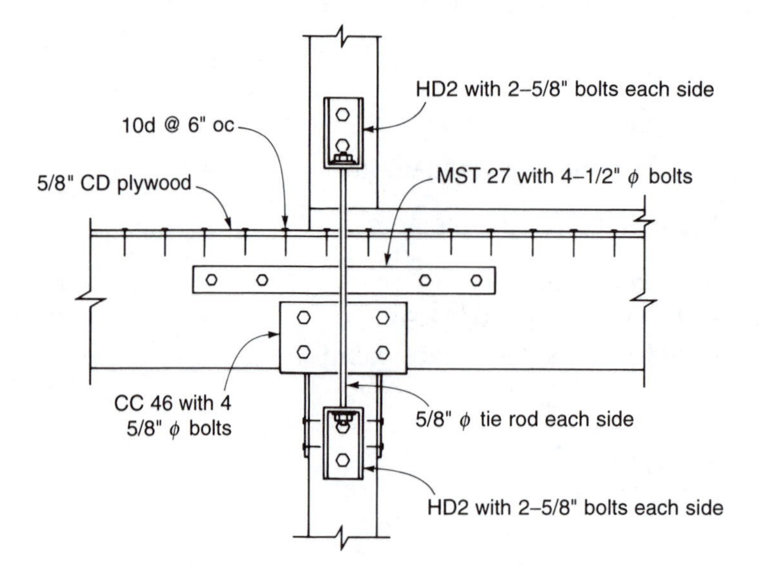

Figure 6–20 Hold-down details for Example 6–4

TOP DRAG STRUT

Provide a double 2 × 4 top plate to shear wall A. The drag force developed at the end of the top plate is

$$F = 10V_A/40$$
$$= 10 \times 8000/40$$
$$= 2000 \text{ pounds}$$

A Simpson type MST136 strap tie with thirty-six 10d × 1½-inch nails has a capacity of

$$F_a = 1945 \times 1.33$$
$$> F$$

The stress in the top plate is

$$f_t = F/A$$
$$= 2000/5.25$$
$$= 381 \text{ pounds per square inch}$$

The allowable tensile stress for Douglas fir–larch number 1 grade lumber is

$$F_t = 675 \text{ pounds per square inch}$$
$$> f_t$$

The drag strut connection details are shown in Figure 6–21.

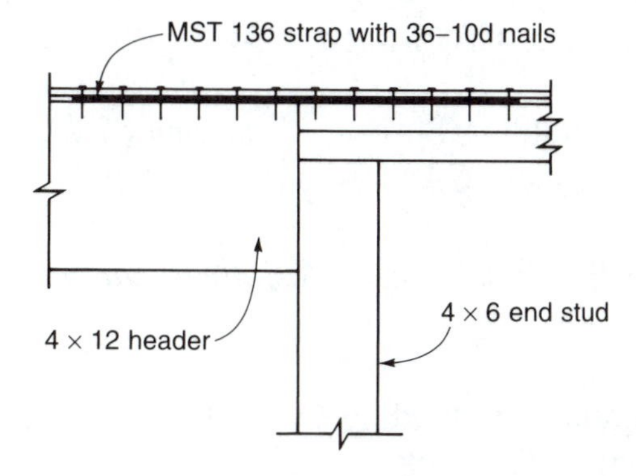

Figure 6–21 Drag strut connection for Example 6–4

BOTTOM JOIST CONNECTION

The drag force transmitted across the column cap to the supporting joists is

$$V_A = 8000 \text{ pounds}$$

A Simpson type CC46 column cap with two ⅝-inch diameter bolts in each joist, after allowing for a ½-inch gap between the joists, provides an end distance of four inches which is less than the specified seven bolt diameters for full design load. The capacity of the bolts in the nominal four inch joists with metal side plates, as given by UBC Table 23–III–N, after allowing for a load duration factor of 1.33 is

$$P_L = 2 \times 2250 \times 1.33 \times 4.0/4.375$$
$$= 5472 \text{ pounds}$$

A Simpson type MST27 strap tie with four ½-inch diameter bolts provides a capacity of

$$F_A = 2205 \times 1.33 = 2932 \text{ pounds}$$

Hence, the total capacity of the column cap and the strap tie is

$$P_L + F_A = 5472 + 2932$$
$$= 8404 \text{ pounds}$$
$$> V_A$$

Details of the bottom joist connection are shown in Figure 6–20.

References

1. Harder, R. Wood frame construction: multi–family code study preliminary findings. *Proceedings of the 1994 seminar of the Structural Engineers Association of Southern California*. Los Angeles, CA, March 1994.

2. City of Los Angeles Building Bureau. *Emergency enforcement measures – wood frame construction.* Los Angeles, CA, 1994.

3. City of Los Angeles Building Bureau. *Nailing requirements for wood frame construction.* Los Angeles, CA, 1992.

4. International Conference of Building Officials. *Plywood specialty products.* Research Report 1952. Whittier, CA, 1981.

5. Tissell, J.R. and Elliot, J.R. *Plywood diaphragms.* Research Report 138. American Plywood Association, Tacoma, WA, 1986.

6. American Plywood Association. *Diaphragms.* APA Design/Construction guide. Tacoma, WA, 1989.

7. Sheedy, P. Anchorage of concrete and masonry walls. *Building Standards,* October 1983 and April 1984. International Conference of Building Officials, Whittier, CA.

8. American Plywood Association. *Anchorage of concrete and masonry walls.* Commentary on reference 7. Tacoma, WA, 1985.

9. Tissel, J.R. *Horizontal plywood diaphragm tests.* Laboratory Report 106. American Plywood Association, Tacoma, WA, 1967.

10. American Forest and Paper Association. *Commentary on the National Design Specification for Wood Construction, Eleventh Edition (ANSI/NFoPA NDS–1991).* Washington, DC, 1991.

11. American Forest and Paper Association. *National Design Specification for Wood Construction, Eleventh Edition (ANSI/NFoPA NDS–1991).* Washington, DC, 1991.

7

Seismic design of masonry structures

7.1 SHEAR WALLS

7.1.1 Design criteria

For the strength design of masonry shear walls, the design assumptions are defined in UBC Section 2108.2.1.2 and illustrated in Figure 7-1. The principal design requirement is to obtain ductile behaviors of the shear wall with adequate warning of impending failure. This may be achieved by ensuring that the mode of failure is by yielding of the vertical tension steel followed by crushing of the masonry at the toe of the wall. Initially, as a bending moment is applied to a shear wall, the section is uncracked and the full section of the wall is effective with the tensile stress in the masonry less than the modulus of rupture given in UBC Formula (8-40) as

f_r = $4.0(f'_m)^{0.5}$ for a fully grouted masonry wall
$\leq$ 235 pounds per square inch

or f_r = $2.5(f'_m)^{0.5}$ for a partially grouted masonry wall
$\leq$ 125 pounds per square inch

where f'_m = masonry compressive stress

The cracking moment is given by UBC Formula (8-39) as

M_{cr} = Sf_r

where S = section modulus of the net wall section.

Sufficient vertical reinforcement must be provided to ensure that the flexural strength exceeds the cracking moment by an adequate margin otherwise cracking of the section may result in a sudden, brittle failure. The required nominal flexural strength is specified in UBC Section 2108.2.5.2 as

$$M_n \geq 1.8M_{cr} \text{ for a fully grouted masonry wall}$$
$$\geq 3.0M_{cr} \text{ for a partially grouted masonry wall}$$

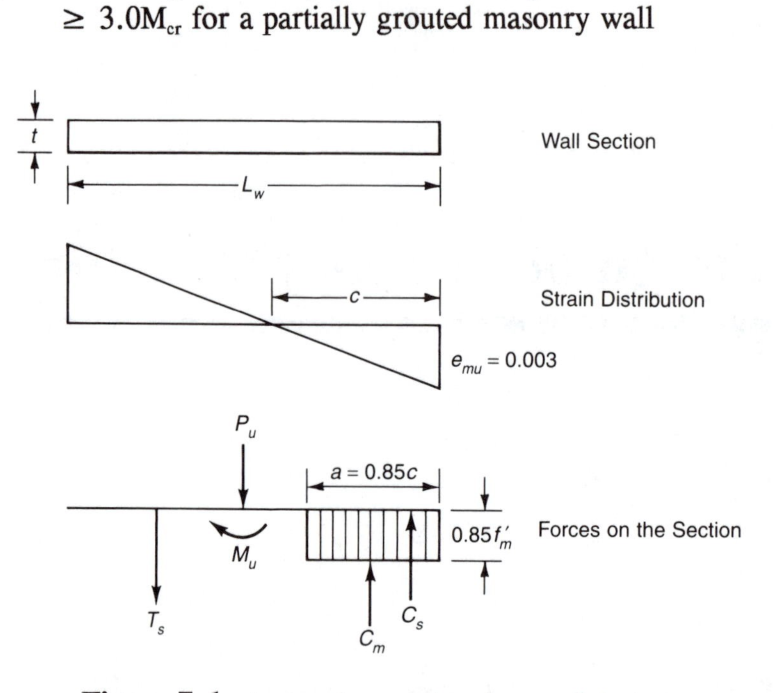

Figure 7–1 Assumptions used in shear wall design

It is also considered desirable[1,2,3,4] to limit the maximum amount of vertical reinforcement to a specified percentage of the balanced reinforcement ratio. An excess of vertical reinforcement combined with a large axial load may produce a sudden brittle failure due to crushing of the masonry prior to yielding of the reinforcement. The balanced condition occurs, in accordance with UBC Section 2108.2.3.3, when the strain in the masonry at the extreme compression fiber of the wall reaches its maximum usable value simultaneously with yielding of the tensile reinforcement at the extreme tension fiber. The appropriate axial load combination associated with the balanced condition is

$$P_{ub} = D + L + 1.4E$$

where
D = dead load
L = live load
E = seismic load

The maximum usable strain is defined in UBC Section 2108.2.1.2 as

$$e_{nu} = 0.003$$

The yield strain in the tension reinforcement is

$$e_s = F_y/E_s$$
$$= 60/29{,}000$$
$$= 0.00207$$

In determining the balanced axial strength of the section, reinforcement in compression may be neglected. The NEHRP Recommended Provisions[3] Section 12A.6.2.3 specifies the maximum reinforcement ratio for walls with a design axial load strength of

$P_n < 0.10f'_mA_n$

as $\rho_{max} = 0.35\rho_b$

where ρ_b = reinforcement ratio producing balanced strain conditions

A_n = net wall area

ϕP_n = design axial load strength

$= 0.80\phi[0.85f'_m(A_n - A_s) + f_yA_s]$ from UBC Formula (8–24)

P_n = nominal axial load strength

A_s = reinforcement area

ϕ = strength reduction factor for axial load with flexure

$= 0.85 - 2\phi P_n/A_nf'_m$

or $= 0.85 - 0.80\phi P_n/P_b$

≥ 0.65

P_b = nominal balanced axial design strength

$= 0.85f'_mta_b$ from UBC Formula (8–10)

t = effective thickness of the wall

$a_b = 0.85d[e_{mu}/(e_{mu} + f_y/E_s)]$ from UBC Formula (8–11)

d = distance from compression fiber to centroid of tensile reinforcement

To prevent failure due to axial compression, the maximum axial load on the wall is limited by UBC Formulas (8–43) and (8–44) to

$P_u \leq \phi P_n$

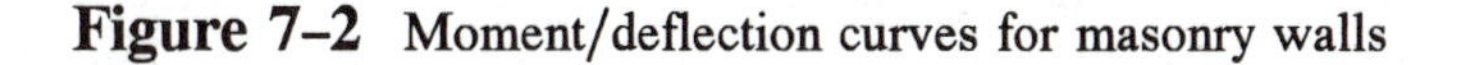

Figure 7–2 Moment/deflection curves for masonry walls

When masonry strain exceeds 0.0015 the wall performance deteriorates rapidly, as shown in Figure 7-2, unless confinement reinforcement is provided to maintain the integrity of the masonry and prevent buckling of the vertical reinforcement. When factored forces determined using a response modification factor R_W of 1.5 produce a compressive strain exceeding 0.0015, UBC Section 2108.2.5.6 requires that confinement reinforcement be provided. Using a response modification factor of 1.5 gives an applied loading which approximates the actual seismic load.

A diagonal shear failure of a shear wall is undesirable as this results in a sudden, brittle collapse of the wall. When the nominal shear strength of the wall exceeds the shear corresponding to the nominal flexural strength, UBC Section 2108.1.4.3.2 specifies a shear strength reduction factor of

$$\phi = 0.80$$

When the nominal shear strength of the wall is less than the shear corresponding to the nominal flexural strength a more severe shear strength reduction factor is imposed of

$$\phi = 0.60$$

To ensure satisfactory seismic performance, minimum reinforcement requirements for shear walls are detailed in UBC Sections 2106.1.12.4 and 2108.2.5.2 and are illustrated in Figure 7-3. Shear reinforcement shall be terminated with a standard hook.

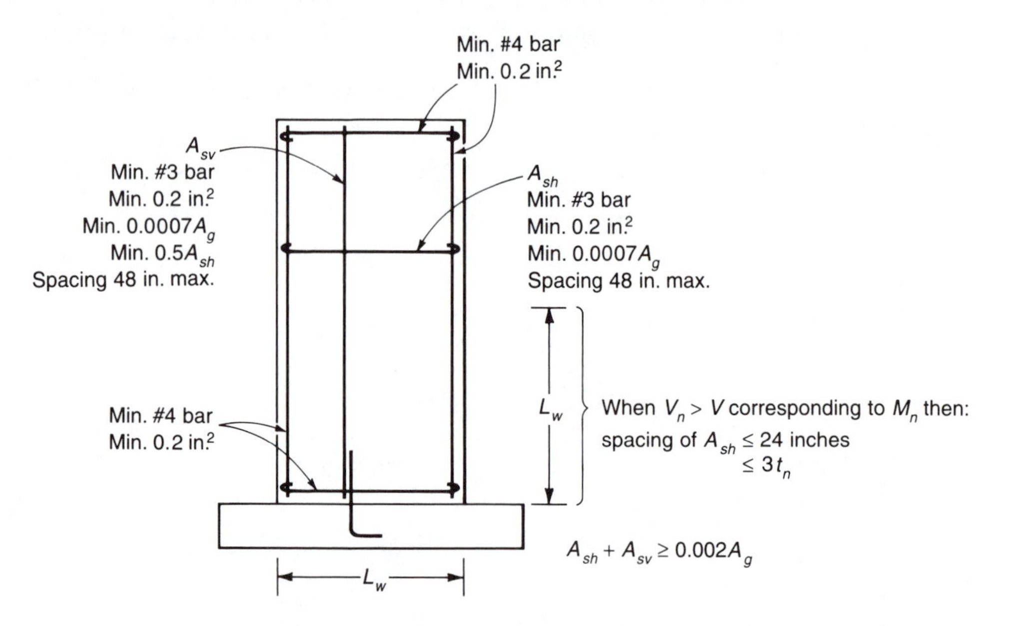

Figure 7-3 Minimum shear wall reinforcement

Maximum and minimum values of the masonry compressive strength are specified in UBC Section 2108.2.5.1 as

$$1500 \leq f'_m \leq 4000 \text{ pounds per square inch}$$

The value of the compressive strength shall be verified by prism tests, in accordance with UBC Section 2105, and continuous special inspection shall be provided during construction as specified in UBC Section 1701.5. The effective width of flanged sections contributing to the compressive strength of the section is specified in UBC Section 2106.2.6 as six times the thickness of the flange as shown in Figure 7-4.

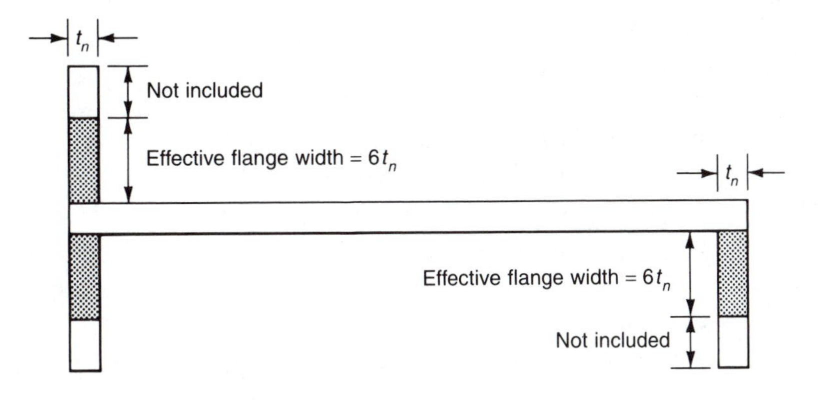

Figure 7-4 Effective flange widths

For the strength design of masonry shear walls, the applicable loading combinations, consisting of service level loads multiplied by the appropriate load factors, are specified in UBC Section 2108.1.3 as

	U	= 1.4(D + L + E)
and	U	= 0.9D ± 1.4E
where	D	= dead load
	L	= live load
	E	= seismic load
	U	= required strength

For dead load plus live load the appropriate load combination is

U = 1.4D + 1.7L

For wind loading the appropriate load combinations are

	U	= 0.75(1.4D + 1.7L + 1.7W)
or	U	= 0.75 (1.4D + 1.7W)
or	U	= 0.9D + 1.3W

7.1.2 Shear capacity

The nominal shear strength of a shear wall may be determined as specified in UBC Section 2108.2.5.5 and given in UBC Formulas (8-45), (8-46), (8-47), and (8-48). When the failure mode of the shear wall is in shear, and the nominal shear strength of the wall is less than the shear corresponding to the nominal flexural strength, both masonry and reinforcement contribute to the shear strength of the wall. The nominal shear strength is given by

	V_n	$= V_m + V_s$
with	V_n/A_{mv}	$\leq 6(f'_m)^{0.5} \quad \leq 380$ pounds per square inch for $M/Vd \leq 0.25$
and	V_n/A_{mv}	$\leq 4(f'_m)^{0.5} \quad \leq 250$ pounds per square inch for $M/Vd \geq 1.00$
where	V_m	= nominal shear strength provided by masonry
		$= C_d A_{mv} (f'_m)^{0.5}$
	A_{mv}	= net shear area
	C_d	= 2.4 for $M/Vd \leq 0.25$
		= 1.2 for $M/Vd \geq 1.00$
	V	= total factored design shear force
	M	= bending moment associated with shear force V
	d	= distance from compression fiber to centroid of tensile reinforcement
	V_s	= nominal shear strength provided by reinforcement
		$= A_{mv} \rho_n f_y$
	ρ_n	= ratio of shear reinforcement on a plane perpendicular to A_{mv}

When the failure mode of the shear wall is in flexure, and the nominal shear strength of the wall exceeds the shear corresponding to the nominal flexural strength, it is assumed that a plastic hinge forms at the base of the wall and extends vertically a distance equal to the length of the wall. The masonry within this region provides no shear resistance and the nominal shear strength is given by

$$V_n = V_s = A_{mv} \rho_n f_y$$

The total factored design shear force for this region is calculated at a height above the base of

$$h_v = L_w/2 \leq h_s/2$$

where	h_s	= story height
	L_w	= length of wall

The nominal shear strength above the plastic hinge region is the sum of the contributions from the masonry and the reinforcement and is given by

$$V_n = V_m + V_s$$

7.1.3 Boundary members

Boundary members are required in a shear wall when the compressive strains exceeds 0.0015. The factored forces producing the compressive strain are determined using a response modification factor R_W of 1.5. The horizontal length of the boundary member is given by UBC Section 2108.2.5.6 as

L_z = zone in which compressive strain exceeds 0.0015 $\geq 3t_n$

where	t_n	= nominal wall thickness

The vertical height of the boundary member is

$$h_z = \text{zone in which compressive strain exceeds } 0.0015 \geq L_w$$

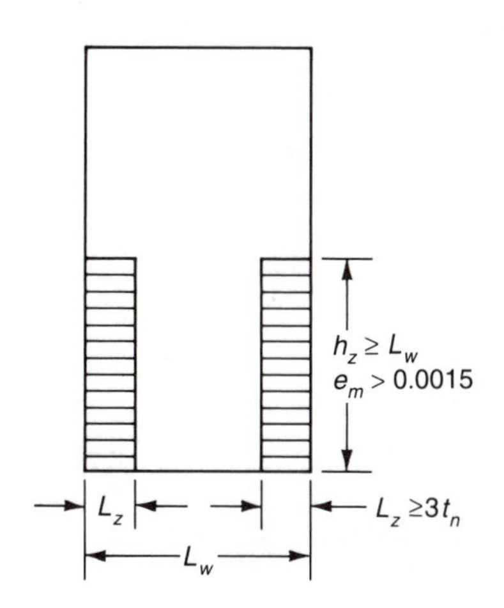

Figure 7–5 Boundary zone dimensions

The dimensions of the boundary member are shown in Figure 7–5. Confinement reinforcement consisting of closed hoops and crossties shall be provided for all vertical reinforcement in the boundary zone. The confinement reinforcement shall be a minimum of number 3 bars at a maximum spacing of eight inches. Equivalent confinement may be used which can develop an ultimate compressive masonry strain of 0.006 and a number of proprietary types have been developed[5].

Example 7–1 (Structural Engineering Examination 1989, Section D, weight 6.0 points)*

GIVEN: A two–story building, as shown in Figure 7–6, consisting of plywood floor and roof diaphragms supported by 8" concrete masonry (CMU) walls.

CRITERIA:
- Seismic Zone 3
- Seismic loads using R_W = 6 are as shown in Figure 7–7
- Total uniform dead load at wall footing = 2200 pounds per linear foot
- Uniform live load from second floor, at wall footing = 200 pounds per linear foot

Materials:
- 8–inch CMU wall, solid grouted with special inspection and laid in running bond.
- CMU compressive strength = 3000 pounds per square inch
- Reinforcing Steel: f_y = 60,000 pounds per square inch

*To conform to current code requirements, the data has been modified.

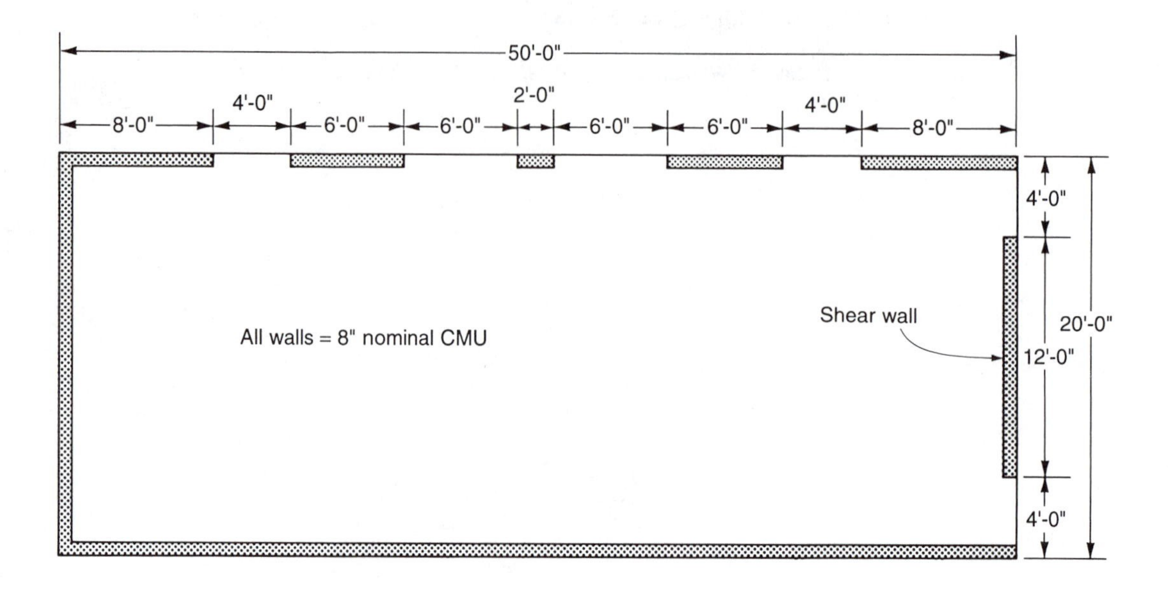

Figure 7–6 Building for Example 7–1

Assumptions:

- Flexible roof and floor diaphragms
- The wall is adequate for flexure plus axial loading
- The nominal shear strength of the wall is less than the shear corresponding to the nominal flexural strength.
- d = 0.8 × wall length for shear stress computations

<u>REQUIRED</u>: 1. Determine the maximum design forces in wall "W1". Consideration of out of plane loads is not required.
2. Determine if the reinforcement shown in Figure 7–8 is adequate for shear.
3. Showing all calculations, determine if confinement is required.

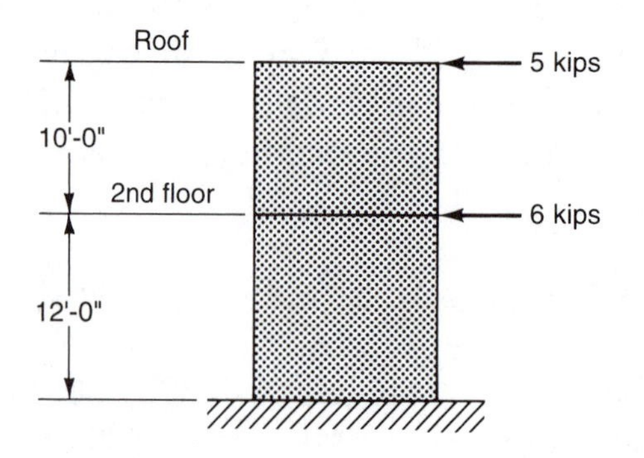

Figure 7–7 Seismic loads on shear wall

SOLUTION

1. DESIGN FORCES

In the bottom story, the factored shear force is given by

$$V_u = 1.4(5 + 6)$$
$$= 15.4 \text{ kips}$$

The factored bending moment at the bottom of the wall due to a response modification factor R_W of 6 is

$$M_u = 1.4(5 \times 22 + 6 \times 12)$$
$$= 255 \text{ kip feet}$$

Using a response modification factor of 1.5 gives a factored bending moment of

$$M'_u = 255 \times 6/1.5$$
$$= 1020 \text{ kip feet}$$

In accordance with the problem statement, the wall is adequate for axial loading. To determine the confinement requirements, the factored axial load is required and this is given by

$$P_u = 1.4(2.2 \times 12 + 0.2 \times 12)$$
$$= 40.32 \text{ kips}$$

2. SHEAR REQUIREMENT

$$M/Vd = 255/(15.4 \times 0.8 \times 12)$$
$$= 1.72$$
$$> 1.0$$

Then from UBC Table 21-K the value of the masonry shear strength coefficient C_d is

$$C_d = 1.2$$

Since the nominal shear strength of the wall is less than the shear corresponding to the nominal flexural strength, the shear capacity of the masonry may be utilized. The nominal shear strength provided by the masonry is given by UBC Formula (8-46) as

$$V_m = C_d A_{mv} (f'_m)^{0.5}$$
$$= 1.2 \times 144 \times 7.63(3000)^{0.5}/1000$$
$$= 72.2 \text{ kips}$$

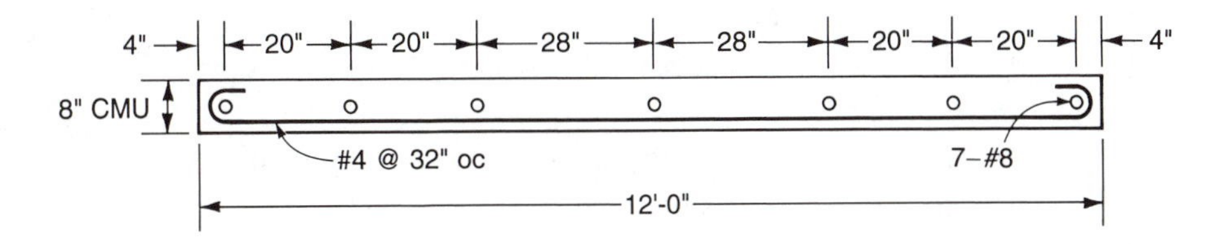

Figure 7–8 Shear wall reinforcement

The design shear strength of the masonry is

$$\begin{aligned} \phi V_m &= 0.6 \times 72.2 \\ &= 43.3 \text{ kips} \\ &> V_u \end{aligned}$$

Hence the masonry is adequate and nominal horizontal reinforcement, only, is required. The minimum allowable horizontal reinforcement area is specified in UBC Section 2106.1.12.4 as

$$\begin{aligned} A_{sh} &= 0.0007A_g \\ &= 0.0007 \times 7.63 \times 12 \\ &= 0.064 \text{ square inches per foot} \end{aligned}$$

Number 4 horizontal bars at 32 inches on center provides a reinforcement area of

$$\begin{aligned} A'_{sh} &= 12 \times 0.2/32 \\ &= 0.075 \text{ square inches per foot} \\ &> A_{sh} \end{aligned}$$

The maximum allowable spacing of the horizontal reinforcement is given by UBC Section 2106.1.12.4 as

$$s = 48 \text{ inches}$$

The spacing provided is

$$\begin{aligned} s' &= 32 \text{ inches} \\ &< s \end{aligned}$$

3. BOUNDARY ZONE REQUIREMENTS

Confinement reinforcement is required in a shear wall, in accordance with UBC Section 2108.2.5.6, when the compressive strains exceed 0.0015. At this level of strain, stress in the masonry may be assumed essentially elastic and the strain distribution across the section and the forces developed are shown in Figure 7-9. The masonry stress corresponding to a strain of 0.001 is given[6] as

$$f_m = 0.40f'_m$$

For a masonry strain of 0.0015 the corresponding stress may be taken as

$$f_m = 0.60f'_m$$

The force in a reinforcing bar is given by

$$\begin{aligned} F &= eA_sE_s \times 0.0015/c \\ &= e \times 0.79 \times 29{,}000 \times 0.0015/c \\ &= 34.365e/c \end{aligned}$$

where e = distance of a reinforcing bar from the neutral axis
A_s = area of a reinforcing bar
c = depth of neutral axis

and

$$\begin{aligned} F_{max} &= 60A_s \\ &= 47.4 \text{ kips} \end{aligned}$$

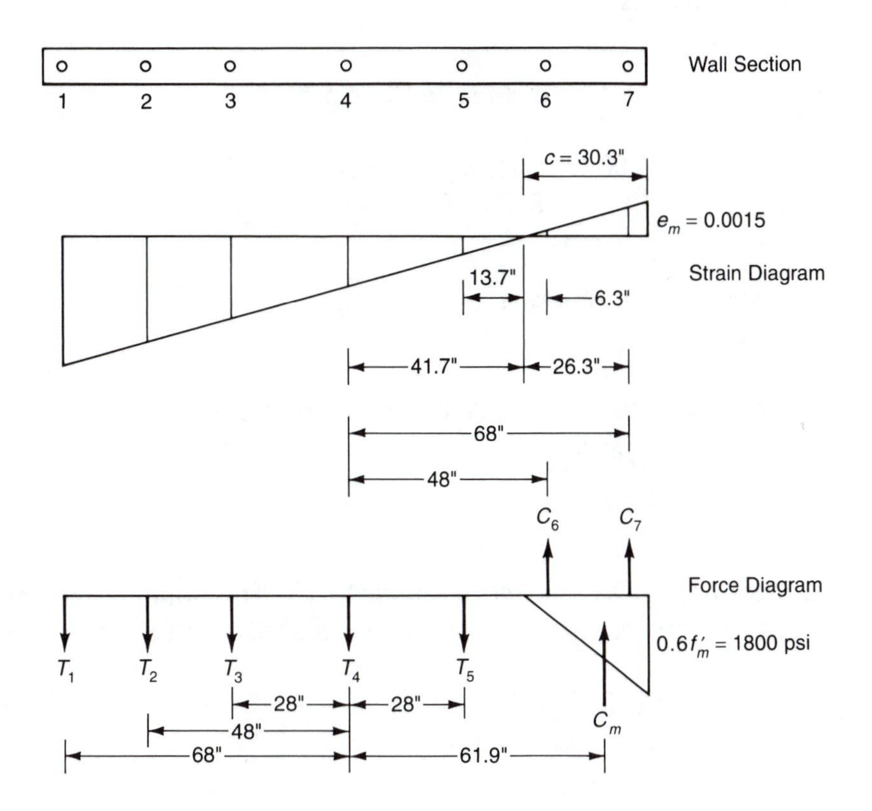

Figure 7–9 Strain distribution for Example 7–1

Assuming the depth to the neutral axis is

	c	= 30.3 inches
then	F	= 1.134e

The tensile forces in the reinforcement are

T_1		= 47.4 kips
T_2		= 47.4 kips
T_3		= 47.4 kips
T_4	$= 1.13 \times 41.7$	= 47.3 kips
T_5	$= 1.13 \times 13.7$	= 15.5 kips

The sum of the tensile forces is

$$\Sigma T = 205.0$$

The compressive force in the reinforcement is

C_6	$= 1.13 \times 6.3$	= 7.2 kips
C_7	$= 1.13 \times 26.3$	= 29.8 kips

The force in the masonry is

$$C_m = 0.5 f_m tc$$

where f_m = maximum stress in the masonry

$$= 0.60 f'_m$$

$$= 1800 \text{ kips per square inch}$$

and

$$C_m = 0.5 \times 1.8 \times 7.63 \times 30.3$$

$$= 208 \text{ kips}$$

The sum of the compressive forces is

$$\Sigma C = 245.0$$

The nominal axial load capacity for this strain distribution is

$$P = \Sigma C - \Sigma T$$

$$= 40.0$$

$$\approx P_u$$

Hence the assumed neutral axis depth is satisfactory. The nominal moment capacity for this strain distribution is obtained by summing moments about the mid-depth of the section and is given by

$$M = 68(T_1 + C_7) + 48(T_2 + C_6) + 28(T_3 - T_5) + 61.9C_m$$

$$= 21639 \text{ kip inches}$$

$$= 1803 \text{ kip feet}$$

$$> M'_u$$

Hence confinement reinforcement is not necessary.

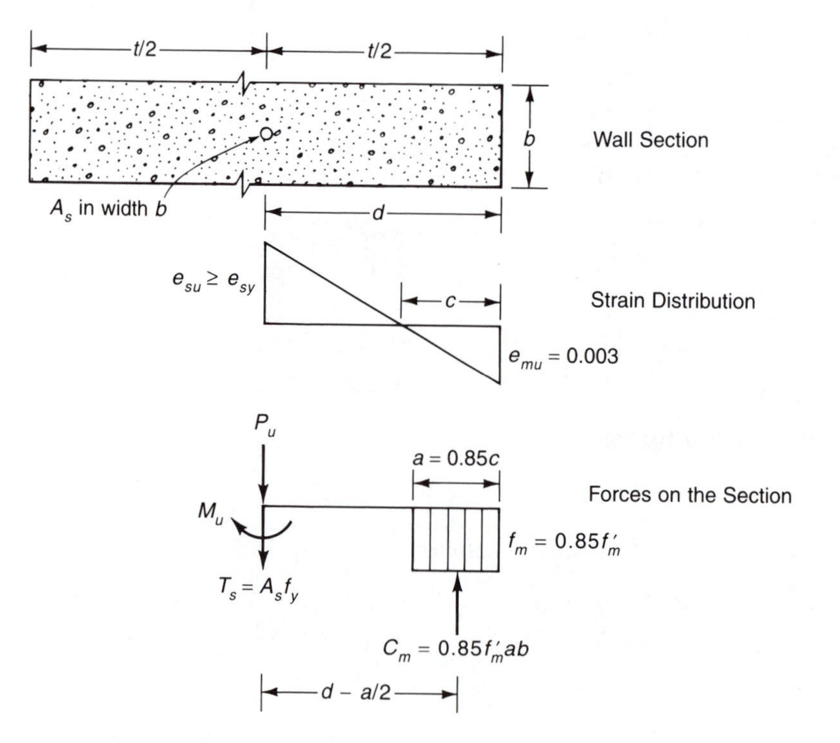

Figure 7–10 Strength design assumptions for masonry

7.2 WALLS WITH OUT-OF-PLANE LOADING

7.2.1 General requirements

The design of slender masonry walls, with a ductile failure mode, is covered in UBC Section 2108.2.4. Alternative methods[7,8] are available and these will not be dealt with in this text. The UBC design method specifies the following criteria:

- The design assumptions are those utilized for strength design as detailed in UBC Section 2108.2.1.2 and shown in Figure 7-10.
- To ensure the ductile failure of the wall with adequate warning of impending failure, the reinforcement ratio is limited by UBC Section 2108.2.4.2 to

$\rho \leq 0.50\rho_b$

where ρ_b = reinforcement ratio producing balanced strain conditions for flexure without axial load as shown in Figure 7-11
$= 87{,}000 \times 0.85 \times 0.85f'_m/f_y(87{,}000 + f_y)$
$= 62{,}858f'_m/f_y(87{,}000 + f_y)$

$\rho = A_s/bd$

A_s = area of reinforcement in the wall of width b

d = distance from compression fiber to centroid of the tensile reinforcement
= t/2 for reinforcement at wall center

t = effective thickness of the wall

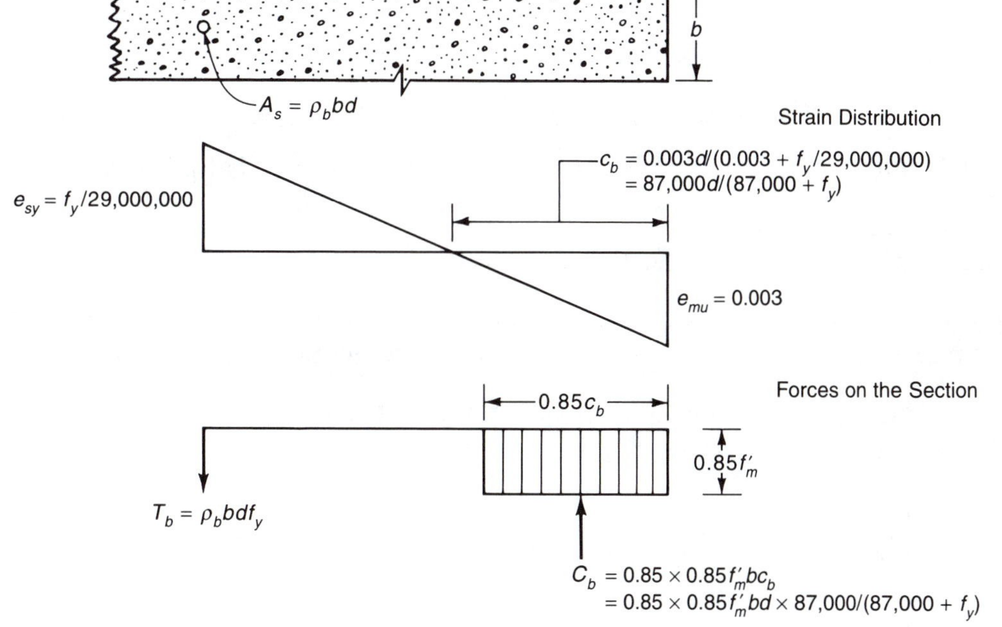

Figure 7-11 Balanced strain conditions for flexure only

- To minimize the P–delta effect, Section 2108.2.4.4 limits the axial load at the location of the maximum moment to

  $P \leq 0.04f'_m A_g$
  where $P = P_w + P_f$
  P_f = gravity load from tributary floor or roof loads
  P_w = weight of wall tributary to the section under consideration
  $A_g = tL_w$
  L_w = length of wall

- To ensure that the walls are representative of the experimental data available, the masonry compressive strength is limited to

  $f'_m \leq 6000$ pounds per square inch

- Due to the excessive influence misplaced reinforcement may have on the deflection and nominal moment capacity of thin sections, UBC Section 2108.2.4.4 limits the nominal thickness of the wall to a value of

  $t_n \geq 6$ inches

- To ensure satisfactory serviceability with full recovery after removal of the applied loads, UBC Section 2108.2.4.6 requires the midheight deflection under service loads to be limited to

  $\Delta_s = 0.007h$
  where h = height of wall between points of support

When the axial load exceeds the value of $0.04f'_m A_g$ additional criteria are imposed by UBC Section 2108.2.4.5:

- The axial load at the location of maximum moment is limited to

  $P < 0.2f'_m A_g$

- To limit the P–delta effect, the slenderness ratio of the wall is limited to

  $h'/t \leq 30$
  where h' = effective height of the wall

- The performance of the wall is determined from an interaction diagram as shown in Figure 7–12. Three points may be adequate to define the diagram and these are

  (i) ϕP_o = design axial load strength at zero eccentricity
  $= \phi \times 0.85f'_m(A_g - A_s) + f_y A_s$
  where $\phi = 0.80$
  $A_g = tL_w$
  (ii) ϕM_o = design flexural capacity without axial load
  $= \phi \times 0.85f'_m a L_w(d - a/2)$
  where $\phi = 0.80$
  a = depth of rectangular stress block
  $= A_s f_y / 0.85f'_m L_w$

(iii) ϕP_b = design axial load under balanced strain conditions
= $\phi \times 0.8f'_m a_b L_w - A_s f_y$
where ϕ = 0.80
a_b = $0.85d \times 87{,}000/(87{,}000 + f_y)$
and ϕM_b = design moment under balanced strain conditions
= $\phi \times (P_b + A_s f_y)(d - a_b/2)$
where ϕ = 0.80

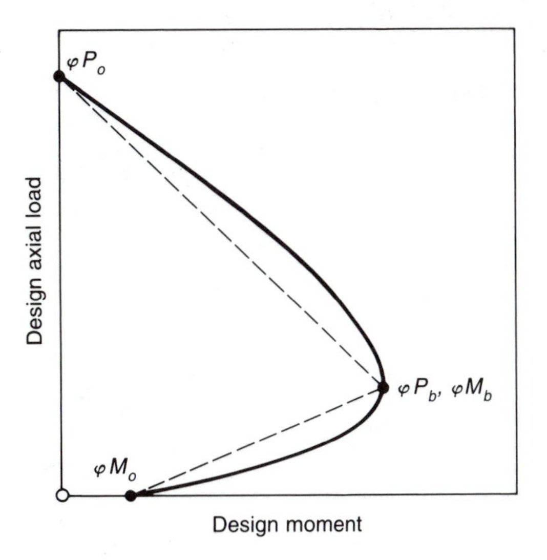

Figure 7–12 Interaction diagram for a slender wall

7.2.2 Design procedure

The factored design moment M_u at midheight of the wall must include the effects of the factored axial loads and eccentricities, the factored lateral load, and the P–delta effect. As shown in Figure 7–13 the ultimate moment is given by UBC Formula (8–29) as

M_u = $w_u h^2/8 + P_{uf}e/2 + P_u\Delta_u$
where w_u = factored lateral load
h = wall height between supports
P_{uf} = factored load from tributary floor and roof loads
P_{uw} = factored weight of wall tributary to section considered
P_u = $P_{uw} + P_{uf}$
e = eccentricity of applied axial load
Δ_u = deflection at midheight of wall due to factored loads and including P–delta effects
= $\Delta_{cr} + 5h^2(M_u - M_{cr})/48E_m I_{cr}$
Δ_{cr} = $5M_{cr}h^2/48E_m I_g$
M_{cr} = cracking moment defined by UBC Formula (8–39)
= $I_g f_r/y_t$

I_g	= moment of inertia of gross wall section
	= $bt^3/12$
y_t	= distance from centroidal axis to extreme fibre in tension
	= $t/2$
t	= effective thickness of wall
f_r	= modulus of rupture of masonry
	= $4.0(f'_m)^{0.5}$ for fully grouted concrete block masonry
	≤ 235 pounds per square inch from UBC Formula (8-40)
E_m	= modulus of elasticity of masonry
	= $750f'_m$ from UBC Formula (6-4)
	≤ 3,000,000 pounds per square inch

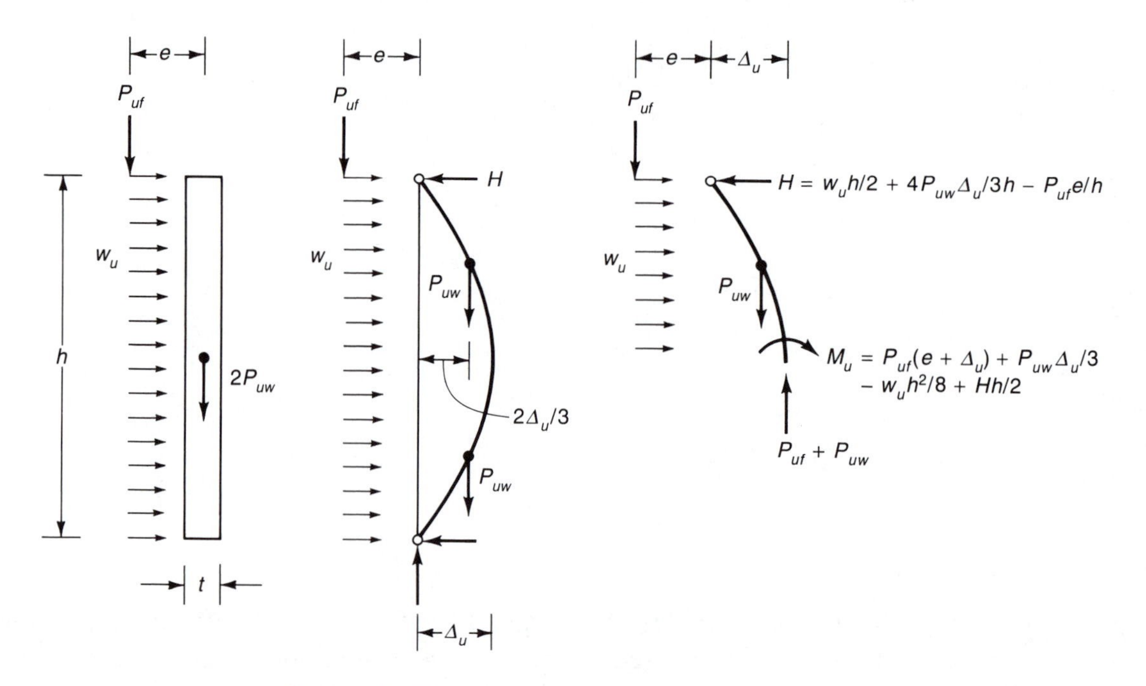

Figure 7–13 Analysis of slender masonry wall

An iterative process is required[9] until the values for M_u converge and the values for Δ_u converge. In accordance with UBC Formula (8-31)

	M_u	$\leq \phi M_n$
where	M_n	= nominal moment strength determined by adding the factored axial load to the force in the reinforcement
		= $A_{se}f_y(d - a/2)$
	A_{se}	= $(P_u + A_s f_y)/f_y$
	a	= depth of the equivalent rectangular stress block defined in UBC Section 2108.2.1.2
		= $(P_u + A_s f_y)/0.85f'_m b$
	ϕ	= strength-reduction factor
		= 0.80

The maximum permissible deflection at midheight of the wall due to service vertical and lateral loads is given by UBC Formula (8-36) as

$$\Delta_s \leq 0.007h$$

When the applied service moment M_{ser} exceeds the cracking moment M_{cr} the service deflection is given by UBC Formula (8-38) as

Δ_s $= \Delta_{cr} + 5h^2(M_{ser} - M_{cr})/48E_mI_{cr}$

where M_{ser} = service moment at midheight of wall, including P-delta effects
$= wh^2/8 + P_fe/2 + P\Delta_s$

w = lateral load

P_f = load from tributary floor and roof loads

P_w = weight of wall tributary to the section considered

P $= P_f + P_w$

I_{cr} = cracked moment of inertia of the wall section assuming the stress in the masonry is essentially elastic
$= bc^3/3 + nA'_{se}(d - c)^2$

c = depth to neutral axis
$= kd$

k $= [2np_e + (np_e)^2]^{0.5} - np_e$

n $= E_s/E_c$

p_e $= A'_{se}/bd$

A'_{se} = equivalent reinforcement area at working load
$= (P + A_sf_y)/f_y$

When the applied service moment is less than the cracking moment the service deflection is given by UBC Formula (8-37) as

Δ_s $= 5M_sh^2/48E_mI_g$

where M_s = the moment of the tensile force in the reinforcement about the centroid of the compressive force in the masonry assuming the stress in the masonry is essentially elastic
$= f_sA_s(d - c/3)$

An iterative process is required until the values for Δ_s and the values for M_{ser} converge.

Example 7-2 (Structural Engineering Examination 1992, Section D, weight 7-points)

GIVEN: The exterior wall of a one-story building is constructed of eight inch concrete masonry units (CMU), as shown in Figure 7-14.

CRITERIA:
- Seismic zone 4

Materials:
- Masonry
  f'_m = 1500 pounds per square inch
  $E_m = 750f'_m$

8 inch medium-weight CMU, weight = 76 pounds per square foot
8 inch nominal masonry = 7.625 inches thick

- Reinforcing steel:
  f_y = 40 kips per square inch
  E_s = 29,000 kips per square inch

Assumptions:

- Wall is pinned at top and bottom.
- The wall is fully grouted.
- Special inspection of masonry is provided.
- Roof framing and connections are adequate to resist wall lateral forces.
- Wind load does not govern.
- $ZI_pC_p = 0.3$

REQUIRED: Use the UBC Slender Wall Design Procedure to:

1a. Check vertical load stresses at midheight for conformance to UBC limitations.
1b. Check minimum and maximum reinforcement requirements.
2. Check wall moment strength for adequacy against seismic loads.
3. Check wall deflection at midheight for adequacy using service seismic and vertical loads (Note: Compute the service load moment M_{ser} by assuming that the lateral wall deflection is equal to the maximum allowable deflection = 0.007h).

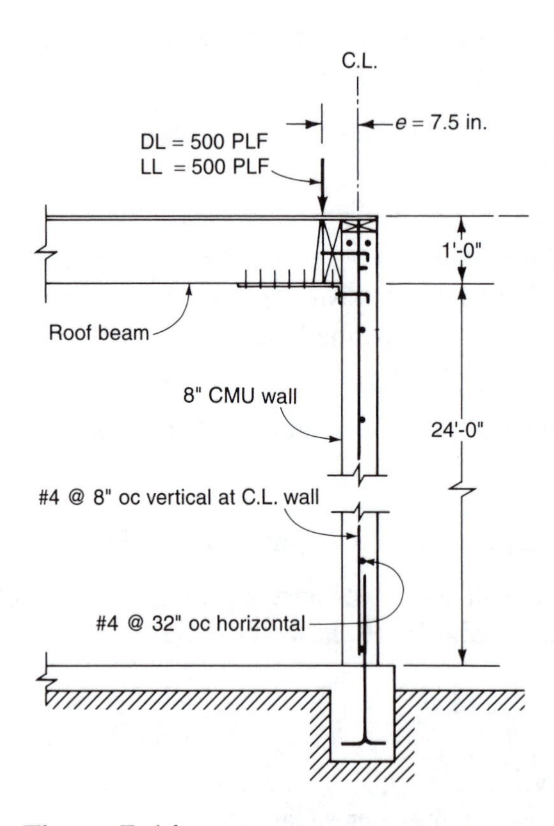

Figure 7–14 Wall section for Example 7–2

SOLUTION

SERVICE LOADS

The weight of wall tributary to the midheight of the wall between supports is

$$
\begin{aligned}
P_w &= 76(h/2 + 1) \\
&= 76(12 + 1) \\
&= 988 \text{ pounds per foot}
\end{aligned}
$$

The dead load of the roof is

$$P_f = 500 \text{ pounds per foot}$$

The total gravity load is

$$
\begin{aligned}
P &= P_w + P_f \\
&= 1.488 \text{ kips per foot}
\end{aligned}
$$

The seismic lateral load is

$$
\begin{aligned}
w &= ZI_pC_p \times 76 \\
&= 0.3 \times 76 \\
&= 22.8 \text{ pounds per square foot}
\end{aligned}
$$

FACTORED LOADS

$$
\begin{aligned}
P_{uf} &= 1.4 \times P_f \\
&= 0.70 \text{ kips per foot} \\
P_u &= 1.4 \times P \\
&= 2.083 \text{ kips per foot} \\
w_u &= 1.4 \times w \\
&= 31.92 \text{ pounds per foot}
\end{aligned}
$$

WALL PROPERTIES

Vertical load eccentricity is

$$e = 7.5 \text{ inches}$$

Effective depth is

$$
\begin{aligned}
d &= t/2 \\
&= 7.625/2 \\
&= 3.813 \text{ inches}
\end{aligned}
$$

Modulus of elasticity of the masonry is

$$
\begin{aligned}
E_m &= 750f'_m \\
&= 1125 \text{ kips per square inch.}
\end{aligned}
$$

The modular ratio is

$$
\begin{aligned}
n &= E_s/E_m \\
&= 25.78
\end{aligned}
$$

The moment of inertia of the gross wall section is

$$I_g = bt^3/12 = 12 \times 7.625^3/12 = 443 \text{ inches}^4 \text{ per foot}$$

The modulus of rupture of the masonry is

$$f_r = 4.0(f'_m)^{0.5} = 4.0 \times (1500)^{0.5} = 155 \text{ pounds per square inch}$$

The cracking moment given by UBC Formula (8–39) is

$$M_{cr} = I_g f_r/y_t = 443 \times 155/3.813 = 18011 \text{ pound inches per foot}$$

The reinforcement area in the wall is

$$A_s = 0.20 \times 12/8 = 0.30 \text{ square inches per foot}$$

For factored loads, the equivalent reinforcement area is

$$A_{se} = (P_u + A_s f_y)/f_y = (2.083 + 0.30 \times 40)/40 = 0.352 \text{ square inches per foot}$$

The nominal moment strength is

$$M_n = A_{se} f_y (d - a/2)$$

where

$$a = (P_u + A_s f_y)/0.85 f'_m b = (2.083 + 0.30 \times 40)/(0.85 \times 1.50 \times 12) = 0.920 \text{ inches}$$

and

$$M_n = 0.352 \times 40 \times (3.813 - 0.920/2) = 47.21 \text{ kip inches}$$

The deflection of the wall corresponding to the cracking moment is

$$\Delta_{cr} = 5M_{cr}h^2/48E_m I_g = 5 \times 18.011 \times 24^2 \times 144/(48 \times 1125 \times 443) = 0.312 \text{ inches}$$

Due to the factored loads, the stress distribution in the section is as shown in Figure 7–10 with a rectangular stress block of depth a in the masonry. The depth to the neutral axis is

$$c = a/0.85 = 0.920/0.85 = 1.082 \text{ inches}$$

The cracked moment of inertia is

$$I_{cr} = bc^3/3 + nA_{se}(d - c)^2$$
$$= 12 \times 1.082^3/3 + 25.78 \times 0.352(3.813 - 1.082)^2$$
$$= 5.072 + 67.681$$
$$= 72.75 \text{ inches}^4 \text{ per foot.}$$

The balanced reinforcement ratio is given by

$$\rho_b = 62.858f'_m/f_y(87 + f_y)$$
$$= 62.858 \times 1.5/40(87 + 40)$$
$$= 0.0186$$

1a. VERTICAL LOAD REQUIREMENTS

The limitation on the axial load is

$$0.04f'_mA_g = 0.04 \times 1.50 \times 7.625 \times 12$$
$$= 5.49 \text{ kips per foot}$$
$$> P$$

Hence the requirements of UBC Section 2108.2.4.4 are satisfied.

1b. REINFORCEMENT REQUIREMENTS

The reinforcement ratio based on the gross section is

$$\rho_g = A_s/A_g$$
$$= 0.30/7.625 \times 12$$
$$= 0.00328$$

For a shear wall in seismic zone 4, UBC Section 2106.1.12.4 specifies a minimum reinforcement ratio of

$$\rho_{min} = 0.0007$$
$$< \rho_g$$

Hence the minimum reinforcement requirement is satisfied.

The reinforcement ratio based on the effective depth is

$$\rho = A_s/bd$$
$$= 0.30/12 \times 3.813$$
$$= 0.00656$$
$$< 0.50\rho_b$$

Hence the maximum reinforcement specified in UBC Section 2108.2.4.2 is satisfied.

2. FLEXURAL CAPACITY

Assume a deflection at midheight due to factored loads of

$$\Delta_{u1} = 2.0 \text{ inches}$$

The ultimate applied moment is

$$M_{u1} = w_u h^2/8 + P_{uf}e/2 + P_u\Delta_{u1}$$
$$= 0.03192 \times 24^2 \times 12/8 + 0.70 \times 7.5/2 + 2.083 \times 2$$
$$= 27.58 + 2.63 + 4.17$$
$$= 34.38 \text{ kip inches}$$
$$> M_{cr}$$

The midheight deflection corresponding to this factored moment is

$$\Delta_{u2} = \Delta_{cr} + 5h^2(M_{u1} - M_{cr})/48E_mI_{cr}$$
$$= 0.312 + 5 \times 24^2 \times 144(34.38 - 18.01)/(48 \times 1125 \times 72.86)$$
$$= 0.312 + 1.726$$
$$= 2.038 \text{ inches}$$
$$\approx 2.0 \text{ inches}$$

Hence, the factored applied moment is

$$M_{u1} = 34.38 \text{ kip inches}$$

The design moment capacity of the section is

$$\phi M_n = 0.8 \times 47.21$$
$$= 37.77 \text{ kip inches}$$
$$> M_{u1}$$

The flexural capacity is adequate.

3. SERVICE DEFLECTION

The maximum allowable deflection under service loads is given by UBC Formula (8-36) as

$$\Delta_{s1} = 0.007h$$
$$= 0.007 \times 24 \times 12$$
$$= 2.016 \text{ inches}$$

The service moment corresponding to this deflection is

$$M_{ser} = wh^2/8 + P_fe/2 + P\Delta_s$$
$$= 0.0228 \times 24^2 \times 12/8 + 0.5 \times 7.5/2 + 1.488 \times 2.016$$
$$= 19.70 + 1.88 + 3.00$$
$$= 24.57 \text{ kip inches}$$
$$> M_{cr}$$

The midheight deflection corresponding to this service moment is given by UBC Formula (8-38) as

$$\Delta_{s2} = \Delta_{cr} + 5h^2(M_{ser} - M_{cr})/48E_mI_{cr}$$
$$= 0.312 + 5 \times 24^2 \times 144(24.57 - 18.01)/(48 \times 1125 \times 72.86)$$
$$= 0.312 + 0.691$$
$$= 1.003 \text{ inches}$$
$$< \Delta_{s1}$$

Hence, the midheight deflection corresponding to the correct service moment is less than the maximum allowed and the section is satisfactory.

7.3 MOMENT-RESISTING WALL FRAMES

7.3.1 Dimensional limitations

The basic requirement in the design of a masonry moment-resisting wall frame is to produce a ductile response to the imposed lateral forces. To ensure this, reinforcement ratios in the members are controlled and limits are imposed on the member dimensions[10]. The strong-pier/weak-beam concept is utilized to prevent instability at the limit state, with plastic hinge formation restricted to the ends of beams and to the bottom of the columns at the base of the structure. As the lateral loading on the frame increases, cracking of the members reduces their effective stiffness. Because of this, the analysis of the frame may be an iterative process in which the effective moment of inertia of a member at a specific loading stage is given by UBC Formula (8-50) as

$$I_e = I_g M_{cr}/M_a + [1 - (M_{cr}/M_a)^3]I_{cr}$$

where
- I_g = gross moment of inertia of the member = $th^3_{eff}/12$
- h_{eff} = effective depth of the member
- t = effective thickness of the member
- I_{cr} = cracked moment of inertia of the member = $tc^3/3 + nA_{se}(d - c)^2$
- c = depth to neutral axis = $a/0.85$
- a = depth of equivalent rectangular stress block = $(P + A_s f_y)/0.85f'_m t$
- n = modular ratio = E_s/E_m
- E_s = 29,000 kips per square inch
- E_m = $750f'_m$
- A_{se} = effective reinforcement area = $(P + A_s f_y)/f_y$
- d = effective depth
- M_{cr} = cracking moment = $2I_g f_r/h_{eff}$
- f_r = modulus of rupture of masonry = $4.0(f'_m)^{0.5}$ for fully grouted concrete block masonry
- M_a = maximum moment on the member at a specific loading stage

The dimensional requirements for the members of a moment-resisting wall frame are detailed in UBC Section 2108.2.6.1.2 and are shown in Figure 7-15. The proportions are selected so as to ensure that flexural yielding occurs in the beams whilst the piers and joints remain essentially elastic.

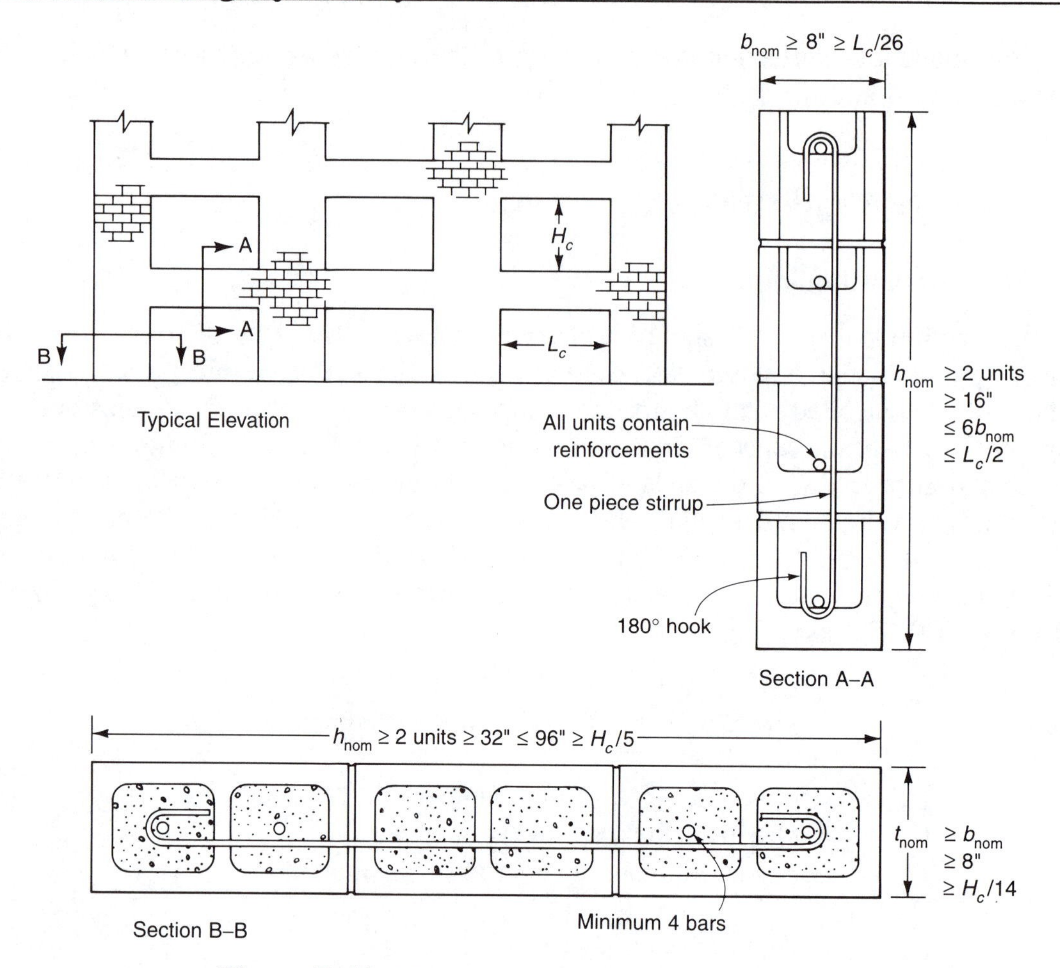

Figure 7–15 Moment-resisting wall frame details

The masonry compressive strength is limited by UBC Section 2108.2.6.2.3 to

$$1500 \leq f'_m \leq 4000 \text{ pounds per square inch}$$

Reinforcement shall be provided in all members such that the flexural strength at any section is not less than one quarter of the flexural strength at the two ends of the members. Longitudinal reinforcement shall be essentially uniformly distributed along the depth of a member with a maximum variation of 50 percent in steel area between units or between reinforced cells. Lap splices are permitted only at the center of the member clear length and the reinforcement ratio based on the gross section is limited to

$$0.022 \leq \rho_g \leq 0.15\, f'_m/f_y$$

where $\rho_g = A_s/A_g$

$A_g = th_{eff}$

Transverse reinforcement shall be provided in all members with a minimum area of

$$A_{v(min)} = 0.0015ts$$

where s = spacing of transverse reinforcement

Transverse reinforcement, consisting of single piece bars, shall be hooked around the extreme longitudinal bars with 180 degree hooks. Within a distance extending one member depth from the member ends, and at any location where a plastic hinge may form, the spacing of the transverse reinforcement shall not exceed one quarter of the nominal member depth. At all other locations the spacing shall not exceed one half the nominal member depth.

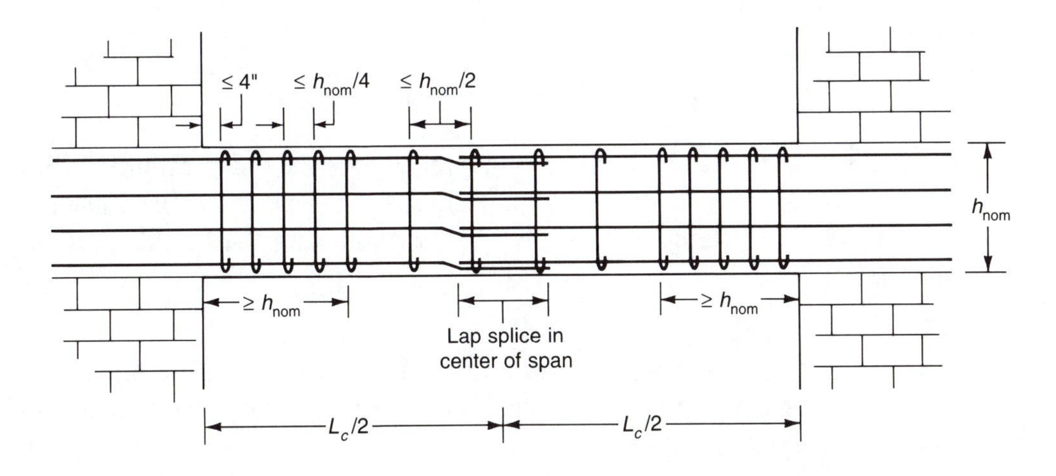

Figure 7–16 Masonry beam details

7.3.2 Beam details

Flexural members are defined in UBC Section 2108.2.6.2.5 as elements having a clear span not less than twice the nominal depth and with a factored axial compressive force of

$$P_u \leq 0.10A_n f'_m$$

where A_n = net area
= actual surface area at a section
= gross area minus all ungrouted areas

Each masonry unit through the beam depth shall contain longitudinal reinforcement. Details of the UBC requirements are shown in Figure 7–16. The factored applied moment on a beam is given by

$$M_u \leq \phi M_n$$

where ϕ = strength reduction factor for flexure given by UBC Formula (8–12)
= $0.85 - 2P_u/A_n f'_m$
≥ 0.65
≤ 0.85
P_u = factored axial load
A_n = net area

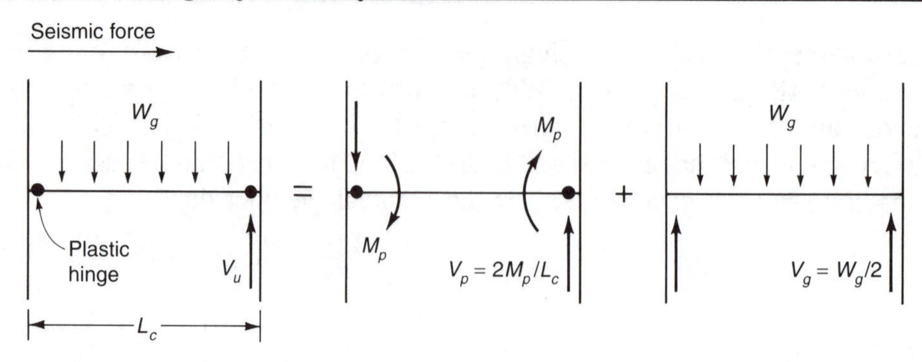

Figure 7–17 Masonry beam shear

To ensure ductile flexural failure of a beam and prevent brittle shear failure, UBC Section 2108.2.6.2.8 requires the nominal shear strength of the section to be determined using 1.4 times the beam nominal flexural strength to allow for variation in the yield strength of the reinforcement. Then

	V_n	$\geq 1.4V_u$
where	V_u	= factored shear force due to the formation of plastic hinges at the ends of the beam plus the shear due to the tributary unfactored gravity load along the span
		$= V_p + V_g$
		$= 2M_p/L_c + W_g/2$ as shown in Figure 7–17
	V_n	$= V_m + V_s$ as given by UBC Formula (8–51)
where	V_n	= nominal shear strength
		$\leq 4A_{mv}(f'_m)^{0.5}$ as given by UBC Formula (8–55)
	A_{mv}	= net shear area
	V_m	= nominal shear strength provided by masonry
		$= 1.2A_{mv}(f'_m)^{0.5}$ as given by UBC Formula (8–54)
		= 0 within a distance extending one beam depth from the ends of the beam and at any location where a plastic hinge may form
	V_s	= nominal shear strength provided by shear reinforcement
		$= A_{mv}\rho_n f_y$ as given by UBC Formula (8–53)
	ρ	= ratio of shear reinforcement on a plane perpendicular to A_{mv}

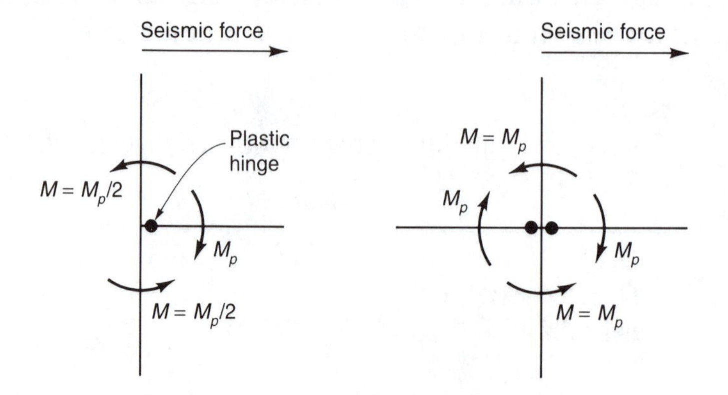

Figure 7–18 Masonry pier moments

7.3.3 Pier details

Piers are defined in UBC Section 2108.2.6.2.6 as members having a clear height not more than five times the nominal depth and with a factored compressive force limited to

$$0.10A_nf'_m \quad \leq P_u \leq 0.15A_nf'_m$$

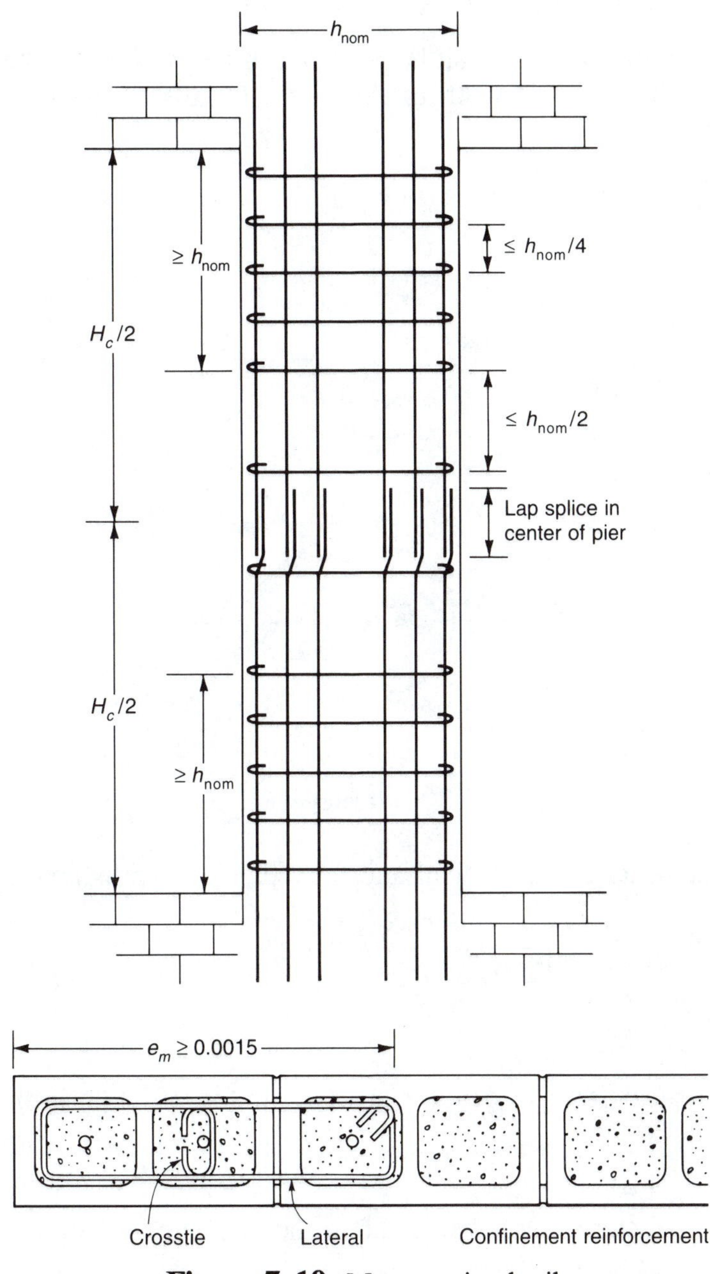

Figure 7–19 Masonry pier details

To allow for possible variation in the yield strength of the tensile reinforcement and to ensure that plastic hinges occur in the beams and not in the pier, UBC Section 2108.2.6.2.7 requires the nominal flexural strength of the piers to be not less than 1.6 times the pier moment corresponding to the development of the plastic hinges in the beams. As shown in Figure 7-18, and assuming that the moment in a beam is divided equally between the upper pier and the lower pier, the required nominal flexural strength is

$$M_n = 1.6M_p/2 \text{ for a single beam connection}$$
$$= 1.6M_p \text{ for a two beam connection}$$

A minimum of four longitudinal bars shall be provided in every pier and the bar diameter shall not exceed one eighth the nominal width of the pier. Details of the UBC requirements are shown in Figure 7-19.

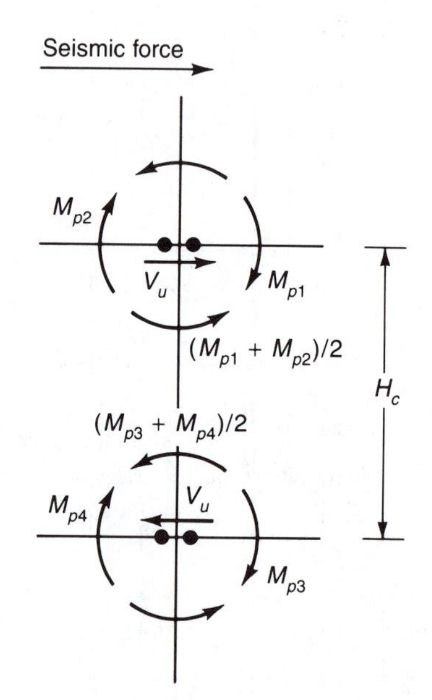

Figure 7-20 Masonry pier shear

The strength reduction factor for axial load and for axial load with flexure is given by UBC Formula (8-12) as

$$\phi = 0.85 - 2P_u/A_n f'_m$$
$$\geq 0.65$$
$$\leq 0.85$$

The specified nominal shear strength of the pier to allow for variation in the yield strength of the beam reinforcement is

$$V_n \geq 1.4V_u$$

where V_n = factored shear force due to the formation of plastic hinges at the ends of the beams

$= (M_{p1} + M_{p2} + M_{p3} + M_{p4})/2H_c$ as shown in Figure 7-20

$V_n = V_m + V_s$ as given by UBC Formula (8-51)

with $V_n/A_{mv} \leq 6(f'_m)^{0.5} \leq 380$ pounds per square inch for $M/Vd \leq 0.25$

and $V_n/A_{mv} \leq 4(f'_m)^{0.5} \leq 250$ pounds per square inch for $M/Vd \geq 1.00$

where V_m = nominal shear strength provided by masonry
= $C_d A_{mv}(f'_m)^{0.5}$

C_d = 2.4 for $M/Vd \leq 0.25$ from UBC Table 21-K
= 1.2 for $M/Vd \geq 1.00$ from UBC Table 21-K

V = total factored design shear force

M = bending moment associated with shear force V

d = distance from compression fibre to centroid of tensile reinforcement

and V_m = 0 within a distance extending one pier depth from the ends of the pier and at piers subjected to net tensile factored forces

V_s = nominal shear strength provided by shear reinforcement
= $A_{mv}\rho_n f_y$ as given by UBC Formula (8-53)

ρ_n = ratio of shear reinforcement on a plane perpendicular to A_{mv}

Confinement reinforcement is required in a pier when the compressive strain exceeds 0.0015. The factored forces producing the compressive strain are determined using a response modification factor R_W of 1.5. The unconfined area of the section, with strain exceeding 0.0015, is neglected in determining the nominal strength of the section. As shown in Figure 7-19 lateral reinforcement consisting of closed hoops and crossties shall be provided for all vertical reinforcement within the area in which strain exceeds 0.0015. The required area of confinement reinforcement is given by UBC Formula (8-49) as

$$A_{sh} \geq 0.09 s h_c f'_m / f_y$$

where s = spacing of confinement reinforcement

h_c = dimension of the grouted core measured center-to-center of confining reinforcement

Equivalent confinement may be used which can develop an ultimate compressive strain of 0.006 and a number of proprietary types have been developed[5].

7.3.4 Beam-pier intersection

Moment-resisting wall frame joints shall be proportioned such that the length of the joint, in both the vertical and horizontal directions, is sufficient to develop the bond strength of the reinforcement passing through the joint. The requirements are given in UBC Formulas (8-56) and (8-57) which are

$$h_p/d_{bb} > 4800/(f'_g)^{0.5}$$

and $$h_b/d_{bp} > 1800/(f'_g)^{0.5}$$

where h_p = effective pier depth in the plane of the wall frame

h_b = effective beam depth

f'_g = compressive strength of grout
$\leq$ 5000 pounds per square inch

d_{bb} = diameter of the largest beam reinforcing bar passing through the joint

d_{bp} = diameter of the largest pier reinforcing bar passing through the joint

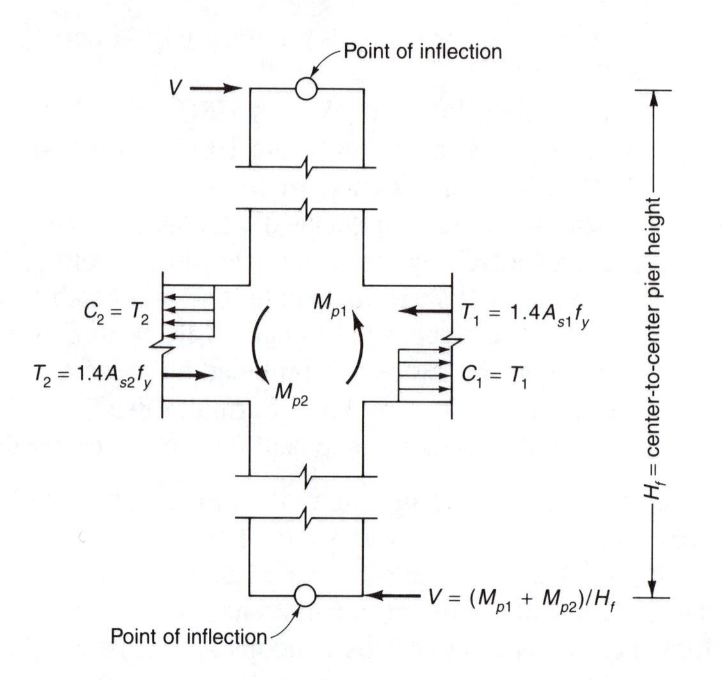

Figure 7–21 Forces acting at a joint

As shown in Figure 7–21 the horizontal shear force at a joint is determined on the assumption that the stress in the beam reinforcement is $1.4f_y$. The net shear acting on the joint is

$$
\begin{aligned}
V_{jh} &= T_1 + T_2 - V \\
&= 1.4(A_{s1} + A_{s2}) - (M_{p1} + M_{p2})/H_f
\end{aligned}
$$

Shear reinforcement, consisting of single piece bars, shall be hooked around the extreme pier reinforcing bars with 180 degree hooks. The total area of shear reinforcement required is given by UBC Formula (8–58) as

$$A_{jh} = 0.5V_{jh}/f_y$$

An additional requirement is that

$$
\begin{aligned}
V_{jh}/A_{mv} &\leq 7(f'_m)^{0.5} \\
&\leq 350 \text{ pounds per square inch}
\end{aligned}
$$

References

1. Kariotis, J.C. and Waqfi, O.M. *Trial designs made in accordance with tentative limit states design standards for reinforced masonry buildings.* TCCMAR Report 9.1-2. National Science Foundation, Washington, DC, 1992.

2. Kariotis, J.C. and Waqfi, O.M. *Recommended procedure for the calculation of the balanced reinforcement ratio.* TCCMAR Report 2.3-7. National Science Foundation, Washington, DC, 1992.

3. Building Seismic Safety Council. *NEHRP recommended provisions for the development of seismic regulations for new buildings: Part 2, Commentary.* Washington, DC, 1991.

4. Building Seismic Safety Council. *NEHRP recommended provisions for the development of seismic regulations for new buildings: Part 1, Provisions.* Washington, DC, 1991.

5. Masonry Institute of America. *Reinforcing steel in masonry.* Los Angeles, CA, 1991.

6. Brandow, G.E. Hart, G.C. and Virdee, A. *Design of masonry structures.* Concrete Masonry Association of California and Nevada, Citrus Heights, CA, 1993.

7. International Conference of Building Officials. *Strength design of one-to-four-story concrete masonry buildings.* Evaluation Report 4115. Whittier, CA, 1984.

8. American Concrete Institute and Structural Engineers Association of Southern California. *Report of the task committee on slender walls.* Los Angeles, CA, 1982.

9. Amrhein, J.E. and Lee, D.E. *Design of reinforced masonry slender walls.* Western States Clay Products Association, San Francisco, CA, 1984.

10. Hart, G.C. Masonry wall frame design and performance. *Proceedings of the Los Angeles Tall Buildings Structural Design Council Seminar.* Los Angeles, CA, May 1993.

This page left blank intentionally.

8

Seismic design of bridges

8.1 DESIGN REQUIREMENTS

8.1.1 Load factor design

The load factor design method, also known as the limit state, or strength design method, utilizes an ultimate limit state analysis to determine the maximum load carrying capacity of the structure. A serviceability limit state analysis is also employed to ensure satisfactory behavior at working load conditions. The objective of ultimate limit state design is to ensure a uniform level of reliability for all structures. This is achieved by multiplying the nominal, or service, loads by load factors to obtain an acceptable probability that the factored loads will be exceeded during the life of the structure. The factored load may be expressed as

$$W_u = \lambda W_n$$

where
W_u = factored load
W_n = nominal load
λ = load factor

The load factor λ is a function of two other partial load factors which are denoted by

γ = factor to compensate for the unfavorable deviation of the loading from the nominal value and uncertainties in strength, methods of analysis and structural behavior.

β = factor to reflect the reduced probability of the full nominal loads, in a combination of loads, being present simultaneously.

Values of γ and β are given in AASHTO[1] Section 3.22. The total effect on the structure of the factored loads shall not exceed the design resistance capacity of the structure and the basic requirement for load factor design may be expressed in terms of resistance capacity as

$$U \leq \phi R_n$$

where U = total effect on the structure of the factored load W_u

ϕR_n = design resistance capacity of the structure

R_n = nominal resistance capacity of the structure

ϕ = strength reduction factor

The factor ϕ is also referred to as the confidence factor, performance factor, resistance factor, and capacity reduction factor. This factor allows for the possibility of adverse variations in material strength, workmanship, and dimensional inaccuracies.

8.1.2 Service load design

The service load, or allowable stress, design method is based on the elastic theory which assumes elastic material properties and a constant modulus of elasticity to predict the material stresses in the structure under the applied working loads. The basic requirement for allowable stress design may be expressed in terms of stresses as

$$D + L + I \leq R$$

where D = stress produced by dead load

L = stress produced by live load

I = stress produced by impact

R = allowable stress

Infrequently occurring overloads are accommodated by permitting a specified increase in the allowable stress and this may be expressed as

$$D + EQ \leq \eta R$$

where EQ = stress produced by seismic effects

In accordance with AASHTOSD[2] Section 4.7.1

η = 1.33 for reinforced concrete elements

= 1.50 for structural steel elements

The allowable stress due to working load is defined as the yield stress or compressive strength divided by a factor of safety. This approach, however, does not ensure a constant factor of safety against failure in different structures.

8.1.3 Load combinations

The different combinations of loads which may act on a structure are represented in AASHTO Section 3.22 by twelve Groups. For seismic loads, the load combination Group which includes seismic loads plus permanent loads is given by AASHTOSD Equation (4–1) as

Seismic Group load $= \gamma(\beta_D D + \beta_B B + \beta_S SF + \beta_E E + \beta_{EQ} EQ)$

For seismic loading, γ and β are defined as unity which gives

Seismic Group load $= 1.0(D + B + SF + E + EQ)$

where
D = dead load
B = buoyancy
SF = stream flow pressure
E = earth pressure
EQ = appropriate elastic seismic force

8.2 DESIGN PRINCIPLES

8.2.1 Design procedure

The objective of the seismic design procedure is to design a bridge structure which may be damaged in an earthquake but which will not collapse and which can quickly be put back into service. The analysis procedure to be adopted is specified in AASHTOSD Section 4.

A detailed seismic analysis is not required for single span bridges. However, minimum support lengths are required, in accordance with AASHTOSD Section 4.9, and the connection of the superstructure to the substructure is designed to resist the product of the dead load reaction and the site acceleration coefficient, as specified in AASHTOSD Section 4.5. The criteria for minimum support lengths are shown in Figure 8-1. For seismic performance category A and B the minimum support length in inches is given by AASHTOSD Formula (4-3)

$$N = 8 + 0.02L + 0.08H$$

For seismic performance category C and D the minimum support length in inches is given by AASHTOSD Formula (4-4) as

$$N = 12 + 0.03L + 0.12H$$

where
L = length in feet of the bridge deck
H = average height in feet of the columns
= 0 for a single span bridge

In addition for special performance categories B, C, and D the support length may not be less than the calculated elastic displacement.

For multi-span structures with a uniform mass distribution and with the stiffness of adjacent supporting members differing by not more than 25 percent, the equivalent static force or single-mode spectral method may be adopted. The method assumes a predominant single mode of vibration which allows a statical method of analysis to be utilized.

The response spectrum or multimode spectral method is used for complex structures with irregular geometry. Several modes of vibration contribute to the overall response of the structure and a space frame analysis program with dynamic capabilities is required.

In order to determine the seismic response of the structure, several factors must be considered and these include the acceleration coefficient, importance classification, performance category, site coefficient and response modification factors. The selection of the design procedure depends on the type of bridge, the magnitude of the acceleration coefficient and on the degree of acceptability of loss of operation. The single-mode spectral method is defined as Procedure 1 and the multimode spectral method is defined as Procedure 2, in AASHTOSD Section 4.2.

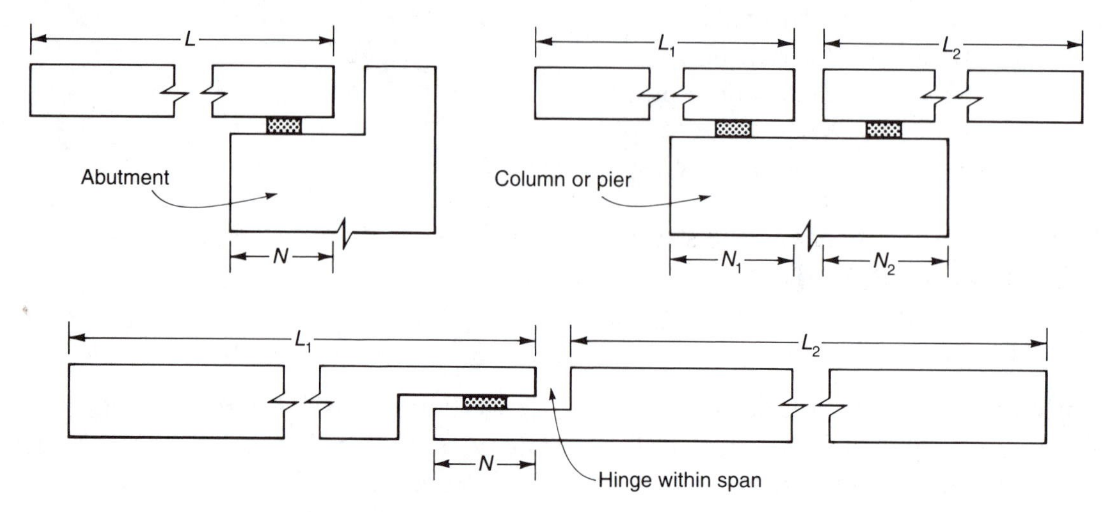

Figure 8-1 Minimum support lengths

8.2.2 Acceleration coefficient

The acceleration coefficient A, given in the contour maps in the Specification, is an estimation of the site dependent design ground acceleration expressed as a fraction of the gravity constant g. The acceleration coefficient corresponds to ground acceleration values with a recurrence interval of 475 years which gives a ten percent probability of being exceeded in a fifty year period. The acceleration coefficients correspond to the effective peak accelerations[3] in bedrock, which are based on historical records and geological data.

8.2.3 Elastic seismic response coefficient

The elastic seismic response coefficient, or lateral design force coefficient C_s, is a function of the seismic zone, the fundamental period of the bridge and the site soil conditions. The value of the lateral design force coefficient is given by AASHTOSD Formula (5-1) as

$$C_s = 1.2AS/T^{2/3}$$
$$\leq 2.5A$$

where S = site coefficient or amplification factor for a specific soil profile
T = fundamental period of the bridge.

The specific site soil profile considerably influences the ground motion characteristics and three profile types and corresponding site coefficients are defined in AASHTOSD Section 3.5.1.

- Soil profile type I consists of rock or rock with an overlying layer of stiff soil less than 200 feet deep. The applicable value for the site coefficient is
  S = 1.0

- Soil profile type II consists of a stiff clay layer exceeding 200 feet in depth and has a site coefficient of
  $S = 1.2$
- Soil profile type III consists of a soft to medium clay layer, at least 30 feet deep, and has a site coefficient of
  $S = 1.5$

The equivalent static force analysis procedure assumes that the first mode of vibration predominates during the seismic response of the structure, and this is the case for regular structures. The mode shape may be represented by the elastic curve produced by the application of a uniform unit virtual load to the structure. From consideration of the kinetic and potential energies in the system, values of the fundamental period of vibration and of the generalized seismic force are obtained[4,5].

The fundamental period for seismic response along the longitudinal axis of the bridge may be reduced to the expression.

$$T = 2\pi(m/k)^{1/2}$$
$$= 0.32(\Delta_W)^{1/2}$$

where m = mass of the system
k = stiffness of the system.
Δ_W = longitudinal displacement in inches due to the total dead weight acting longitudinally

The analysis for seismic response in the transverse direction requires the use of numerical integration techniques[6] to evaluate the relevant expressions. The fundamental period T and the equivalent static force are obtained by using the technique detailed in AASHTOSD Section 3.5 which involves the determination of the integrals given in the AASHTOSD expressions (5-5), (5-6) and (5-7) where the limits of the integrals extend over the whole length of the bridge superstructure. For seismic response in the longitudinal direction these expressions simplify when the dead weight per unit length of the superstructure and tributary substructure is constant, and when the displacement profile is constant, as shown in Figure 8-2. Hence, for longitudinal seismic force, the expressions for α, β, and γ reduce to

$$\alpha = \int v_s(x)\,dx$$
$$= v_s \int dx$$
$$= v_s L$$
$$\beta = \int w(x)\,v_s(x)\,dx$$
$$= wv_s \int dx$$
$$= wv_s L$$
$$\gamma = \int w(x)\,v_s(x)^2\,dx$$
$$= wv_s^2 \int dx$$
$$= wv_s^2 L$$

where $v_s(x)$ = displacement profile due to p_o
v_s = total longitudinal displacement of the structure due to p_o
p_o = uniform unit virtual load
$w(x)$ = distribution of dead weight per unit length of the superstructure and tributary substructure
= w for a constant dead weight

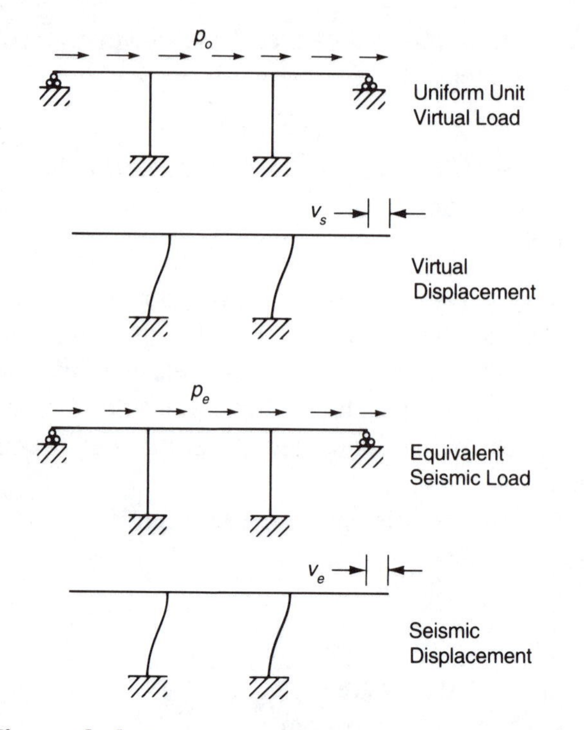

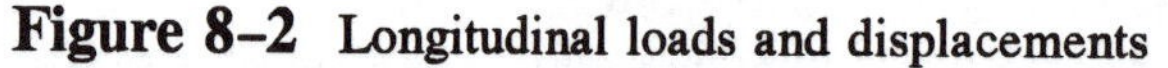

Figure 8–2 Longitudinal loads and displacements

The fundamental period is given by AASHTOSD Expression (5–8) as

$$
\begin{aligned}
T &= 2\pi(\gamma/p_o g\alpha)^{1/2} \\
&= 2\pi(wv_s/p_o g)^{1/2} \\
&= 0.32(Wv_s/P_o)^{1/2} \\
&= 0.32(W/k)^{1/2} \\
&= 0.32(\Delta_W)^{1/2}
\end{aligned}
$$

where W = wL = total weight of superstructure and tributary substructure
P_o = p_oL = total applied virtual load
k = total stiffness of the structure
Δ_W = longitudinal displacement in inches due to the total dead weight acting longitudinally

Hence, the elastic seismic response coefficient C_s may be obtained from AASHTOSD Formula (5–1) and the equivalent static seismic loading is given by AASHTOSD Expression (5–9) as

$$
\begin{aligned}
P_e(x) &= p_e = \beta C_s w(x) v_s(x)/\gamma \\
&= wC_s
\end{aligned}
$$

and this produces the longitudinal displacement v_e as shown in Figure 8–2.

The total elastic seismic shear is given by

$$
\begin{aligned}
V &= p_e L \\
&= wLC_s \\
&= WC_s
\end{aligned}
$$

The procedure for determining the transverse seismic response is shown in Figure 8–3. The abutments are assumed to be rigid and to provide a pinned end restraint at each end of the superstructure. A transverse uniform unit virtual load is applied as shown and is resisted by the lateral stiffness of the superstructure and by the stiffness of the central column bent. As shown in Figure 8–4 the displacements produced are the sum of the displacements in the cut-back structure due to the unit virtual load plus the displacement in the cut back structure due to the reaction in the column bent. The value of the reaction in the column bent is obtained by equating the displacements of node 2 in cases (i), (ii), and (iii). The displacement of node 2 in the original structure, as shown at (i), is

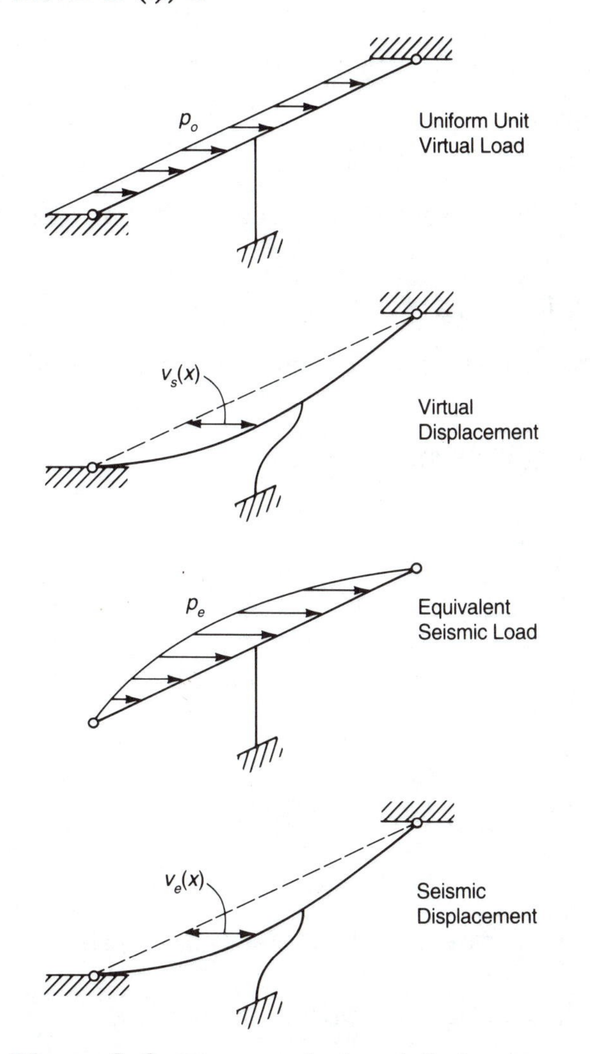

Figure 8–3 Transverse loads and displacements

	δ_2	$= k_C R$
where	k_C	= stiffness of the column bent
		$= 12EI_C/H^3$ for a fixed ended column
	R	= reaction in the column bent.
	I_C	= moment of inertia of column
	H	= height of column

The displacement of node 2 in the cut–back structure due to the uniform unit virtual load, as shown at (ii), is

	δ_2'	$= 5L^4/384EI$
where	E	= modulus of elasticity of the superstructure
	I	= lateral moment of inertia of the superstructure.

The displacement of node 2 in the cut–back structure due to the reaction in the column bent, as shown at (iii), is

$$\delta_2'' = -RL^3/48EI$$

Equating these displacements gives

$$\delta_2 = \delta_2' + \delta_2''$$

Hence $\quad R = 5L^4/(8L^3 + 384EIk_C)$

The displacement profile in the cut–back structure due to the uniform unit virtual load, as shown at (ii), is

$$v_s'(x) = x(L^3 - 2Lx^2 + x^3)/24EI$$

The displacement profile in the cut–back structure due to the reaction in the column bent, as shown at (iii), is

$$v_s''(x) = -Rx(3L^2 - 4x^2)/48EI$$

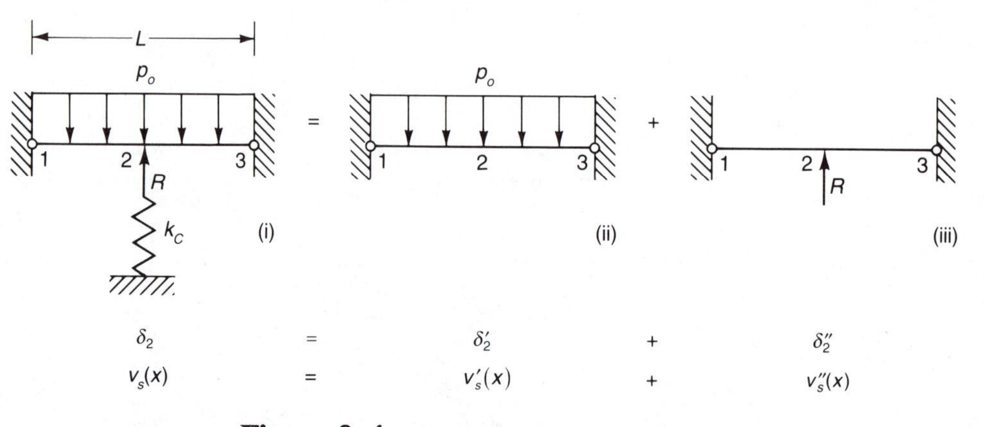

Figure 8–4 Transverse displacements

The displacement profile in the original structure, as shown at (i), is

$$v_s(x) = v_s'(x) + v_s''(x)$$
$$= [2x^4 + 4Rx^3 - 4Lx^2 + (2L - 3R)L^2x]/48EI$$

The values for α, β and γ are obtained from AASHTOSD Expressions (5–5), (5–6) and (5–7)

$$\alpha = \int v_s(x)\,dx$$
$$\beta = \int w(x)v_s(x)\,dx$$
$$\gamma = \int w(x)v_s(x)^2\,dx$$

The limits of integration extend over the whole length of the bridge and the numerical values of the integrals may be obtained by means of a calculator[6]. Alternatively, values of the integrands may be computed at discrete intervals over the length of the bridge and the numerical integration performed manually.

The fundamental period is obtained from AASHTOSD Expression (5-8) as

$$T = 2\pi(\gamma/p_o g\alpha)^{1/2}$$

The seismic response coefficient is obtained from AASHTOSD Formula (5-1) as

$$C_s = 1.2AS/T^{2/3}$$

The equivalent seismic loading is defined by AASHTOSD Formula (5-9) as

$$p_e(x) = \beta C_s w(x) v_s(x)/\gamma$$

This equivalent static load may now be applied to the structure and the resultant forces calculated.

8.2.4 Importance classification

The importance classification is defined in AASHTOSD Section 3.3 and two categories are specified. An importance classification of I is assigned to essential bridges which for social or security considerations must remain functional after an earthquake. For this situation a design is required which will ensure the continued operation of the facility. An importance classification of II is assigned to non-essential bridges.

8.2.5 Seismic performance category

The seismic performance category is a function of the acceleration coefficient and the importance classification and is defined in AASHTOSD Section 3.4. The four categories are shown in Table 8-1 and these determine the necessary requirements for selection of the design procedure, minimum support lengths, and substructure design details.

Acceleration Coefficient	Essential Bridges: Importance Classification I	Other Bridges: Importance Classification II
$A \leq 0.09$	A	A
$0.09 < A \leq 0.19$	B	B
$0.19 < A \leq 0.29$	C	C
$0.29 < A$	D	C

Table 8-1 Seismic performance category

Example 8.1 (Minimum support lengths)
A single span bridge with a span of 75 feet is located in San Diego County. The bridge is considered a non-essential structure. Determine the minimum required support length.

Solution
The applicable acceleration coefficient for San Diego County is obtained from AASHTOSD Figure 1-5 as

$$A \quad > 0.29$$

The importance classification for a non-essential bridge is given by AASHTOSD Section 3.3 as

$$IC \quad = II$$

From AASHTOSD Section 3.4 for a value of the acceleration coefficient exceeding 0.29 and an importance classification of II, the relevant seismic performance category is

$$SPC \quad = C$$

For seismic performance category C, the minimum support length is given by AASHTOSD Formula (4-4) as

$$N = 12 + 0.03L + 0.12H$$

where L = length of bridge deck
= 75 feet
H = height of columns
= 0

then $N = 12 + 0.03 \times 75 + 0$
= 14.25 inches

8.2.6 Analysis procedure

Seismic Performance Category	Bridges with 2 or more Spans	
	Regular	Irregular
A	N/A	N/A
B	1	1
C	1	2
D	1	2

Table 8-2 Analysis procedure

AASHTOSD Section 4.2 defines two analysis procedures. Procedure 1 is the single mode spectral analysis technique and Procedure 2 is the multimode spectral analysis technique. The procedure selected depends on the seismic performance category and on the bridge classification and is summarized in Table 8-2. A multi-span bridge with a uniform mass distribution and with the stiffness of adjacent supporting members differing by not more than 25 percent, is classified as a regular structure. In this type of bridge, the fundamental mode of vibration predominates during the seismic response of the structure and higher modes of vibration do not significantly effect the distribution of seismic forces. An irregular bridge is one that does not satisfy the definition of a regular bridge and, in this type of structure, the higher modes of vibration significantly effect the seismic response.

A detailed seismic analysis is not required for single span bridges or for bridges classified as seismic performance category A. However, minimum support lengths are required to accommodate the maximum inelastic displacement, in accordance with AASHTOSD Section 4.9, and the connection of the superstructure to the substructure is designed to resist the product of the dead load reaction and the site acceleration coefficient, as specified in AASHTOSD Section 4.5.

8.2.7 Response modification factors

To design a bridge structure to remain within its elastic range during a severe earthquake is uneconomical. Limited structural damage is acceptable provided that total collapse is prevented and public safety is not endangered. Any damage produced in a severe earthquake should be readily detectable, accessible and repairable.

Element	R–factor
Wall Type Pier:	
Strong axis	2
Weak axis	3
Reinforced Concrete Pile Bents:	
Vertical Piles Only	3
One or More Batter Piles	2
Single Columns:	3
Steel or Composite Pile Bents:	
Vertical Piles Only	5
One or More Batter Piles	3
Multiple Column Bent:	5

Table 8–3 Response modification factors for substructures

To achieve this end, the response modification factor R, is specified in AASHTOSD Section 3.6 and this represents the ratio of the force in a component which would develop in a linearly elastic system to the prescribed design force. The response modification factors are selected to ensure that columns will yield during a severe earthquake while connections and foundations will have little, if any, damage. The response modification values for substructure components reflect the non-linear energy dissipation capability, the increase in natural period and damping, and the ductility and redundancy of the component. Values of the R-factors are shown in Tables 8-3 and 8-4. Thus, the R-factor for a single column is 3 and for a multiple column bent is 5 which is an indication of the redundancy provided by the multiple column bent. The R-factor is applied to moments, only. Elastic design values are adopted for axial force and shear force unless the values corresponding to plastic hinging of the columns are smaller, in which case, the smaller values are used.

For foundations, the value of the R-factor is taken as half the value of the R-factor for the substructure element to which it is attached. For pile bents, the value of the R-factor is identical to that for the substructure.

Connection type	R-factor	Design force
Superstructure to abutment:		
Single span	N/A	A × DL
SPC A	N/A	0.2 × DL
SPC B,C,D	0.8	Elastic/R
Expansion joints:	0.8	Elastic/R
Pinned columns:		
SPC A	N/A	0.2 × DL
SPC B, C, D	1.0	Elastic/R
Fixed columns:		
SPC A	N/A	0.2 × DL
SPC B	1.0	Elastic/R
SPC C, D	N/A	Plastic hinge forces

Table 8-4 Response modification factors for connections

The R-factors of 1.0 and 0.8 assigned to connectors requires connectors to be designed for 100 percent or 125 percent of the elastic force. This is to ensure enhanced overall integrity of the structure at strategic locations with little increase in construction costs. However, the connector design forces need not exceed the values determined using the maximum probable plastic hinge moment capacities developed in the columns. The maximum probable capacity or overstrength capacity results from the actual material strength exceeding the minimum specified strength. In accordance with AASHTOSD Section 4.8.2, the overstrength capacities are calculated using strength reduction factors of

ϕ = 1.3 ... for reinforced concrete columns
ϕ = 1.25 ... for structural steel columns.

These values may be compared with the normal strength reduction factors of

ϕ = 0.9 ... for reinforced concrete members
ϕ = 1.0 ... for structured steel members.

Example 8-2 (Longitudinal seismic response)

GIVEN: The two span bridge with a single central column support shown in Figure 8-5 is located in the vicinity of San Louis Obispo on an important strategic route. The soil profile at the site consists of a 45 feet layer of soft clay. The column is fixed at the top and bottom.

CRITERIA: Column moment of inertia = 40 feet^4
Column modulus of elasticity = 432,000 kips per square foot
Weight of the superstructure and tributary substructure = 7 kips per foot

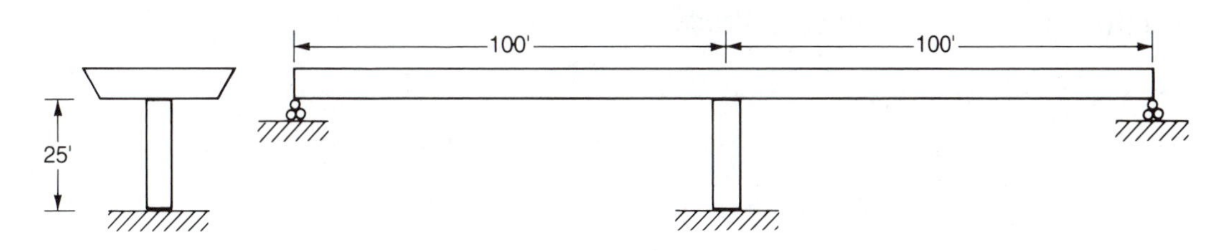

Figure 8-5 Details for Example 8-2

REQUIRED:
1. The applicable acceleration coefficient.
2. The importance classification for the structure.
3. The seismic performance category for the structure
4. The value of the site coefficient.
5. The analysis procedure to be used.
6. The stiffness of the column.
7. The fundamental period of the bridge in the longitudinal direction.
8. The value of the elastic seismic response factor.
9. The elastic seismic moment in the column due to the longitudinal seismic force.
10. The reduced design moment in the column.

SOLUTION

1. ACCELERATION COEFFICIENT

From AASHTOSD Section 3.2, the applicable acceleration coefficient for the San Louis Obispo area is

A = 0.4

2. IMPORTANCE CLASSIFICATION

From AASHTOSD Section 3.3, for a bridge located on an important strategic route, the importance classification is

$$IC = I$$

3. PERFORMANCE CATEGORY

From AASHTOSD Section 3.4, for a value of the acceleration coefficient exceeding 0.29 and an importance classification of I, the relevant seismic performance category is

$$SPC = D$$

4. SITE COEFFICIENT

From AASHTOSD Section 3.5, the relevant site coefficient for a soft clay layer exceeding 30 feet is

$$S = 1.5$$

5. ANALYSIS PROCEDURE

From AASHTOSD Section 4.2, for a regular bridge with a seismic performance category of D, the required analysis procedure is Procedure 1.

6. COLUMN STIFFNESS

The stiffness of a column fixed at the top and bottom is given by

$$\begin{aligned} k_C &= 12EI/H^3 \\ &= 12 \times 432{,}000 \times 40/25^3 \\ &= 13{,}271 \text{ kips per foot} \end{aligned}$$

7. FUNDAMENTAL PERIOD

The total weight of the superstructure and tributary substructure is

$$\begin{aligned} W &= wL \\ &= 7 \times 200 \\ &= 1400 \text{ kips} \end{aligned}$$

The longitudinal stiffness of the bridge is

$$\begin{aligned} k_C &= 13{,}271/12 \\ &= 1106 \text{ kips per inch} \end{aligned}$$

The fundamental period of the bridge in the longitudinal direction is given by

$$\begin{aligned} T &= 0.32(W/k_C)^{1/2} \\ &= 0.32(1400/1106)^{1/2} \\ &= 0.36 \text{ seconds} \end{aligned}$$

8. ELASTIC SEISMIC RESPONSE

From AASHTOSD Section 5.2, the value of the elastic seismic response coefficient is given by Formula (5-1) as

$$C_s = 1.2AS/T^{2/3}$$
$$= 1.2 \times 0.40 \times 1.5/(0.36)^{2/3}$$
$$= 1.42$$

The maximum allowable value of C_s is

$$C_s = 2.5A$$
$$= 2.5 \times 0.4$$
$$= 1.0 \text{ ... governs}$$

9. ELASTIC SEISMIC MOMENT
The total elastic seismic shear is given by

$$V = WC_s$$
$$= 1400 \times 1.0$$
$$= 1400 \text{ kips}$$

The elastic moment in the column is

$$M_E = VH/2$$
$$= 1400 \times 25/2$$
$$= 17{,}500 \text{ kip feet}$$

10. REDUCED DESIGN MOMENT
The response modification factor for a single column is given in Table 3-5 as

$$R = 3$$

Hence, the reduced design moment in the column is

$$M_R = M_E/3$$
$$= 17{,}500/3$$
$$= 5833 \text{ kip feet}$$

8.2.8 Combination of orthogonal forces

AASHTOSD Section 4.4 requires the combination of orthogonal seismic forces to account for the directional uncertainty of the earthquake motion and for the possible simultaneous occurrence of earthquake motions in two perpendicular horizontal directions. The combinations specified are:

Load Case 1: 100 percent of the forces due to a seismic event in the longitudinal direction plus 30 percent of the forces due to a seismic event in the transverse direction.

Load Case 2: 100 percent of the forces due to a seismic event in the transverse direction plus 30 percent of the forces due to a seismic event in the longitudinal direction.

8.2.9 Column plastic hinges

The determination of shear and axial forces in column bents, due to plastic hinging, is shown in Figure 8-6 for transverse seismic loading.

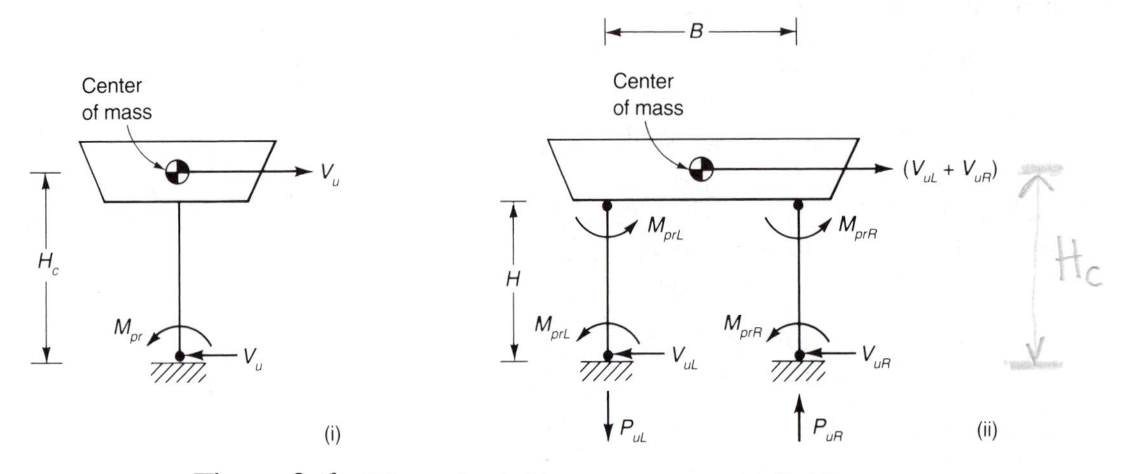

Figure 8-6 Column plastic hinges: transverse earthquake

For a single column, as shown at (i), the maximum probable or overstrength moment capacity at the foot of the column is

$$M_{pr} = \phi M_n$$

where M_n = nominal moment capacity.

The shear developed in the column is given by

$$V_u = M_{pr}/H_C$$

where H_C = height of center of mass

For a double column bent with both columns fixed at top and bottom, as shown at (ii), the shears developed in the left and right columns are given by

$$V_{uL} = 2M_{prL}/H$$
$$V_{uR} = 2M_{prR}/H$$

where H = height of column

The axial force developed in the columns is obtained by equating moments of external forces about the base of one column and is given by

$$P_{uL} = -P_{uR} = [H_C(V_{uL} + V_{uR}) - (M_{prL} + M_{prR})]/B$$

Shear and axial forces, due to plastic hinging, for longitudinal seismic loading are shown in Figure 8-7.For a single column, pinned at the top, as shown at (i), the shear developed in the column is given by

$$V_u = M_{pr}/H$$

For a single column fixed top and bottom, as shown at (ii), the shear developed in the column is given by

$$V_u = 2M_{pr}/H$$

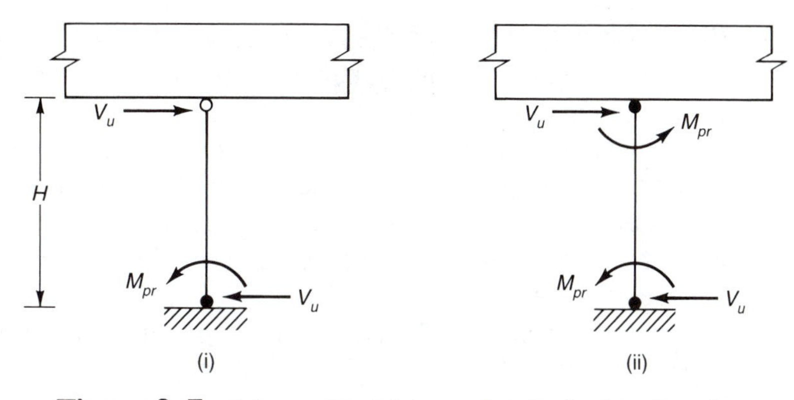

Figure 8–7 Column plastic hinges: longitudinal earthquake

Example 8–3 (Column plastic hinges)

GIVEN: Figure 8–8 shows the central bent in a regular two span bridge located in the vicinity of Los Angeles on a strategic route. The 4 feet diameter columns may be considered fixed at the top and bottom and the axial force, due to dead load, at the bottom of each column is 800 kips. Each column is reinforced with 24 number 14 deformed bars, grade 60, and the concrete strength is 3250 pounds per square inch. The relevant column interaction diagram is shown in the Figure.

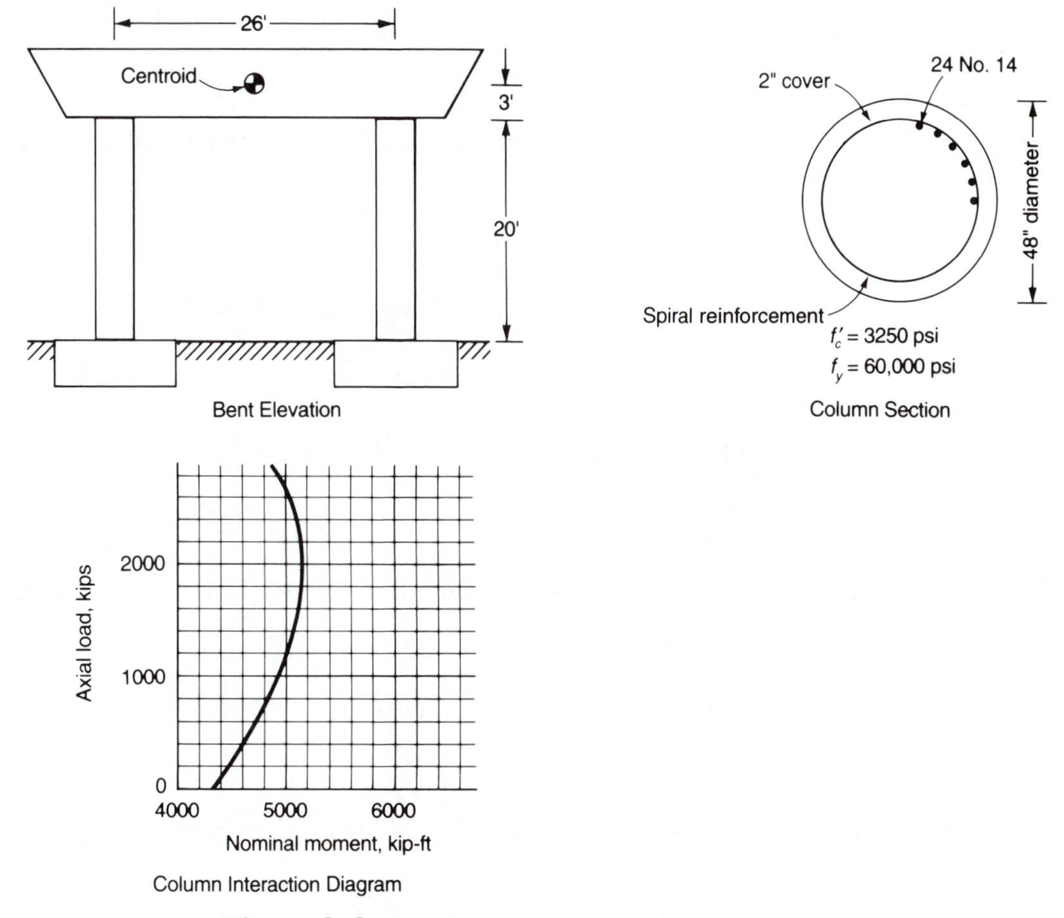

Figure 8–8 Details for Example 8–3

REQUIRED:
1. Do slenderness effects have to be considered in designing the columns?
2. Determine the maximum probable plastic hinging moment at the base of each column.
3. If the overstrength plastic hinge capacities at the top and bottom of the column may be assumed equal, determine the maximum transverse shear force developed in a column in the two column bent.
4. Determine the design shear strength provided by the concrete section outside of the end regions.
5. Determine the design shear strength required from shear reinforcement outside of the end regions.
6. Determine the pitch required for a spiral of number 6 reinforcement outside of the end regions.
7. Determine the length of the end regions, over which special confinement reinforcement is required.
8. Determine the minimum design shear strength provided by the concrete within the end regions of the column.
9. Determine the design shear strength required from shear reinforcement in the end regions of the left column.
10. Determine the pitch required for a spiral of number 7 reinforcement in the end regions of the left column.

SOLUTION

1. SLENDERNESS RATIO

From AASHTO Section 8.16.5.2, the slenderness ratio of a circular column is given by

slenderness ratio $= kl_u/r$

where
- k = effective length factor
- l_u = unsupported column length = 20 feet
- r = radius of gyration = 0.25 × diameter = 1 foot

For an unbraced frame, with both ends of the column fixed, the effective length factor is[7]

$$k = 1.0$$

$$\text{and, slenderness ratio} = 1.0 \times 20/1.0 = 20 < 22$$

Hence, the column is classified as a short column and slenderness effects may be neglected.

2. PLASTIC HINGING MOMENT

From the interaction diagram, for an axial load of 800 kips, the nominal plastic hinging moment is

$$M_n = 4900 \text{ kip feet}$$

In accordance with AASHTOSD Section 4.8.2 the overstrength plastic hinge capacity is

$$M_{pr} = 1.3M_n = 1.3 \times 4900 = 6370 \text{ kip feet}$$

3. MAXIMUM TRANSVERSE SHEAR FORCE

The shear forces produced in the columns by the plastic hinges, for a seismic force acting to the right, are given by

$$V_{uL} = 2M_{prL}/H$$
$$= 2 \times 6370/20$$
$$= 637 \text{ kips}$$
$$V_{uR} = 2M_{prR}/H$$
$$= 2 \times 6370/20$$
$$= 637 \text{ kips}$$

The total shear force in the bent is

$$V_1 = V_{uL} + V_{uR}$$
$$= 637 + 637$$
$$= 1274 \text{ kips}$$

The axial forces produced in the columns by the plastic hinges are given by

$$P_{uL} = -[H_C(V_{uL} + V_{uR}) - (M_{prL} + M_{prR})]/B$$
$$= -[23(637 + 637) - (6370 + 6370)]/26$$
$$= -1127 + 490$$
$$= -637 \text{ kips}$$
$$P_{uR} = +[H_C(V_{uL} + V_{uR}) - (M_{prL} + M_{prR})]/B$$
$$= 1127 - 490$$
$$= 637 \text{ kips}$$

The axial forces produced in the columns by the dead load plus the plastic hinges are given by

$$P_L = P_D + P_{uL}$$
$$= 800 - 637$$
$$= 163 \text{ kips}$$
$$P_R = P_D + P_{uR}$$
$$= 800 + 637$$
$$= 1437 \text{ kips}$$

Using these revised axial forces, the nominal plastic hinging moments in the columns are obtained from the interaction diagram as

$$M_{nL} = 4450 \text{ kip feet}$$
$$M_{nR} = 5100 \text{ kip feet}$$

The corresponding overstrength plastic hinge capacities are given by

$$M_{prL} = 1.3 \times 4450$$
$$= 5785 \text{ kip feet}$$
$$M_{prR} = 1.3 \times 5100$$
$$= 6630 \text{ kip feet}$$

The shear forces produced in the columns by the revised plastic hinge capacities are given by

$$\begin{aligned} V_{uL} &= 2M_{prL}/H \\ &= 2 \times 5785/20 \\ &= 579 \text{ kip feet} \\ V_{uR} &= 2M_{prR}/H \\ &= 2 \times 6630/20 \\ &= 663 \text{ kips feet} \end{aligned}$$

The total shear force in the bent is

$$\begin{aligned} V_2 &= V_{uL} + V_{uR} \\ &= 579 + 663 \\ &= 1242 \text{ kips} \end{aligned}$$

The percentage change in total shear between cycle 1 and cycle 2 is

$$\begin{aligned} \Delta V &= 100\,(V_1 - V_2)/V_1 \\ &= 100\,(1274 - 1242)/1274 \\ &= 2.5\% \\ &< 10\% \end{aligned}$$

Hence, in accordance with AASHTOSD Section 4.8.2 no further iterations are necessary and the maximum probable shear force in the right column is

$$V_{mas} = 663 \text{ kips}$$

4. CONCRETE SHEAR STRENGTH OUTSIDE OF THE END REGIONS

In accordance with AASHTO Section 8.16.6.2 the design shear strength provided by the concrete, outside of the end regions, is given by Equations (8-46) and (8-49) as

$$\phi V_c = 2\phi bd(f'_c)^{1/2}$$

where ϕ = strength reduction factor = 0.85 from Section 8.16.1.2

b = column diameter = 48 inches

d = distance from compressive fibre to centroid of reinforcement in opposite side of member = 37 inches

f'_c = concrete compressive strength = 3250 pounds per square inch

Hence

$$\begin{aligned} \phi V_c &= 2 \times 0.85 \times 48 \times 37 \times (3250/1000)^{1/2} \\ &= 172 \text{ kips} \end{aligned}$$

5. SHEAR STRENGTH REQUIRED FROM REINFORCEMENT OUTSIDE END REGIONS

The design shear strength required from the shear reinforcement is given by AASHTO Equation (8-46) and (8-47) as

$$\begin{aligned} \phi V_s &= V_u - \phi V_c \\ &= 663 - 172 \\ &= 491 \text{ kips} \\ &< 8\phi V_c \end{aligned}$$

Hence, in accordance with AASHTO Section 8.16.6.3, the required shear reinforcement strength is within allowable limits.

6. PITCH OF SPIRAL REINFORCEMENT OUTSIDE END REGIONS

To satisfy the requirements for lateral reinforcement in a compression member, in accordance with AASHTO Section 8.18.2, the clear distance between spirals shall not exceed 3 inches and the minimum volumetric ratio of the spiral reinforcement to the concrete core is given by Equation (8-63) as

$\rho_s = 0.45\,(A_g/A_c - 1)f'_c/f_y$

$= A_v\pi(D_c - D_s)/sA_c$

where

A_g = gross area of column = 1810 square inches

A_c = area of core measured to outside of spiral = 1521 square inches

D_c = diameter of core measured to outside of spiral = 44 inches

D_s = diameter of spiral reinforcement = 0.75 inches

A_v = area of spiral reinforcement = 0.44 square inches

s = pitch of spiral reinforcement

Hence, the required pitch is given by

$s = A_vf_y\pi(D_c - D_s)/0.45A_cf'_c(A_g/A_c - 1)$

$= 0.44 \times 60{,}000\pi(44 - 0.75)/[0.45 \times 1521 \times 3250\,(1810/1521 - 1)]$

$= 8.5$ inches

$s = 3.75$ inches maximum ... governs

To satisfy the requirements for shear strength, in accordance with AASHTO Section 8.16.6.3, the required spiral pitch is given by Equation (8-53) as

$s = \phi A_vf_yd/\phi V_s$

$= 0.85 \times 2 \times 0.44 \times 60 \times 37/491$

$= 3.38$ inches

< 3.75 inches

Hence, the required maximum pitch is

$s = 3.38$ inches

7. LENGTH OF THE END REGIONS

In accordance with AASHTOSD Section 8.4.1 the length of the end regions is the larger of

- 18 inches
- $H/6 = 20 \times 12/6 = 40$ inches
- Column diameter = 48 inches ... governs

8. CONCRETE SHEAR STRENGTH IN THE END REGIONS

The final axial forces produced in the columns by the final values of the overstrength plastic hinge moments are given by

$$\begin{aligned}
P_{uL} &= -[H_C(V_{uL} + V_{uR}) - (M_{prL} + M_{prR})]/B \\
&= -\ [23(579 + 663) - (5785 + 6630)]/26 \\
&= -\ 1098 + 478 \\
&= -\ 621 \text{ kips} \\
P_{uR} &= +\ [H_C(V_{uL} + V_{uR}) - (M_{prL} + M_{prR})]/B \\
&= 1098 - 478 \\
&= 621 \text{ kips}
\end{aligned}$$

The final axial forces produced in the columns by the dead load plus the plastic hinges are

$$\begin{aligned}
P_L &= P_D + P_{uL} \\
&= 800 - 621 \\
&= 179 \text{ kips} \\
P_R &= P_D + P_{uR} \\
&= 800 + 621 \\
&= 1421 \text{ kips}
\end{aligned}$$

The axial force value given by

$$\begin{aligned}
A_g f'_c/10 &= 1810 \times 3.25/10 \\
&= 588 \text{ kips} \\
&> P_L
\end{aligned}$$

Hence, in accordance with AASHTOSD Section 8.4.1, the design shear strength of the concrete in the end region of the left column must be neglected because of cracking under load reversals.

9. SHEAR STRENGTH REQUIRED FROM SHEAR REINFORCEMENT IN END REGIONS

For the left column, the design shear strength required from the shear reinforcement is given by AASHTO Equations (8–46) and (8–47) as

$$\begin{aligned}
\phi V_s &= V_{uL} - \phi V_c \\
&= V_{uL} - 0 \\
&= 579 \text{ kips}
\end{aligned}$$

10. PITCH OF SPIRAL REINFORCEMENT IN THE END REGIONS

The required pitch of confinement reinforcement is given by the smaller value obtained from AASHTOSD Equations (8–1) and (8–2). Then, for number 7 reinforcement,

$$\begin{aligned}
s &= A_v f_y \pi (D_c - D_s)/0.45 A_c f'_c (A_g/A_c - 1) \\
&= 0.60 \times 60{,}000\pi\ (44 - 0.875)/[0.45 \times 1521 \times 3250\ (1810/1521 - 1)] \\
&= 11.6 \text{ inches}
\end{aligned}$$

or

$$\begin{aligned}
s &= A_v f_y \pi (D_c - D_s)/0.12\ A_c f'_c \\
&= 0.60 \times 60{,}000\ \pi(44 - 0.875)/[0.12 \times 1521 \times 3250] \\
&= 8.22 \text{ inches}
\end{aligned}$$

To satisfy the requirements for shear strength, in accordance with AASHTO Section 8.16.6.3, the required spiral pitch is given by Equation (8–53) as

$$
\begin{aligned}
s &= \phi A_v f_y d/\phi V_s \\
&= 0.85 \times 2 \times 0.60 \times 60 \times 37/579 \\
&= 3.91 \text{ inches}
\end{aligned}
$$

Since the clear distance between spirals shall not exceed 3 inches, in accordance with AASHTO Section 8.18.2, the maximum pitch is

$$
\begin{aligned}
s &= 3 + D_s \\
&= 3 + 0.875 \\
&= 3.875
\end{aligned}
$$

References

1. American Association of State Highway and Transportation Officials. *Standard specifications for highway bridges, Fifteenth edition.* Washington, D.C. 1992.

2. American Association of State Highway and Transportation Officials. *Standard specifications for highway bridges, Fifteenth edition: Division I-A: Seismic Design.* Washington, D.C. 1992.

3. Building Science Safety Council. *NEHRP recommended provisions for the development of seismic regulations for new buildings: Part 2, Commentary.* Washington, D.C. 1992.

4. Paz, M. Structural dynamics. Van Nostrand Reinhold, New York, 1991.

5. Federal Highway Administration. *Seismic design and retrofit manual for highway bridges.* Washington, D.C. 1987.

6. Hewlett Packard Company. *HP-28S calculator reference manual.* Corvallis, OR 1987.

7. Portland Cement Association. *Notes on ACI 318-89: Building code requirements for reinforced concrete.* Skokie, IL, 1990.

APPENDIX I

List of useful equations

CHAPTER I:

The circular natural frequency or angular velocity is

$$\omega = (k/m)^{1/2}$$
$$= 2\pi/T$$
$$= 2\pi f$$

where
- T = natural period
- f = natural frequency of vibration
- k = stiffness

The stiffness of a fixed ended column is

$$k_F = 12EI/l^3$$

The stiffness of a cantilever column is

$$k_C = 3EI/l^3$$

The critical damping coefficient is

$$c_c = 2m\omega$$

The damped circular frequency is

$$\omega_D = [\omega^2 - (c/2m)^2]^{1/2}$$
$$= \omega[1 - (c/2m\omega)^2]^{1/2}$$
$$= \omega(1 -(c/c_c)^2]^{1/2}$$
$$= \omega(1 - \xi^2)^{1/2}$$

where c = damping coefficient.
ξ = damping ratio

The damped natural period is

$$T_D = 2\pi/\omega_D$$
$$= 2\pi/\omega(1 - \xi^2)^{1/2}$$
$$= T/(1 - \xi^2)^{1/2}$$

The logarithmic decrement is

$$\delta = cT_D/2m$$
$$= \xi\omega T_D$$
$$= 2\pi\xi/(1-\xi^2)^{1/2}$$
$$= 2\pi\xi$$

The matrix of participation factors is

$$\{P\} = [\Phi\}^T[M\}\{1\}/[\Phi]^T[M][\Phi]$$

where $\{P\}$ = column vector of partition factors for all modes considered
$[\Phi]$ = mode shape matrix or eigenvectors
$\{1\}$ = column vector of ones
$[M]$ = mass matrix, the diagonal matrix of lumped masses
$[\Phi]^T[M][\Phi]$ = modal mass matrix

and $\Sigma P_j\phi_{1j}$ = 1.0

where P_j = the participation factor associated with the specific mode j
ϕ_{1j} = the component, for the first node of the system, of the eigenvector associated with the specific mode j

The matrix of maximum node displacements is

$$[x] = [\Phi][P][S_d]$$
$$= [\Phi][P][S_v][\omega]^{-1}$$
$$= [\Phi][P][S_a][\omega^2]^{-1}$$

where $[P]$ = diagonal matrix of participation factors
$[S_d]$ = diagonal matrix of spectral displacements

$[S_v]$ = diagonal matrix of spectral velocities
$[S_a]$ = diagonal matrix of spectral accelerations
$[\omega]$ = diagonal matrix of modal frequencies
$[\omega^2]$ = diagonal matrix of squared modal frequencies

The matrix of peak acceleration response is

$$[\ddot{x}] = [\Phi][P][S_a] = [\Phi][P][S_v][\omega] = [\Phi][P][S_d][\omega^2]$$

The matrix of lateral forces at each node is

$$\begin{aligned}[F] &= [M][\ddot{x}] \\ &= [M][\Phi][P][S_a] \\ &= [M][\Phi][P][S_v][\omega] \\ &= [M][\Phi][P][S_d][\omega^2] \\ &= [K][\Phi][P][S_d] \\ &= [K][\Phi][P][S_v][\omega]^{-1} \\ &= [K][\Phi][P][S_d][\omega^2]^{-1} \\ &= [K][x\ ]\end{aligned}$$

The column vector of total base shear forces is

$$\{V\} = [F]^T\{1\} = [P][S_a][\Phi]^T[M]\{1\}$$

For normalized eigenvectors:

$$[\Phi[^T[M][\Phi] = [I]$$

where $[I]$ = identity matrix
$[\Phi]^T[M][\Phi]$ = modal mass matrix
and $\{P\} = [\Phi]^T[M]\{1\}$

The participation factor is

$$P = \Sigma M_i\phi_i/M$$

where M_i = mass at floor level i
ϕ_i = mode shape component for node point i for the given mode
M = modal mass
$= \Sigma M_i\phi_i^2$

The effective mass is

$$
\begin{aligned}
M^E &= (\Sigma M_i\phi_i)^2/\Sigma M_i\phi_i^2 \\
&= P\Sigma M_i\phi_i \\
&= (\Sigma M_i\phi_i)^2/M \\
&= P^2M
\end{aligned}
$$

The effective weight is

$$W^E = (\Sigma W_i\phi_i)^2/\Sigma W_i\phi_i^2$$

where W_i = weight at floor level i

The peak acceleration at a node is

$$\ddot{x} = \phi_i P S_a$$

where S_a = spectral acceleration for the given mode

The maximum displacement at a node is

$$x_i = \phi_i P S_d$$

where S_d = spectral displacement for the given node

The lateral force at a node is

$$
\begin{aligned}
F_i &= M_i\ddot{x} \\
&= M_i\phi_i P S_a \\
&= M_i\omega^2 x_i
\end{aligned}
$$

The total base shear is

$$
\begin{aligned}
V &= \Sigma F_i \\
&= PS_a\Sigma M_i\phi_i \\
&= P^2MS_a \\
&= M^ES_a \\
&= W^ES_a/g
\end{aligned}
$$

where g = acceleration due to gravity
= 386 inches per second²

For normalized eigenvectors:

M	=	modal mass
	=	$\Sigma M_i \phi_i^2$
	=	1.0
P	=	participation factor
	=	$\Sigma M_i \phi_i$
M^E	=	effective mass
	=	$(\Sigma M_i \phi_i)^2$
W^E	=	effective weight
	=	$(\Sigma W_i \phi_i)^2/g$

The spectral acceleration is

	S_a	=	$\omega S_v \quad = \omega^2 S_d$
where	S_a	=	spectral acceleration
	S_v	=	spectral velocity
	S_d	=	spectral displacement.

CHAPTER 2:

The seismic force coefficient C is

	C	=	$1.25S/T^{2/3}$
where	S	=	site coefficient, for a specific soil type, from UBC Table 16–J
	T	=	fundamental period of vibration.

The natural period of vibration using method A is

	T	=	$C_t(h_n)^{3/4}$
where	h_n	=	height in feet of the roof above the base, not including the height of penthouses or parapets
	C_t	=	0.035 for steel moment–resisting frames
		=	0.030 for reinforced concrete moment–resisting frames and eccentric braced steel frames
		=	0.020 for all other buildings.

The natural period of vibration using method B is

	T	=	$2\pi(\Sigma w_i \delta_i^2 / g \Sigma f_i \delta_i)^{1/2}$
where	δ_i	=	elastic deflection at level i
	f_i	=	lateral force at level i
	w_i	=	seismic dead load located at level i
	g	=	acceleration due to gravity

The seismic base shear is

$V = (ZIC/R_W)W$

where
Z = seismic zone factor from UBC Table 16-I
I = importance factor from UBC Table 16-K
C = seismic force coefficient
R_W = response modification factor from UBC Table 16-N
W = seismic dead load

In a multi-story structure the design lateral force at level x is

$F_x = (V - F_t)w_x h_x / \Sigma w_i h_i$

where
V = base shear
F_t = 0.07TV
= 0.0 for $T \leq 0.7$ seconds
T = fundamental period of vibration.
w_x = seismic dead load located at level x
h_x = height above the base to level x
w_i = seismic dead load located at level i
h_i = height above the base to level i.

In a multi-story structure the force acting on a horizontal diaphragm at level x is

$F_{px} = (F_t + \Sigma F_i)w_{px}/\Sigma w_i$
$\not< 0.35ZIw_{px}$
$\not> 0.75ZIw_{px}$

where
F_t = concentrated lateral force at roof level
F_i = lateral force at level i
w_i = total seismic dead load located at level i
w_{px} = seismic dead load tributary to the diaphragm at level x, not including walls parallel to the direction of the seismic load.

For a structural element the design seismic force is

$F_p = ZI_pC_pW_p$ where,
Z = zone coefficient from UBC Table 16-I
I_p = importance factor from UBC Table 16-K
C_p = horizontal force factor from UBC Table 16-O
W_p = weight of element or component.

For a rigid structure the design seismic force is

$V = 0.5ZIW$

The amplification factor for accidental eccentricity is

$A_x = (\delta_{max}/1.2\delta_{avg})^2$
≤ 3.0

where δ_{max} = maximum displacement at level x
δ_{avg} = average of displacements at extreme points of the structure at level x

The deflection of a pier due to a unit applied load is given by

$\delta = \delta_F + \delta_S$

where δ_F = deflection due to flexure
$= 4(H/L)^3/Et$ for a cantilever pier
$= (H/L)^3/Et$ for a pier fixed at top and bottom
H = height of pier
L = length of pier
E = modulus of elasticity of pier
t = thickness of pier
δ_S = deflection due to shear
$= 1.2H/GA$
$= 3(H/L)/Et$
G = rigidity modulus of pier
$= 0.4E$
A = cross sectional area of pier
$= tL$

The rigidity, or stiffness, of a pier is

$R = 1/\delta$

CHAPTER 8:

The lateral design force coefficient is

$C_s = 1.2AS/T^{2/3}$
$\leq 2.5A$

where S = site coefficient or amplification factor for a specific soil profile
T = fundamental period of the bridge.
A = acceleration coefficient

The fundamental period for seismic response along the longitudinal axis of a bridge is

$$T = 2\pi(m/k)^{1/2}$$
$$= 0.32(\Delta_W)^{1/2}$$

where
- m = mass of the system
- k = stiffness of the system.
- Δ_W = longitudinal displacement in inches due to the total dead weight acting longitudinally

The fundamental period for seismic response along the transverse axis of a bridge is

$$T = 2\pi(\gamma/p_o g\alpha)^{1/2}$$

where
- $\alpha = \int v_s(x)\, dx$
- $\gamma = \int w(x)v_s(x)^2\, dx$

The equivalent seismic loading is

$$p_e(x) = \beta C_s w(x) v_s(x)/\gamma$$

where
- $\beta = \int w(x)v_s(x)\, dx$

APPENDIX II

Multiple choice questions for the special seismic examination

QUESTIONS

For the following questions, choose the answer which is most nearly correct.

1. What stress increase is allowed when considering seismic forces?

   (A) 0.25
   (B) 0.33
   (C) 0.40
   (D) 0.50
   (E) 0.75

2. What are the load combinations which include seismic loads?

   (A) DL + Seismic
   (B) Floor LL + Seismic
   (C) DL + Floor LL + Seismic
   (D) DL + Snow + Seismic
   (E) DL + Floor LL + Snow + Seismic

3. Does a snow load of 30 pounds per square foot need to be combined with seismic load?

   (A) Yes
   (B) No

4. By what amount may snow loads be reduced when combined with seismic load?

   (A) 0.25
   (B) 0.33
   (C) 0.40
   (D) 0.50
   (E) 0.75

5. For concrete design in seismic zones 3 and 4, what is the required ultimate strength of a member when seismic loads are considered?

   (A) $0.9D \pm 1.4E$
   (B) $0.9D + 1.7L \pm 1.4E$
   (C) $1.4(D + L + E)$
   (D) $1.4D + 1.7L + 1.1E$
   (E) $1.4D + 1.7L + 1.4E$

6. For a structure with vertical irregularity Type D in seismic zones 3 and 4, what is the required axial design force in supporting columns?

   (A) $0.9D + 1.4E$
   (B) $0.85D \pm 3ER_w/8$
   (C) $1.0D + 0.8L + 3ER_w/8$
   (D) $1.4D + 1.7L + 1.1E$
   (E) $1.4D + 1.7L + 1.4E$

7. What allowable stress increase may be used when designing the columns in Question number 6.

   (A) 1.25
   (B) 1.33
   (C) 1.50
   (D) 1.70
   (E) 1.75

8. When designing for overturning, what factor must be applied to the dead load?

   (A) 0.50
   (B) 0.75
   (C) 0.85
   (D) 1.00
   (E) 1.25

9. What factor of safety against overturning is required for seismic loads?

   (A) 0.75
   (B) 1.00
   (C) 1.25
   (D) 1.33
   (E) 1.50

10. What factor of safety against overturning is required for wind loads?

    (A) 0.75
    (B) 1.00
    (C) 1.25
    (D) 1.33
    (E) 1.50

11. In calculating accidental torsion, what displacement is assumed for the mass of the structure?

    (A) 0.02L
    (B) 0.03L
    (C) 0.04L
    (D) 0.05L
    (E) 0.06L

12. What minimum design force must be used for the anchorage of masonry walls to a roof diaphragm?

    (A) 100 pounds per linear foot
    (B) 125 pounds per linear foot
    (C) 150 pounds per linear foot
    (D) 175 pounds per linear foot
    (E) 200 pounds per linear foot

13. Masonry walls must be designed to resist bending when the anchor spacing exceeds what distance?

    (A) 2 feet
    (B) 3 feet
    (C) 4 feet
    (D) 5 feet
    (E) 6 feet

14. What constitutes seismic dead load, W?

    (A) Structure dead load
    (B) Permanent equipment
    (C) 25 percent warehouse floor load
    (D) 10 pounds per square foot partition load
    (E) Snow load exceeding 30 pounds per square foot

15. A structure consists of a bearing wall system with concrete shear walls in the north–south direction and a special moment-resisting steel frame in the east–west direction. What is the applicable R_W value in each direction?

    (A) 4
    (B) 6
    (C) 8
    (D) 10
    (E) 12

16. In seismic zone 3, which structural systems must be used for buildings exceeding 240 feet in height?

    (A) Steel eccentrically braced frame
    (B) Special moment-resisting frame
    (C) Special moment-resisting frame/concrete shear wall
    (D) Special moment-resisting frame/steel eccentrically braced frame
    (E) Special moment-resisting frame/steel concentrically braced frame

17. In seismic zone 4, what factor must be applied to seismic forces when designing chevron bracing?

    (A) 0.75
    (B) 1.00
    (C) 1.25
    (D) 1.33
    (E) 1.50

18. In seismic zone 3, what percentage of the tributary gravity load is supported by chevron bracing?

    (A) 0%
    (B) 10%
    (C) 20%
    (D) 30%
    (E) 40%

19. In which seismic zones may K bracing be utilized in a three story building?

    (A) 1
    (B) 2A
    (C) 2B
    (D) 3
    (E) 4

20. In seismic zone 4, what limitation is placed on the L/R ratio for the bracing members in a three story structure?

    (A) $65/\sqrt{F_y}$
    (B) $95/\sqrt{F_y}$
    (C) $190/\sqrt{F_y}$
    (D) $317/\sqrt{F_y}$
    (E) $720/\sqrt{F_y}$

21. What is the fundamental period of a ten story building frame system structure with a story height of twelve feet?

    (A) 0.65
    (B) 0.68
    (C) 0.73
    (D) 0.75
    (E) 0.83

22. A dual structural system with a steel moment–resisting frame and a braced frame is subjected to a seismic load of 200 kips. The relative stiffnesses of the braced frame and the moment frame are 100:30. For what lateral force must the frames be designed?

    (A) 50
    (B) 77
    (C) 113
    (D) 154
    (E) 182

23. If the relative stiffnesses of the frames in the above structure are 100:40, what are the design forces?

    (A) 38
    (B) 57
    (C) 115
    (D) 143
    (E) 196

24. The structure in Question number 21 is constructed on a site with soil profile type S_4. What is the value of the seismic coefficient C?

    (A) 1.75
    (B) 2.17
    (C) 2.56
    (D) 2.75
    (E) 3.10

25. A special moment-resisting concrete frame with a fundamental period of two seconds is constructed on a site with soil profile type S_1. What is the value of the seismic coefficient C?

    (A) 0.67
    (B) 0.75
    (C) 0.79
    (D) 0.85
    (E) 0.90

26. The structure in Question number 21 is constructed on a site with an undetermined soil profile. What is the value of the seismic coefficient C?

    (A) 1.5
    (B) 2.0
    (C) 2.3
    (D) 2.5
    (E) 2.7

27. A single-story fire station in Los Angeles is constructed with masonry shear walls which support the timber roof structure. What percentage is the base shear of the total seismic dead load?

    (A) 19 percent
    (B) 21 percent
    (C) 23 percent
    (D) 25 percent
    (E) 27 percent

28. In seismic zone 4, what is the maximum height of a structure which may be designed by static force procedures?

    (A) 65 feet
    (B) 160 feet
    (C) 240 feet
    (D) 320 feet
    (E) No Limit

29. The natural period of a building, located on a site in zone 4 with an undetermined soil profile, is determined by Method A to be, T_A = 0.6 seconds and by Method B to be, T_B = 1.0 seconds. What is the value of the seismic coefficient C?

    (A) 1.8
    (B) 2.0
    (C) 2.2
    (D) 2.4
    (E) 2.6

30. A three-story timber framed bearing wall structure has a story height of h_s = 12 feet and a natural period of T = 0.8 seconds. What is the maximum allowable story drift ratio?

    (A) 0.00350
    (B) 0.00375
    (C) 0.00400
    (D) 0.00425
    (E) 0.00450

31. The structure in Question number 30 has a calculated drift in the top story, resulting from the required lateral seismic forces, of Δ = 0.15 inches. For what movement must exterior, nonbearing wall panels be designed?

    (A) 0.40 inches
    (B) 0.45 inches
    (C) 0.50 inches
    (D) 0.55 inches
    (E) 0.60 inches

32. The twelve feet high columns of a special moment-resisting steel frame have a lateral displacement of 0.5 inches when subjected to a lateral seismic force of 3 kips and an axial load of 100 kips. Is it necessary to consider P-Δ effects in zone 4 or zone 2B?

    (A) Yes
    (B) No

33. A reinforced concrete cantilever, in zone 4, is twenty feet long and weighs 0.5 kips per linear foot. For what moment must it be designed to resist vertical upward seismic effects?

    (A) 11 kips feet
    (B) 14 kips feet
    (C) 17 kips feet
    (D) 20 kips feet
    (E) 23 kips feet

34. A prestressed concrete girder, in seismic zone 3, spans twenty feet, weighs 0.4 kips per linear foot and has a section modulus about the bottom fibre of S = 4000 in^3. When considering seismic effects, what is the maximum allowable design stress in the bottom of the girder due to self weight?

    (A) 10 pounds per square inch
    (B) 15 pounds per square inch
    (C) 20 pounds per square inch
    (D) 25 pounds per square inch
    (E) 30 pounds per square inch

35. A regular building is analyzed by the dynamic lateral force procedure and the calculated base shear is V_D = 1500 kips. A static lateral force procedure carried out on the same structure gives a value for the base shear of V_A = 2500 kips using Method A, and a value of V_B = 2000 kips using Method B. What scaling factor must be applied to the seismic forces determined by the dynamic method?

    (A) 1.11
    (B) 1.25
    (C) 1.33
    (D) 1.40
    (E) 1.50

36. An emergency generator in the basement of a hospital in seismic zone 4 weighs ten kips and has a fundamental period of 0.055 seconds. Calculate the design lateral seismic force.

    (A) 1 kips
    (B) 2 kips
    (C) 3 kips
    (D) 4 kips
    (E) 5 kips

37. A water tank supported on the roof of an office building in seismic zone 4 weighs 1 kips and has a fundamental period of 0.065 seconds. Calculate the design lateral seismic force.

    (A) 0.2 kips
    (B) 0.3 kips
    (C) 0.4 kips
    (D) 0.5 kips
    (E) 0.6 kips

38. What is the minimum force for which a plywood diaphragm may be designed?

    (A) $0.25ZIw_p$
    (B) $0.30ZIw_p$
    (C) $0.35ZIw_p$
    (D) $0.40ZIw_p$
    (E) $0.45ZIw_p$

39. What increase is required in the value of C_p when calculating the anchorage force of a masonry wall to the midspan of a plywood diaphragm?

    (A) 10 percent
    (B) 25 percent
    (C) 33 percent
    (D) 50 percent
    (E) 67 percent

40. What stress increase is allowed in seismic zone 3 when designing the connection of a drag strut to a shear wall if the structure has a plan irregularity type B?

    (A) 0 percent
    (B) 5 percent
    (C) 10 percent
    (D) 15 percent
    (E) 20 percent

41. What is the maximum allowable span:width ratio for a horizontal plywood diaphragm?

    (A) 2:1
    (B) 2.5:1
    (C) 3:1
    (D) 3.5:1
    (E) 4:1

42. A hospital located in seismic zone 3 has a six inch thick concrete parapet which is five feet high. What is the maximum bending moment produced in the parapet by seismic forces?

    (A) 603 pounds feet per foot
    (B) 653 pounds feet per foot
    (C) 703 pounds feet per foot
    (D) 753 pounds feet per foot
    (E) 803 pounds feet per foot

43. A billboard in seismic zone 4 with a weight of 1 kips and a calculated natural period of 0.73 seconds is grade supported. Determine the seismic lateral load.

    (A) 0.10 kips
    (B) 0.15 kips
    (C) 0.20 kips
    (D) 0.25 kips
    (E) 0.30 kips

44. An emergency water tank for a hospital, weighing 60 kips, is mounted on a braced tower as shown in Figure A–1. The site specific response curve for the locality is shown. Determine the UBC design base shear if the natural period of the tower is 0.3 seconds.

    (A) 15 kips
    (B) 20 kips
    (C) 25 kips
    (D) 30 kips
    (E) 35 kips

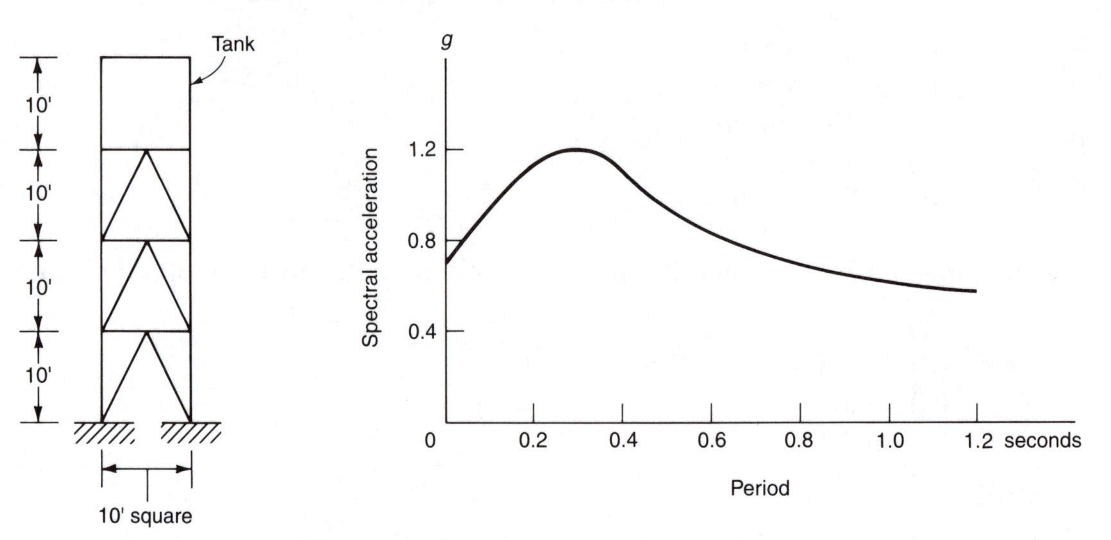

Figure A–1 Details for Question A–44

45. Determine the design uplift in one column of the braced tower shown in Figure A–1.

    (A) 50.5 kips
    (B) 55.5 kips
    (C) 60.5 kips
    (D) 65.5 kips
    (E) 70.5 kips

46. What is the ratio of the shear forces resisted by the masonry cantilever piers A and B shown in Figure A–2?

    (A) 28.3
    (B) 33.3
    (C) 38.3
    (D) 43.3
    (E) 48.3

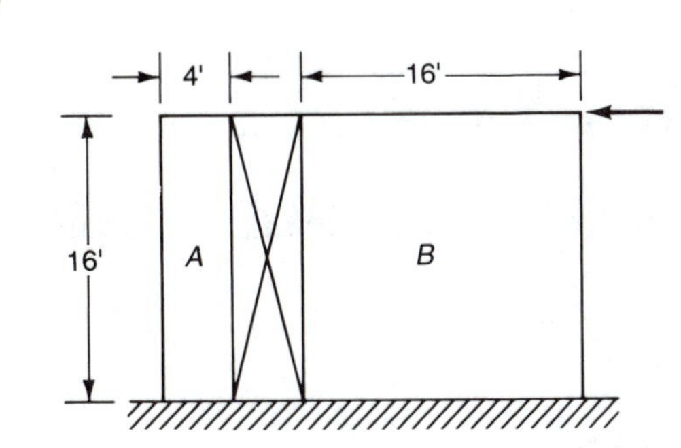

Figure A–2 Details for Question 46

47. What is the ratio of the shear forces resisted by the wood shear walls A and B shown in Figure A–2?

    (A) 1
    (B) 2
    (C) 3
    (D) 4
    (E) 5

48. Calculate the distance of the center of mass from the west wall of the building shown in Figure A–3. The weight of the roof is 50 pounds per square foot, the shear walls 60 pounds per square foot and the air conditioner 50 kips.

    (A) 38.9 feet
    (B) 41.9 feet
    (C) 44.9 feet
    (D) 47.9 feet
    (E) 50.9 feet

49. Calculate the distance of the center of rigidity from the west wall of the building shown in Figure A–3.

    (A) 32 feet
    (B) 34 feet
    (C) 36 feet
    (D) 38 feet
    (E) 40 feet

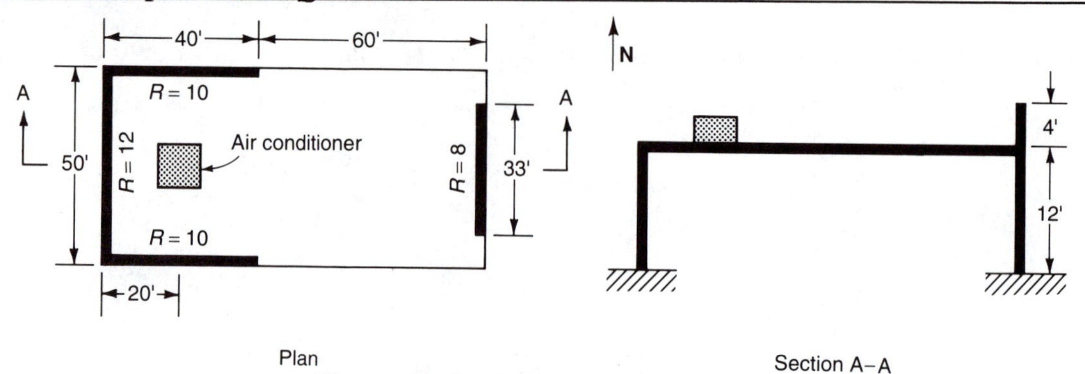

Figure A–3 Building for Question 48

50. For a seismic base shear of 100 kips determine the maximum UBC design torsion for the building shown in Figure A–3.

    (A) 650 kips feet
    (B) 670 kips feet
    (C) 690 kips feet
    (D) 710 kips feet
    (E) 730 kips feet

SOLUTIONS

	UBC REFERENCE
1. One third **THE CORRECT ANSWER IS (B)**	1603.5
2. DL + Floor LL + Seismic DL + Floor LL + Snow + Seismic **THE CORRECT ANSWER IS (C) + (E)**	1603.6
3. No **THE CORRECT ANSWER IS (B)**	1603.6
4. 75% **THE CORRECT ANSWER IS (E)**	1603.6
5. U = 1.4 (D+L+E) U = 0.9D ± 1.4E **THE CORRECT ANSWER IS (A) + (C)**	1921.2.7

6. $P = 1.0D + 0.8L + 3ER_w/8$ 1628.7.2
$P = 0.85D \pm 3ER_w/8$
THE CORRECT ANSWER IS (B) + (C)

7. 1.7 1628.7.2
THE CORRECT ANSWER IS (D)

8. 0.85 1631.1
THE CORRECT ANSWER IS (C)

9. 1.0 1628.7.1
THE CORRECT ANSWER IS (B)

10. 1.0 for structures with $H<60'$ and $H/B<0.5$ otherwise 1.5. 1619.1
THE CORRECT ANSWER IS (B) + (E)

11. 0.05L 1628.5
THE CORRECT ANSWER IS (D)

12. 200 pounds per linear foot 1611
THE CORRECT ANSWER IS (E)

13. 4 feet 1611
THE CORRECT ANSWER IS (C)

14. 1628.5
(i) structure dead load
(ii) weight of permanent equipment
(iii) 25% warehouse floor loading
= 31.25 pounds per square foot (light storage)
= 62.5 pounds per square foot (heavy storage)
(iv) 10 pounds per square foot partition load
(v) snow load if greater than 30 pounds per square foot
THE CORRECT ANSWER IS (A) + (B) + (C) + (D) + (E)

15. R_W = 6 in both directions — 1628.3.3, Table 16-N

THE CORRECT ANSWER IS (B)

16. (i) Special moment-resisting frame in steel or concrete — Table 16-N
    (ii) Dual system with special moment-resisting frame and concrete shear walls
    (iii) Dual system with steel special moment-resisting frame and steel eccentrically braced frame
    (iv) Dual system with steel special moment-resisting frame and steel concentrically braced frame

THE CORRECT ANSWER IS (B) + (C) + (D) + (E)

17. 1.5 — 2211.8.4.1

THE CORRECT ANSWER IS (E)

18. 0% — 2211.8.4.1

THE CORRECT ANSWER IS (A)

19. Zones 1 and 2 — 2211.8.4.2, 2212.6.4

THE CORRECT ANSWER IS (A) + (B) + (C)

20. $720/\sqrt{F_y}$ — 2211.8.2.1

THE CORRECT ANSWER IS (E)

21. $T = 0.02\ (120)^{3/4}$ — (28-3)
    $= 0.73$ seconds

THE CORRECT ANSWER IS (C)

22. Braced frame: — 1627.6.5
    $V = 200 \times 100/130$
    $= 154$ kips

THE CORRECT ANSWER IS (D)

Moment frame:
    $V = 200 \times 25/100$
    $= 50$ kips

THE CORRECT ANSWER IS (A)

23. Braced frame: 1627.6.5

$V = 200 \times 100/140$
$= 143$ kips

THE CORRECT ANSWER IS (D)

Moment frame:

$V = 200 - 143$
$= 57$ kips

THE CORRECT ANSWER IS (B)

24. $C = 1.25S/T^{2/3}$ 1628.2

$= 1.25 \times 2/(0.73)^{2/3}$ (28–2)

$= 3.1$ Table 16–J

Use $C = 2.75$ maximum

THE CORRECT ANSWER IS (D)

25. $C = 1.25\ S/T^{2/3}$

$= 1.25 \times 1/2^{2/3}$ (28–2)

$= 0.79$ Table 16–J

Use $C = 0.075R_W$ Table 16–N

$= 0.075 \times 12$

$= 0.9$ minimum

THE CORRECT ANSWER IS (E)

26. $C = 1.25S/T^{2/3}$ (28–2)

$= 1.25 \times 1.5/(0.73)^{2/3}$ Table 16–J

$= 2.3$

THE CORRECT ANSWER IS (C)

27. $V = (ZIC/R_W)W$ (28–1)

$= (0.4 \times 1.25 \times 2.75/6)W$ Table 16–I

$= 0.23W$ Table 16–K

$= 23\%\ W$ Table 16–N

THE CORRECT ANSWER IS (C)

28. Regular structure: $h_n < 240$ feet 1627.8.2

Irregular structure: $h_n \leq 65$ feet 1627.8.3.4

Provided that, if the structure is located on soil profile type S_4, its period does not exceed 0.7 seconds.

THE CORRECT ANSWER IS (A) + (C)

29. In zone 4, the natural period determined by Method B is limited to — 1628.2.2, 1628.2.1

$T = 1.3T_A$
$= 1.3 \times 0.6$
$= 0.78$ seconds ... governs
$< T_B$

The seismic coefficient is given by

$C = 1.25S/T^{2/3}$
$= 1.25 \times 1.5/(0.78)^{2/3}$
$= 2.21$

THE CORRECT ANSWER IS (C)

30. Maximum allowable drift is given by — 1628.8.2, 1625

$\Delta = 0.004 \times 12 \times 12 = 0.58$ inches
or $\Delta = 0.03 \times 12 \times 12/8 = 0.54$ inches ... governs

Maximum allowable story drift ratio is given by

$\Delta_R = \Delta/h_s$
$= 0.54/(12 \times 12)$
$= 0.00375$

THE CORRECT ANSWER IS (B)

31. Required movement is — 1631.2.4.2

$\delta = \Delta \times 3R_W/8$
$= 0.15 \times 3 \times 8/8$
$= 0.45$ inches
Use $\delta = 0.5$ inches, minimum

THE CORRECT ANSWER IS (C)

32. For zone 4:

Story drift ratio is — 1628.9

$\Delta_R = \Delta/h_s$
$= 0.5/(12 \times 12)$
$= 0.00347$
$0.02/R_W = 0.02/12$
$= 0.00167$
$< \Delta_R$

Hence consider $P - \Delta$ effects for Zone 4.

THE CORRECT ANSWER IS (A)

For Zone 2B:
Primary moment is
$M_p = V \times h_s$
$= 3 \times 12 \times 12$
$= 432$ kips inches
Secondary moment is
$M_s = P \times \Delta$
$= 100 \times 0.5$
$= 50$ kips inches
Moment ratio is
$M_s/M_p = 50/432$
$= 0.12$
> 0.1
Hence, consider $P - \Delta$ effects for zone 2B.

THE CORRECT ANSWER IS (A)

33. Required moment is — 1628.10
$M = 0.5ZW_pL^2/2$
$= 0.5 \times 0.4 \times 0.5 \times 20^2/2$
$= 20$ kips feet

THE CORRECT ANSWER IS (D)

34. Maximum allowable self weight moment is — 1628
$M_s = 0.5wL^2/8$ — 1628.10
$= 0.5 \times 400 \times 20^2 \times 12/8$
$= 120{,}000$ pound inches
Maximum allowable bottom fibre stress is
$f_b = M_s/S$
$= 120{,}000/4{,}000$
$= 30$ pounds per square inch

THE CORRECT ANSWER IS (E)

35. 90% $V_B = 0.9 \times 2000$ — 1629.5.3
$= 1800$ kips
80% $V_A = 0.8 \times 2500$
$= 2000$ ... governs
$> V_B$
Scaling Factor is
$R = 0.8V_A/V_D$
$= 2000/1500$
$= 1.33$

THE CORRECT ANSWER IS (C)

36.	$F_p = ZI_p(2C_p/3)W_p$	1630.2
	$= 0.4 \times 1.5 \times (2 \times 0.75/3) \times 10$	(30–1)
	$= 3$ kips	Table 16–O
	Check as an independent structure:	
	$V = 0.5ZIW$	1632.2
	$= 0.5 \times 0.4 \times 1.5 \times 10$	(32–1)
	$= 3$ kips	
	Hence, design lateral force = 3 kips	

THE CORRECT ANSWER IS (C)

37.	$F_p = ZI_p(2C_p)W_p$	1630.2
	$= 0.4 \times 1 \times (2 \times 0.75) \times 1$	(30–1)
	$= 0.6$ kips	Table 16–O

THE CORRECT ANSWER IS (E)

38. $F_p = 0.35ZIw_p$ minimum — 1631.2.9.2

THE CORRECT ANSWER IS (C)

39. 50% increase — Table 16–O

THE CORRECT ANSWER IS (D)

40. No stress increase is allowed — 1631.2.9.6

THE CORRECT ANSWER IS (A)

41. 4:1 — Table 23-I-I

THE CORRECT ANSWER IS (E)

42.	Parapet weight is	1630.2
	$W_p = 150 \times 5/2$	(30–1)
	$= 375$ pounds per linear foot	Table 16–O
	Seismic force is	
	$F_p = ZI_pC_pW_p$	
	$= 0.3 \times 1.25 \times 2 \times 375$	
	$= 281$ pounds per linear foot	
	Bending moment is	
	$M = 2.5$	
	$F_p = 2.5 \times 281$	
	$= 703$ pounds feet per linear foot	

THE CORRECT ANSWER IS (C)

43. Seismic coefficient is — 1632.5, Table 16-P, (28-1), (28-2)

$C = 1.25S/T^{2/3}$
$= 1.25 \times 1.5/(0.73)^{2/3}$
$= 2.3$
< 2.75 ... OK

Use minimum

$C = 0.5R_W$
$= 0.5 \times 5$
$= 2.5$

Lateral force is

$V = (ZIC/R_W)W$
$= (0.4 \times 1 \times 2.5/5) \times 1$
$= 0.2$ kips

THE CORRECT ANSWER IS (C)

44. Base shear is — Table 16-K, Table 16-P

$V = WS_aI/R_W$
$= 60 \times 1.2 \times 1.25/3$
$= 30$ kips

THE CORRECT ANSWER IS (D)

45. Each column forms part of two intersecting lateral force systems. Allowing for orthogonal effects, the overturning moment on one column is — 1631.1

$M_O = 1.3VH/2$
$= 1.3 \times 30 \times 35/2$
$= 682.5$ kips feet

Reduced dead load is

$W_E = 0.85W$
$= 0.85 \times 60$
$= 51$ kips

The restoring moment on one column is

$M_R = W_E \times B/4$
$= 51 \times 10/4$
$= 127.5$ kips feet

Net uplift is

$T = (M_o - M_R)/B$
$= (682.5 - 127.5)/10$
$= 55.5$ kips

THE CORRECT ANSWER IS (B)

46. Unit load produces the deflection

$\delta = [4(H/L)^3 + 3(H/L)]/Et$

$\delta_A = [4(16/4)^3 + 3(16/4)]/Et$
$= 268/Et$

$\delta_B = [4(16/16)^3 + 3(16/16]/Et$
$= 7/Et$

Ratio of shear forces is

$V_B/V_A = \delta_A/\delta_B$
$= 268/7$
$= 38.3$

THE CORRECT ANSWER IS (C)

47. The ratio of the shear forces is

$V_B/V_A = L_B/L_A$
$= 16/4$
$= 4$

THE CORRECT ANSWER IS (D)

48. Roof weight is

$W_R = 50 \times 100 \times 50/1000$
$= 250$ kips

East wall weight is

$W_E = 60 \times 33 \times 16/1000$
$= 31.7$ kips

West wall weight is

$W_W = 60 \times 50 \times 12/1000$
$= 36$ kips

North plus south wall weight is

$W_{NS} = 2 \times 60 \times 40 \times 12/1000$
$= 57.6$

Total weight is

$W_T = 250 + 31.7 + 36 + 57.6 + 50$
$= 425.3$ kips

Distance of center of mass from the west wall is

$\overline{x} = (250 \times 50 + 31.7 \times 100 + 36 \times 0 + 57.6 \times 20 + 50 \times 20)/425.3$
$= 41.9$ feet

THE CORRECT ANSWER IS (B)

49. Distance of center of rigidity from west wall is

$x_R = R_E \times 100/(R_E + R_W)$
$= 8 \times 100/(8 + 12)$
$= 40$ feet

THE CORRECT ANSWER IS (E)

50. Accidental eccentricity is

1628.5

$e_a = 0.05 \times L$
$= 0.05 \times 100$
$= 5$ feet

Total eccentricity is

$e_T = e_x + e_a$
$= \bar{x} - x_R + e_a$
$= 41.9 - 40 + 5$
$= 6.9$ feet

The design torsion is

$M = Ve_T$
$= 100 \times 6.9$
$= 690$ kips feet

THE CORRECT ANSWER IS (C)

This page left blank intentionally.

APPENDIX III

Structural design programs for the HP-28S calculator

SECTION A.1:	**CONCRETE DESIGN PROGRAMS**
A.1.1: TRD:	Design reinforced concrete beam with tension reinforcement only.
A.1.2: TRA:	Analysis of reinforced concrete beam with tension reinforcement only.
A.1.3: CRD:	Design reinforced concrete beam with compression reinforcement.
A.1.4: CRA:	Analysis of reinforced concrete beam with compression reinforcement.
A.1.5: TBD:	Design of reinforced concrete T-beam.
A.1.6: TBA:	Analysis of reinforced concrete T-beam.
SECTION A.2:	**MASONRY DESIGN PROGRAMS**
A.2.1: EAS:	Analysis of masonry beam to determine stresses.
A.2.2: EAM:	Analysis of masonry beam to determine allowable moment.
A.2.3: EDC:	Design of masonry beam with compression reinforcement.
SECTION A.3:	**WOOD DESIGN PROGRAMS**
A.3.1: HANK:	Hankinson's formula.
A.3.2: BEAM:	Allowable bending design stress.
A.3.3: COL:	Allowable compression design stress.
SECTION A.4:	**STEEL DESIGN PROGRAMS**
A.4.1: FB:	Allowable stress in slender beam.
A.4.2: CBX:	Combined axial compression and bending.
A.4.3: MPCO:	Plastic moment with combined axial compression and bending.
A.4.4: STPL:	Column stiffener plates.
SECTION A.5:	**PROPERTIES OF SECTIONS**
A.5.1: IXX:	Moment of inertia of composite sections.
A.5.2: RIG:	Rigidity of walls.

SECTION A.1: CONCRETE DESIGN PROGRAMS

A.1.1: TRD: DESIGN REINFORCED CONCRETE BEAM WITH TENSION REINFORCEMENT ONLY

PROGRAM LISTING:

KEYSTROKES	COMMENTARY
MEMORY HOME ' CONC CRDIR	Create parent directory, CONC.
USER CONC ' TRD MEMORY CRDIR	Create subdirectory, TRD.
USER TRD	Open current directory, TRD.
≪ ' 12 * M/(B * D^2) ' EVAL ENTER ' K STO	Define the parameter, K. Store K.
≪ ' (1-√(1-2 * K/(.9 * .85 * FC))) * .85 * FC/FY ' EVAL ENTER ' R STO	Define the reinforcement ratio, R. (ie. ρ of reference 1). Store R.
≪ IF FC 4 < THEN .85 ELSE IF FC 8 > THEN .65 ELSE ' .85-(FC-4)/20 ' EVAL END END ENTER ' B1 STO	Define the factor, B1 (i.e. β_1 of reference 2). Store B1.
≪ ' .75 * .85 * 87 * B1 * FC/(FY * (87+FY)) ' EVAL ENTER ' RMAX STO	Define the maximum reinforcement ratio, RMAX (i.e. ρ_{max} of reference 3). Store RMAX.
≪ ' .2/FY ' EVAL ENTER ' RMIN STO	Define the minimum reinforcement ratio, RMIN (i.e. ρ_{min} of reference 4). Store RMIN.
≪ ' R * B * D ' EVAL ENTER ' AS STO	Define reinforcement area, AS. Store AS.
≪ ' .85 * 2 * B * D * √(1000 * FC)/1000 ' EVAL ENTER ' PHVC STO	Define concrete design shear strength, PHVC (i.e. ϕV_c of reference 5). Store PHVC.

≪ ' 12 * 50 * B/(1000 * FY) ' EVAL ENTER	Define minimum shear reinforcement AVS (ie. A_v/s of reference 6).
' AVS STO	Store AVS.
≫ {STO M FY FC B D} MENU ENTER	Set up an input menu for data entry
' DAT STO	Store the data menu

EXAMPLE

MEMORY HOME USER CONC TRD	Recall program TRD
DAT	Prepare data entry.
53.5 M	Required factored moment, kip ft.
60 FY	Reinforcement yield strength, ksi
3 FC	Concrete compressive strength, ksi.
12 B	Beam width, in.
14.4 D	Effective depth of reinforcement, in.
USER NEXT NEXT	Display output functions.
K	= 0.258 ksi
R	Reinforcement ratio, ρ = 0.0051
NEXT NEXT	
B1	Compression zone factor β_1 = 0.85
RMAX	Maximum reinforcement ratio, ρ_{max} = 0.016
RMIN	Minimum reinforcement ratio ρ_{min} = 0.0033
AS	Reinforcement area A_s = 0.878 in^2
PHVC	Concrete design shear strength, ϕV_c = 16.09 kip
AVS	Minimum required shear reinforcement A_v/s = 0.12 in^2/ft.

A.1.2: TRA: ANALYSIS OF REINFORCED CONCRETE BEAM WITH TENSION REINFORCEMENT ONLY

PROGRAM LISTING:

KEYSTROKES	COMMENTARY
MEMORY HOME USER CONC	Open parent directory, CONC.
' TRA MEMORY CRDIR	Create subdirectory, TRA.
USER TRA	Open current directory, TRA.
≪ ' AS/(B * D) ' EVAL ENTER ' R STO	Define the reinforcement ratio, R. Store R.
≪'.9 * AS * FY * D * (1-.59 * R * FY/ FC)/12 ' EVAL ENTER ' M STO	Define the design moment strength, M. (i.e ϕM_n of reference 1). Store M.
≪ IF FC 4 < THEN .85 ELSE IF FC 8 > THEN .65 ELSE ' .85-(FC-4)/20 ' EVAL END END ENTER ' B1 STO	Define the factor, B1 (i.e. β_1 of reference 2). Store B1.
≪ ' .75 * .85 * 87 * B1 * FC/(FY * (87+ FY)) ' EVAL ENTER ' RMAX STO	Define the maximum reinforcement ratio, RMAX (i.e. ρ_{max} of reference 3). Store RMAX.
≪ ' .2/FY ' EVAL ENTER ' RMIN STO	Define the minimum reinforcement ratio, RMIN (i.e. ρ_{min} of reference 4). Store RMIN
≪ ' .85 * 2 * B * D * √(1000 * FC)/1000 ' EVAL ENTER ' PHVC STO	Define concrete design shear strength, PHVC (i.e. ϕV_c of reference 5). Store PHVC
≪ ' 12 * 50 * B/(1000 * FY) ' EVAL ENTER ' AVS STO	Define minimum shear reinforcement, AVS (i.e. A_v/s of reference 6). Store AVS
≪ {STO AS B D FY FC} MENU ENTER ' DAT STO	Set up an input menu for data entry. Store the data menu.

EXAMPLE

MEMORY HOME USER CONC TRA	Recall program TRA.
DAT	Prepare for data entry.
.878 AS	Reinforcement area, in^2.
12 B	Beam width, in.
14.4 D	Effective depth of reinforcement, in.
60 FY	Reinforcement yield strength, ksi.
3 FC	Concrete compressive strength, ksi.
USER NEXT NEXT	Display output functions.
R	Reinforcement ratio, ρ = 0.0051.
NEXT NEXT	
M	Design moment strength ϕM_n = 53.5 kip ft.
B1	Compression zone factor, β_1 = 0.85
RMAX	Maximum reinforcement ratio, ρ_{max} = 0.016
RMIN	Minimum reinforcement ratio, ρ_{min} = 0.0033
PHVC	Concrete design shear strength, ϕV_c = 16.09 kip.
AVS	Minimum required shear reinforcement, A_v/s = 0.12 in^2/ft.

The actual reinforcement ratio lies between the maximum and minimum allowable values and is satisfactory.

A.1.3: CRD: DESIGN REINFORCED CONCRETE BEAM WITH COMPRESSION REINFORCEMENT

PROGRAM LISTING:

KEYSTROKES	COMMENTARY
MEMORY HOME USER CONC	Open parent directory, CONC.
' CRD MEMORY CRDIR	Create subdirectory, CRD.
USER CRD	Open current directory, CRD.

```
« IF FC 4 < THEN .85 ELSE IF FC 8
> THEN .65 ELSE ' .85-(FC-4)/20 '
EVAL END END ENTER
' B1 STO
```

Define compression zone factor, B1 (i.e. β_1 of reference 2).
Store B1.

```
« ' .75 * .85 * 87 * B1 * FC/(FY *
(87+FY)) ' EVAL ENTER
' RMAX STO
```

Define the maximum reinforcement ratio for a singly reinforced section, RMAX
Store RMAX.

```
« ' RMAX * B * D ' EVAL ENTER
' AMAX STO
```

Define the maximum reinforcement area for a singly reinforced section, AMAX.
Store AMAX.

```
« ' .9 * AMAX * FY * D * (1-.59 *
RMAX * FY/ FC)/12 ' EVAL ENTER
' MMAX STO
```

Define the maximum design moment strength for a singly reinforced section.
Store MMAX.

```
« ' M-MMAX ' EVAL ENTER
' MR STO
```

Define the residual moment MR (i.e. factored applied moment - MMAX).
Store MR.

```
« IF ' 87-D1 * (87+FY)/D ' EVAL FY
> THEN FY ELSE ' 87-D1 *
(87+FY)/D ' EVAL END ENTER
' FSC STO
```

Define the stress in compression reinforcement at balanced strain conditions (i.e. f'_{sb} of reference 1), FSC.
Store FSC.

```
« ' 12 * MR/(.9 * FSC * (D-D1)) '
EVAL ENTER
' ASC STO
```

Define the required area of compression reinforcement, ASC.
Store ASC.

```
« ' AMX + ASC * FSC/FY ' EVAL
ENTER
' AST STO
```

Define the required area of tension reinforcement, AST.
Store AST

```
« ' .85 * 2 * B * D * √(1000 * FC)/1000 '
EVAL ENTER
' PHVC STO
```

Define concrete design shear strength, PHVC (i.e. ϕV_c of reference 5).
Store PHVC.

```
« ' 12 * 50 * B/(1000 * FY) ' EVAL
ENTER
' AVS STO
```

Define minimum shear reinforcement AVS (i.e. A_v/s of reference 6).
Store AVS.

```
« {STO M FY FC B D D1} MENU
ENTER
' DAT STO
```

Set up an input menu for data entry.
Store the data menu.

EXAMPLE

MEMORY HOME USER CONC CRD	Recall program CRD
NEXT DAT	Prepare for data entry.
190 M	Required factored moment, kip ft.
60 FY	Reinforcement yield strength, ksi.
5 FC	Concrete compressive strength, ksi.
12 B	Beam width, in.
12 D	Effective depth of reinforcement, in.
2.4 D1	Depth to compression reinforcement, in.
USER NEXT NEXT	Display output functions.
B1	Compression zone factor, $\beta_1 = 0.85$.
RMAX	Maximum reinforcement ratio, $\rho_{max} = 0.025$
AMAX	Maximum tension reinforcement area for singly reinforced beam = 3.62 in^2
MMAX	Maximum design moment strength for a singly reinforced section = 160.76 kip ft.
MR	Residual moment = 29.24 kip ft.
NEXT NEXT	
FSC	Stress in compression reinforcement = 57.6 ksi.
ASC	Compression reinforcement area = 0.705 in^2
AST	Tension reinforcement area = 4.30 in^2
PHVC	Concrete design shear strength, ϕV_c = 17.31 kip.
AVS	Minimum required shear reinforcement, $A_v/s = 0.12$ in^2/ft.

A.1.4: CRA: ANALYSIS OF REINFORCED CONCRETE BEAM WITH COMPRESSION REINFORCEMENT

PROGRAM LISTING:

KEYSTROKES	COMMENTARY
MEMORY HOME USER CONC	Open parent directory, CONC.
' CRA MEMORY CRDIR	Create subdirectory, CRA.
USER CRA	Open current directory, CRA.

```
≪ IF FC 4 < THEN .85 ELSE IF FC 8
> THEN .65 ELSE ' .85-(FC-4)/20 '
EVAL END END ENTER
' B1 STO
```

Define compression zone factor, B1 (i.e. β_1 of reference 2)

Store B1

```
≪ IF ' 87-D1 * (87+FY)/D ' EVAL
FY > THEN FY ELSE ' 87-D1 *
(87+FY)/D ' EVAL END ENTER
' FSC STO
```

Define the stress in compression reinforcement at balanced strain conditions, FSC (ie. f'_{sb} of reference 1). Store FSC.

```
≪ ' 75 * .85 * 87 * B1 * FC/(FY * (87+
FY))+ASC * FSC/(B * D * FY) ' EVAL
ENTER
' RMAX STO
```

Define the maximum reinforcement ratio, RMAX (i.e. ρ_{max} of reference 3).

Store RMAX

```
≪ ' AST/(B * D) ' EVAL ENTER
' R STO
```

Define the reinforcement ratio, R. Store R.

```
≪ ' .85 * FC * B * C * B1 ' EVAL
ENTER
' CC STO
```

Define the compressive force in the concrete compression zone, CC. Store CC.

```
≪ ' ASC * 87 * (1-D1/C) ' EVAL
ENTER
' CS STO
```

Define the compressive force in the compression reinforcement, CS. Store CS.

```
≪ ' AST * FY ' EVAL ENTER

' T STO
```

Define the tensile force, T, in the tension reinforcement.
Store T.

```
≪ ' .9 * (CC * (D-B1 * C/2)+CS *
(D-D1))/12 ' EVAL ENTER
' M STO
```

Define the design moment strength, M(i.e. ϕM_n of reference 1). Store M.

```
≪ ' .85 * 2 * B * D * √(1000 * FC)/1000 '
EVAL ENTER
' PHVC STO
```

Define concrete design shear strength, PHVC (i.e. ϕV_c of reference 5) Store PHVC

```
≪ ' 12 * 50 * B/(1000 * FY) ' EVAL
ENTER
' AVS STO
```

Define minimum shear reinforcement, AVS (i.e. A_v/s of reference 6). Store AVS.

```
≪ {STO AST ASC FY FC B D D1 C}
MENU ENTER
' DAT STO
```

Set up an input menu for data entry.

Store the data menu.

EXAMPLE

MEMORY HOME USER CONC CRA	Recall program CRA
NEXT DAT	Prepare for data entry.
9.36 AST	Tension reinforcement area, in^2
6.24 ASC	Compression reinforcement area in^2
60 FY	Reinforcement yield strength, ksi.
4 FC	Concrete compressive strength, ksi.
24 B	Beam width, in.
27 D	Effective depth of reinforcement, in.
NEXT	
3 D1	Depth to compression reinforcement.
4.98 C	Initial estimate of depth to neutral axis, in.
USER NEXT NEXT NEXT	Display output functions.
B1	Compression zone factor, $\beta_1 = 0.85$.
NEXT NEXT NEXT	
FSC	Compression reinforcement stress= 60 ksi
RMAX	Maximum reinforcement ratio,ρ_{max}= 0.031
R	Reinforcement ratio, $\rho = 0.014 < \rho_{max}$...ok
CC	Compressive force in the concrete compressive zone = 345.4 kip.
CS	Compressive force in the compression reinforcement = 215.8 kip.
T	Tensile force in the tension reinforcement = 561.6 kip ≈ CC + CS Hence, the initial estimate for C is correct.
NEXT NEXT NEXT	
M	Design moment strength,ϕM_n=1033 kip ft
PHVC	Concrete design shear strength, ϕV_c = 69.67 kip.
AVS	Minimum required shear reinforcement, $A_v/s = 0.24\ in^2/ft$.

A.1.5: TBD: DESIGN OF REINFORCED CONCRETE T-BEAM

(Note: Assumes flange thickness is less than depth of the equivalent rectangular stress block)

PROGRAM LISTING:

KEYSTROKES	COMMENTARY
MEMORY HOME USER CONC	Open parent directory, CONC.
' TBD MEMORY CRDIR	Create subdirectory, TBD.
USER TBD	Open current directory, TBD.
≪ ' .85 * FC * (B–BW) * HF ' EVAL ENTER	Define the compressive force in the flange, CF (i.e. C_f of reference 1).
' CF STO	Store CF
≪ ' CF/FY ' EVAL ENTER	Define the reinforcement area, ASF, required to balance the flange force (i.e. A_{sf} of reference 1).
' ASF STO	Store ASF.
≪ ' .9 * ASF * FY * (D–.5 * HF)/12 ' EVAL ENTER	Define the design moment strength, MF, due to ASF.
' MF STO	Store MF.
≪ ' M–MF ' EVAL ENTER	Define the residual moment MW (i.e. the factored applied moment – MF)
' MW STO	Store MW.
≪ ' 12 × MW/(BW * D^2) ' EVAL ENTER	Define the parameter, KW.
' KW STO	Store KW.
≪ ' (1–√(1–2 * KW/(.9 * .85 * FC))) * .85 * FC/FY ' EVAL ENTER	Define the web reinforcement ratio, RW.
' RW STO	Store RW.
≪ ' RW * BW * D ' EVAL ENTER	Define the reinforcement area, ASW required to resist the residual moment.
' ASW STO	Store ASW.
≪ ' ASF + ASW ' EVAL ENTER	Define the total required area of tension reinforcement, AS.
' AS STO	Store AS.

« IF FC 4 < THEN .85 ELSE IF FC 8 > THEN .65 ELSE ' .85-(FC-4)/20 ' EVAL END END ENTER	Define the factor, B1 (i.e. β_1 of reference 2).
' B1 STO	Store B1.
« ' .85 * 87 * B1 * FC/(FY * (87+FY)) ' EVAL ENTER	Define the web balanced reinforcement ratio, RB (i.e. $\bar{\rho}_b$ of reference 3).
' RB STO	Store RB.
« ' ASF/(BW * D) ' EVAL ENTER	Define the reinforcement ratio, RF, due to ASF (i.e. ρ_f of reference 1).
' RF STO	Store RF.
« ' .75 * BW * (RB+RF)/B ' EVAL ENTER	Define the maximum reinforcement ratio, RMAX (i.e. ρ_{max} of reference 3).
' RMAX STO	Store RMAX
« ' AS/(B * D) ' EVAL ENTER	Define the reinforcement ratio, R(i.e. ρ of reference 1).
' R STO	Store R.
« ' ASW * FY/(.85 * FC * BW) ' EVAL ENTER	Define the depth of the equivalent rectangular stress block, A (i.e. a of reference 2).
' A STO	Store A.
« ' .85 * 2 * BW * D * √(1000 * FC)/ 1000 ' EVAL ENTER	Define concrete design shear strength, PHVC (i.e. ϕV_c of reference 5).
' PHVC STO	Store PHVC.
« ' 12 * 50 * BW/(1000 * FY) ' EVAL ENTER	Define the minimum shear reinforcement, AVS (i.e. A_v/s of reference 6).
' AVS STO	Store AVS.
« {STO M FY FC B BW D HF} MENU ENTER	Set up an input menu for data entry.
' DAT STO	Store the data menu.

EXAMPLE

MEMORY HOME USER CONC TBD	Recall program TBD.
NEXT DAT	Prepare for data entry.
400 M	Required factored moment, kip ft.
60 FY	Reinforcement yield strength, ksi.
4 FC	Concrete compressive strength, ksi.
30 B	Flange width, in.
10 BW	Web width, in.
19 D	Effective depth of reinforcement, in.
NEXT	
2.5 HF	Flange depth, in.
USER NEXT NEXT NEXT	Display output functions.
CF	Flange compressive force C_f = 170 kip.
ASF	Reinforcement area, A_{sf} = 2.83 in^2
MF	Design moment due to A_{sf} = 226 kip ft.
MW	Residual moment = 174 kip ft.
KW	KW = 0.577
RW	Web reinforcement ratio = 0.012
NEXT NEXT NEXT	
ASW	Reinforcement area, ASW = 2.27 in^2
AS	Total reinforcement area = 5.1 in^2
B1	Compression zone factor, β_1 = 0.85
RB	Reinforcement ratio, $\bar{\rho}_b$ = 0.0285
RF	Reinforcement ratio, ρ_f = 0.0149
RMAX	Reinforcement ratio, ρ_{max} = 0.0109
NEXT NEXT NEXT	
R	Actual reinforcement ratio, ρ = 0.0090 $< \rho_{max}$ .....ok
A	Depth of stress block, a = 4.01" $> h_f$ ...ok
PHVC	Concrete design shear strength ϕV_c = 20.43 kip.
AVS	Minimum required shear reinforcement A_v / s = 0.0996 in^2 / ft.

A.1.6: TBA: ANALYSIS OF REINFORCED CONCRETE T-BEAM

(Note: Assumes flange thickness is less than depth of the equivalent rectangular stress block)

PROGRAM LISTING:

KEYSTROKES	COMMENTARY
MEMORY HOME USER CONC	Open parent directory, CONC.
' TBA MEMORY CRDIR	Create subdirectory, TBA.
USER TBA	Open current directory, TBA
≪ ' .85 * FC * (B–BW) * HF ' EVAL ENTER	Define the compressive force in the flange, CF(i.e. C_f of reference 1).
' CF STO	Store CF.
≪ ' CF/FY ' EVAL ENTER	Define the reinforcement area, ASF, required to balance the flange force (i.e. A_{st} of reference 1).
' ASF STO	Store ASF.
≪ ' .9 * ASF * FY * (D–.5 * HF)/12 ' EVAL ENTER	Define the design moment strength, MF, due to ASF.
' MF STO	Store MF.
≪ ' AS - ASF ' EVAL ENTER	Define the reinforcement area, ASW, required to resist the residual moment.
' ASW STO	Store ASW.
≪ ' ASW/(BW * D) ' EVAL ENTER	Define the web reinforcement ratio, RW, required to resist the residual moment.
' RW STO	Store RW.
≪ ' .9 * ASW * FY * D * (1–.59 * RW * FY/FC)/12 ' EVAL ENTER	Define the residual moment, MW.
' MW STO	Store MW.
≪ ' MF + MW ' EVAL ENTER	Define the design moment strength, M. (i.e.ϕM_n of reference 1).
' M STO	Store M

« IF FC 4 < THEN .85 ELSE IF FC 8 > THEN .65 ELSE ' .85-(FC-4)/20 ' EVAL END END ENTER	Define the factor, B1 (i.e. β_1 of reference 2).
' B1 STO	Store B1.
« ' .85 * 87 * B1 * FC/(FY * (87+FY)) ' EVAL ENTER	Define the web balanced reinforcement ratio, RB(i.e. $\bar{\rho}_b$ of reference 3).
' RB STO	Store RB.
« ' ASF/(BW * D) ' EVAL ENTER	Define the reinforcement ratio, RF, due to ASF (i.e. ρ_f of reference 1).
' RF STO	Store RF.
« ' .75 * BW * (RB+RF)/B ' EVAL ENTER	Define the maximum reinforcement ratio, RMAX (i.e. ρ_{max} of reference 3).
' RMAX STO	Store RMAX.
« ' AS/(B * D) ' EVAL ENTER	Define the reinforcement ratio, R(i.e. ρ of reference 1).
' R STO	Store R.
« ' RW * FY * D/(.85 * FC) ' EVAL ENTER	Define the depth of the equivalent rectangular stress block, A (i.e. a of reference 2).
' A STO	Store A
« ' .85 * 2 * BW * D * √(1000 * FC) ' EVAL ENTER	Define concrete design shear strength, PHVC (i.e. A_v/s of reference 6).
' PHVC STO	Store PHVC
« ' 12 * 50 * BW/(1000 * FC) ' EVAL ENTER	Define the minimum shear reinforcement, AVS (i.e. A_v/s of reference 6).
' AVS STO	Store AVS.
« {STO AS FY FC B BW D HF} MENU ENTER	Set up an input menu for the data entry.
' DAT STO	Store the data menu

EXAMPLE

Keystrokes	Description
MEMORY HOME USER CONC TBA	Recall program TBA.
NEXT DAT	Prepare for data entry.
5 AS	Reinforcement area, in^2.
60 FY	Reinforcement yield strength, ksi.
4 FC	Concrete compressive strength, ksi.
30 B	Flange width, in.
10 BW	Web width, in.
19 D	Effective depth of reinforcement, in.
NEXT	
2.5 HF	Flange depth, in.
USER NEXT NEXT NEXT	Display output functions.
CF	Flange compressive force, C_f = 170 kip.
ASF	Reinforcement area, A_{sf} = 2.83 in^2.
MF	Design moment due to A_{sf} = 226 kip ft.
ASW	Reinforcement area, ASW = 2.17 in^2.
RW	Web reinforcement ratio = 0.011.
MW	Residual moment = 167 kip ft.
NEXT NEXT NEXT	
M	Design moment strength, ϕM_n = 393 kip ft.
B1	Compression zone factor, β_1 = 0.85.
RB	Reinforcement ratio, ρ_b = 0.0285.
RF	Reinforcement ratio, ρ_f = 0.0149.
RMAX	Reinforcement ratio, ρ_{max} = 0.0109
R	Actual reinforcement ratio, ρ = 0.00877 $< \rho_{max}$ .....ok.
NEXT NEXT NEXT	
A	Depth of stress block, a = 3.82 in $> h_f$ ....ok.
PHVC	Concrete design shear strength, ϕV_c = 20.43 kip.
AVS	Minimum required shear reinforcement, A_v/s = 0.0996 in^2/ft

SECTION A.2: MASONRY DESIGN PROGRAMS

A.2.1: EAS: ANALYSIS OF MASONRY BEAM TO DETERMINE STRESSES

PROGRAM LISTING:

KEYSTROKES	COMMENTARY
MEMORY HOME ' MAS CRDIR	Create parent directory, MAS.
USER MAS ' EAS MEMORY CRDIR	Create subdirectory, EAS.
USER EAS	Open current directory, EAS.
≪ ' ES/EM ' EVAL ENTER	Define the modular ratio, N (i.e n of reference 7).
' N STO	Store N.
≪ ' AS/(B * D) ' EVAL ENTER	Define the reinforcement ratio, P (i.e. ρ of reference 7).
' P STO	Store P.
≪ ' √(2 * N * P + P^2 * N^2)-N * P ' EVAL ENTER	Define the neutral axis depth ratio, K (i.e. k of equation (7–33), reference 8).
' K STO	Store K.
≪ ' 1-K/3 ' EVAL ENTER	Define lever arm ratio, J (i.e. j of equation (7–35), reference 8).
' J STO	Store J.
≪ ' 24 * M/(J * K * B * D^2) ' EVAL ENTER	Define the compressive stress in the masonry, FM (i.e. f_b of equation (7–31), reference 8).
' FM STO	Store FM.
≪ ' 12 * M/(AS * J * D) ' EVAL ENTER	Define the tensile stress in the reinforcement, FS (i.e. f_s of equation (7–32), reference 8).
' FS STO	Store FS.
≪ {STO M AS ES EM B D} MENU ENTER	Set up an input menu for data entry.
' DAT STO	Store the data menu.

EXAMPLE

MEMORY HOME USER MAS EAS	Recall program EAS.
NEXT DAT	Prepare data entry.
30 M	Applied moment, kip ft.
.88 AS	Reinforcement area, in^2.
29000 ES	Modulus of elasticity of reinforcement, ksi
1100 EM	Modulus of elasticity of the masonry, ksi.
9.63 B	Beam width in.
34 D	Effective depth of reinforcement, in.
USER NEXT NEXT	Display output functions.
N	Modular ratio, n = 26.36
NEXT NEXT	
P	Reinforcement ratio, p = 0.0027.
K	Neutral axis depth ratio, k = 0.312.
J	Lever arm ratio, j = 0.896.
FM	Masonry compressive stress f_b= 0.23 ksi.
FS	Reinforcement tensile stress f_s= 13.43 ksi

A.2.2: EAM: ANALYSIS OF MASONRY BEAM TO DETERMINE ALLOWABLE MOMENT

PROGRAM LISTING:

KEYSTROKES	COMMENTARY
MEMORY HOME USER MAS	Open parent directory, MAS.
' EAM MEMORY CRDIR	Create subdirectory, EAM.
USER EAM	Open current directory, EAM.
≪ ' ES/EM ' EVAL ENTER	Define the modular ratio, N (i.e. n of reference 7).
' N STO	Store N.
≪ ' AS/(B * D) ' EVAL ENTER	Define the reinforcement ratio, P (i.e. ρ of reference 7).
' P STO	Store P.
≪ ' √(2 * N * P + P^2 * N^2)-N * P ' EVAL ENTER	Define the neutral axis depth ratio, K (i.e. k of equation (7–33), reference 8).
' K STO	Store K.

≪ ' 1-K/3 ' EVAL ENTER	Define the lever arm ratio, J (i.e. j of equation (7-35), reference 8).
' J STO	Store J.
≪ ' FM * J * K * B * D^2/24 ' EVAL ENTER	Define the moment capacity of the masonry, MM (i.e. M_m of reference 7).
' MM STO	Store MM.
≪ ' FS * AS * J * D/12 ' EVAL ENTER	Define the moment capacity of the reinforcement, MS(i.e. M_s of reference 7).
' MS STO	Store MS.
≪ {STO AS FS FM ES EM B D} MENU ENTER	Set up an input menu for data entry.
' DAT STO	Store the data menu.

EXAMPLE:

MEMORY HOME USER MAS EAM	Recall program EAM.
NEXT DAT	Prepare data entry.
.33 AS	Reinforcement area, in^2.
24 FS	Allowable reinforcement stress, ksi.
.3 FM	Allowable masonry stress, ksi.
29000 ES	Modulus of elasticity of reinforcement, ksi
1350 EM	Modulus of elasticity of the masonry, ksi.
12 B	Beam width, in.
NEXT	
5 D	Effective depth of reinforcement, in.
USER NEXT NEXT	Display output functions
N	Modular ratio, n = 21.48.
P	Reinforcement ratio, p = 0.0055.
K	Neutral axis depth ratio, k = 0.382.
NEXT NEXT	
J	Lever arm ratio, j = 0.873.
MM	Moment capacity of the masonry, M_m = 1.250 kip ft....governs.
MS	Moment capacity of the reinforcement, M_s = 2.880 kip ft.

A.2.3: EDC: DESIGN OF MASONRY BEAM WITH COMPRESSION REINFORCEMENT

PROGRAM LISTING:

KEYSTROKES	COMMENTARY
MEMORY HOME USER MAS	Open parent directory, MAS.
' EDC MEMORY CRDIR	Create subdirectory, EDC.
USER EDC	Open current directory, EDC.
« ' ES/EM ' EVAL ENTER	Define the modular ratio, N (i.e. n of reference 7).
' N STO	Store N.
« ' N * FM/(2 * FS * (N+FS/FM)) ' EVAL ENTER	Define the reinforcement ratio, PB, for balanced stress conditions (i.e. p_b of equation (6–7), reference 9).
' PB STO	Store PB.
« ' √(2 * N * PB + PB^2 * N^2)-N * PB ' EVAL ENTER	Define the neutral axis depth ratio, KB, for balanced stress conditions.
' KB STO	Store KB.
« ' 1-KB/3 ' EVAL ENTER	Define the lever arm ratio, JB.
' JB STO	Store JB.
« ' FM * JB * KB * B * D^2/24 ' EVAL ENTER	Define the moment capacity of the masonry, MB, for balanced stress conditions.
' MB STO	Store MB.
« ' 12 * MB/(FS * JB * D) ' EVAL ENTER	Define the tensile reinforcement area, AB, required for balanced stress conditions.
' AB STO	Store AB.
« ' M–MB ' EVAL ENTER	Define the residual moment MR (i.e. the applied moment – MB).
' MR STO	Store MR.

≪ IF ' FS * (KB-D1/D)/(1-KB) ' EVAL FS > THEN FS ELSE ' FS * (KB-D1/D)/(1-KB) ' EVAL END ENTER	Define the stress in the compression reinforcement, FSC, at balanced stress condition.
' FSC STO	Store FSC.
≪ ' 12 * MR/(FSC * (D-D1) * (N-1)/N) ' EVAL ENTER	Define the required area of compression reinforcement, ASC.
' ASC STO	Store ASC
≪ ' AB + 12 * MR/(FS * (D-D1)) ' EVAL ENTER	Define the required area of tension reinforcement, AST.
' AST STO	Store AST.
≪ {STO M FS FM ES EM B D D1} MENU ENTER	Set up an input menu for data entry.
' DAT STO	Store the data menu.

EXAMPLE

MEMORY HOME USER MAS EDC	Recall program EDC.
NEXT DAT	Prepare data entry.
33 M	Applied moment, kip ft.
24 FS	Allowable reinforcement stress, ksi.
.5 FM	Allowable masonry stress, ksi.
29000 ES	Modulus of elasticity of the reinforcement, ksi.
1125 EM	Modulus of elasticity of the masonry, ksi.
7.63 B	Beam width, in.
NEXT	
24.5 D	Effective depth of reinforcement, in.
4.5 D1	Depth to the compression reinforcement, in.
USER NEXT NEXT NEXT	Display output functions.
N	Modular ratio, n = 25.78
NEXT NEXT NEXT	
PB	Balanced reinforcement ratio, p_b = 0.00364
KB	Neutral axis depth ratio = 0.349

JB	Lever arm ratio = 0.884
MB	Moment capacity of the masonry = 29.45 kip ft.
AB	Tension reinforcement area for balanced stress conditions = 0.68 in^2
MR	Residual moment = 3.55
NEXT NEXT NEXT	
FSC	Stress in compression reinforcement = 6.113 ksi.
ASC	Compression reinforcement area = 0.36 in^2
AST	Tension reinforcement area = 0.77 in^2

SECTION A.3: WOOD DESIGN PROGRAMS

A.3.1: HANK: HANKINSON'S FORMULA[10,11]

PROGRAM LISTING:

KEYSTROKES	COMMENTARY
MEMORY HOME ' WOOD CRDIR	Create parent directory, WOOD.
USER WOOD ' HANK MEMORY CRDIR	Create subdirectory, HANK.
USER HANK	Open current directory, HANK.
≪ ' P * Q/(P * (SIN(THET))^2 + Q * (COS(THET))^2) ' EVAL ENTER	Define the dowel bearing strength N at an angle $\theta°$ with the direction of the grain (i.e. $F_{e\theta}$ of reference 10).
' N STO	Store N.
≪ {STO P Q THET} MENU ENTER	Set up an input menu for data entry.
' DAT STO	Store the data menu.

EXAMPLE

MEMORY HOME USER WOOD HANK	Recall program HANK.
DAT	Prepare data entry.
5600 P	Dowel bearing strength parallel to the grain, psi.
2600 Q	Dowel bearing strength perpendicular to the grain, psi.
19 THET	Angle of inclination between the applied load and the direction of the grain, $\theta°$.
USER	Display output functions.
N	Dowel bearing strength $F_{e\theta}$ = 4990 psi.

A.3.2: BEAM: ALLOWABLE BENDING DESIGN STRESS[10,11]

PROGRAM LISTING:

KEYSTROKES	COMMENTARY
MEMORY HOME USER WOOD	Open parent directory, WOOD.
' BEAM MEMORY CRDIR	Create subdirectory, BEAM.
USER BEAM	Open current directory, BEAM.
≪ ' √(12 * LE * D/B^2) ' EVAL ENTER ' RB STO	Define the slenderness ratio. Store R_b
≪ ' FB * CD * CM * CT * CS ' EVAL ENTER ' FAB STO	Define the adjusted tabulated bending value. Store F_b^*
≪ ' 1000000 * E * CM * CT ' EVAL ENTER ' E1 STO	Define the allowable modulus of elasticity. Store E'
≪ ' KBE * E1/RB^2 ' EVAL ENTER ' FBE STO	Define the critical buckling design value. Store F_{bE}
≪ ' FBE/FAB ' EVAL ENTER ' F STO	Define F_{bE}/F_b^* Store this value.
≪ ' (1 + F)/1.9 − √(((1 + F)/1.9)^2 − F/0.95) ' EVAL ENTER ' CL STO	Define the beam slenderness factor. Store C_L
≪ ' KL * (1291.5/(B * D * L))^(1/X) ' EVAL ENTER ' CV STO	Define the volume factor. Store C_V
≪ IF CL VB > THEN ' FAB * VB ' EVAL ELSE ' FAB * CL ' EVAL END ENTER ' F1B STO	Define the allowable bending design stress. Store F_b'
≪ {STO FB E KBE B D LE X KL L CD CM CT CS} MENU ENTER ' DAT STO	Set up an input menu for data entry. Store the data menu.

EXAMPLE

MEMORY HOME USER WOOD BEAM	Recall program BEAM.
NEXT DAT	Prepare data entry.
2400 FB	Tabulated bending stress F_b for a 24F- V5 visually graded DF/HF (western species) glued-laminated beam.
1.5 E	Modulus of elasticity E_{yy} x 10^6, psi.
0.69 KBE	Euler buckling coefficient K_{bE} for glued-laminated beams.
5.125 B	Beam width, in.
36 D	Beam depth, in.
8 * 1.63 + 3 * 3 LE	Effective length l_e for a uniformly loaded, 32 ft span simply supported beam, braced at 8 ft on centers.
NEXT	
10 X	x = 10 for western species glued-laminated beams.
1.0 KL	Loading condition coefficient for distributed loading.
32 L	Length between points of zero moment.
1.25 CD	Load duration factor C_D for roof live loading.
1.0 CM	Wet service factor C_M for interior use.
1.0 CT	Temperature factor C_t for normal temperatures.
NEXT	
1.0 CS	Size factor C_F (not applicable to glued-laminated beams).
USER NEXT NEXT NEXT	Display output functions.
RB	Slenderness ratio R_B = 19.04
FAB	Adjusted tabulated bending stress F_b^* = 3000 psi.
E1	Allowable modulus of elasticity E' = 1.5 $\times 10^6$ psi.
FBE	Critical buckling design value F_{bE} = 2520 psi.
NEXT NEXT NEXT	
CL	Beam stability factor C_L = 0.737
CV	Volume factor C_V = 0.859
F1B	Allowable bending stress F_b' = 2210 psi.

A.3.3: COL: ALLOWABLE COMPRESSION DESIGN STRESS[10,11]

PROGRAM LISTING:

KEYSTROKES	COMMENTARY
MEMORY HOME USER WOOD	Open parent directory, WOOD.
' COL MEMORY CRDIR	Create subdirectory, COL.
USER COL	Open current directory, COL.
≪ ' FC * CD * CM * CT * CS ' EVAL ENTER	Define the adjusted tabulated compression value.
' FAC STO	Store F_c^*
≪ ' 1000000 * E * CM * CT * CTS' EVAL ENTER	Define the allowable modulus of elasticity.
' E1 STO	Store E′
≪ ' KCE * E1/SRA^2 ' EVAL ENTER	Define the critical buckling design value.
' FCE STO	Store F_{cE}
≪ ' FCE/FAC ' EVAL ENTER	Define F_{cE}/F_c^*
' F STO	Store this value.
≪ ' (1 + F)/(2 * C) - √(((1 + F)/(2 * C))^2 - F/C) ' EVAL ENTER	Define the column stability factor.
' CP STO	Store C_P
≪ ' FAC * CP ' EVAL ENTER	Define the allowable compressive design stress.
' F1C STO	Store F_c'
≪ IF 11 SRF > THEN 0 ELSE IF K SRF > THEN ' (SRF-11)/(K-11) ' EVAL ELSE 1 END END ENTER	Define the column stability factor for flexural effects, J(i.e. J of reference 14).
' J STO	Store J.
≪ {STO FC E SRA KCE C CD CM CT CS CTS} MENU ENTER	Set up an input menu for data entry.
' DAT STO	Store the data menu.

EXAMPLE

MEMORY HOME USER WOOD COL	Recall program COL.
DAT	Prepare data entry.
1300 FC	Tabulated compression value F_c Douglas fir-larch no. 2 grade, psi.
1.6 E	Tabulated modulus of elasticity × 10^6, psi.
1 * 12 * 12/7.25 SRA	Slenderness ratio $K_e l/d$ for a 4 × 8 pin ended column, 12 feet long, braced about the weak axis.
0.3 KCE	Euler buckling coefficient K_{cE} for visually graded lumber.
0.8 C	Column parameter c for sawn lumber.
1.0 CD	Load duration factor C_D for normal live load duration.
NEXT	
1.0 CM	Wet service factor C_M for interior use.
1.0 CT	Temperature factor C_t for normal temperatures.
1.05 CS	Size factor C_F
USER NEXT NEXT	Display output functions.
FAC	Adjusted tabulated compression stress F_c^* = 1365 psi.
E1	Allowable modulus of elasticity E′ = 1.6 × 10^6 psi.
FCE	Critical buckling design value F_{cE} = 1217 psi.
CP	Column stability factor C_P = 0.650
NEXT NEXT	
F1C	Allowable compression stress F_c' = 887 psi.

SECTION A.4: STEEL DESIGN PROGRAMS

A.4.1: FB: ALLOWABLE STRESS IN SLENDER BEAM

PROGRAM LISTING:

KEYSTROKES	COMMENTARY
MEMORY HOME ' STEEL CRDIR	Create parent directory, STEEL.
USER STEEL ' FB MEMORY CRDIR	Create subdirectory, FB.
USER FB	Open current directory, FB
≪ ' CB * 12000/(L * DAF) ' EVAL ENTER	Define the allowable bending stress, FB2 (i.e. F_b of equation (F1-8), reference 12).
' FB2 STO	Store FB2
≪ ' L/RT ' EVAL ENTER	Define the factor, LRT (i.e. l/r_T of reference 12).
' LRT STO	Store LRT.
≪ ' √(102000 * CB/FY) ' EVAL ENTER	Define the factor, B0
' B0 STO	Store B0
≪ ' B0 * √5 ' EVAL ENTER	Define the factor, B1.
' B1 STO	Store B1.
≪ IF LRT B0 <THEN ' .6 * FY ' EVAL ELSE IF LRT B1 > THEN ' 170000 * CB/LRT^2 ' EVAL ELSE ' FY * (.667-FY * LRT^2/(1530000 * CB)) ' EVAL END END ENTER	Define the allowable bending stress, FB1 (i.e. F_b of equations (F1-6) & (F1-7), reference 12).
' FB1 STO	Store FB1.
≪ {STO FY L CB DAF RT} MENU ENTER	Set up an input menu for data entry.
' DAT STO	Store the data menu.

EXAMPLE

MEMORY HOME USER STEEL FB	Recall program FB.
DAT	Prepare data entry.
36 FY	Specified yield stress, F_y ksi.
20x12 L	Unbraced length, l, ins.
1 CB	Bending coefficient, C_b.
4.23 DAF	Ratio, depth of beam/area compression flange, d/A_f, in^{-1}.
2.57 RT	Radius of gyration of compression area about the web, r_T.
USER NEXT	Display output functions.
FB2	Allowable bending stress from equation (F1-8), F_b = 11.82 ksi.
LRT	l/r_T = 93.39.
B0	B0 = 53.2 $<$93.39.
B1	B1 = 119.0 $>$ 93.39...equation (F1-6) applies.
FB1	Allowable bending stress from equation (F1-7), F_b = 16.62 ksi....governs

A.4.2: CBX: COMBINED AXIAL COMPRESSION AND BENDING

PROGRAM LISTING:

KEYSTROKES	COMMENTARY
MEMORY HOME USER STEEL	Open parent directory, STEEL.
' CBX MEMORY CRDIR	Create subdirectory, CBX.
USER CBX	Open current directory, CBX.
« ' FA/FPA +CM * FB/(FPB * (1-FA/FE)) ' EVAL ENTER ' E31A STO	Define the value of E31A (i.e. left hand side of equation (H1-1), reference 13). Store E31A.
« ' FA/(.6 * FY) + FB/FPB ' EVAL ENTER ' E31B STO	Define the value of E31B (i.e. left hand side of equation (H1-2), reference 13). Store E31b.

≪ ' FA/FPA + FB/FPB ' EVAL ENTER	Define the value of E32 (i.e. the left hand side of equation (H1-3), reference 13).
' E32 STO	Store E32
≪ {STO FA FPA CM FB FPB FE FY} MENU ENTER	Set up an input menu for data entry.
' DAT STO	Store the data menu.

EXAMPLE

MEMORY HOME USER STEEL CBX	Recall program CBX.
NEXT DAT	Prepare data entry.
4.82 FA	Computed axial stress, f_a, ksi.
15.1 FPA	Axial compressive stress permitted in the absence of bending moment, F_a, ksi...$f_a/F_a > 0.15$; equation (H1-3) is not applicable.
.85 CM	Coefficient applied to bending term, C_m.
17.68 FB	Computed bending stress, f_b ksi.
22 FPB	Bending stress permitted in the absence of axial force, F_b, ksi.
22.2 FE	Euler stress divided by factor of safety, F'_e, ksi.
NEXT	
36 FY	Specified yield stress, F_y, ksi.
USER NEXT	Display output functions.
E31A	Left hand side of equation (H1-1) = 1.19.
E31B	Left hand side of equation (H1-2) = 1.03.

A.4.3: MPCO: PLASTIC MOMENT OF MEMBER SUBJECT TO COMBINED AXIAL COMPRESSION AND BENDING.

PROGRAM LISTING:

KEYSTROKES	COMMENTARY
MEMORY HOME USER STEEL	Open parent directory, STEEL.
' MPCO MEMORY CRDIR	Create subdirectory, MPCO.
USER MPCO	Open current directory, MPCO.
« ' (1-P/PCR) * (1-P/PE)/CM ' EVAL ENTER	Define the value of M2, the ratio of the maximum factored moment to the critical moment which can be resisted by the member in the absence of axial load (i.e. M/M_m of equation (N4-2), reference 14).
' M2 STO	Store M2.
« ' (1-P/PY) * 1.18 ' EVAL ENTER	Define the value of M3, the ratio of the maximum factored moment to the fully plastic moment (i.e, M/M_p of equation (N4-3), reference 14).
' M3 STO	Store M3.
« {STO P PCR PE PY CM} MENU ENTER	Set up an input menu for data entry.
' DAT STO	Store the data menu.

EXAMPLE

MEMORY HOME USER STEEL MPCO	Recall program MPCO.
DAT	Prepare data entry.
300 P	Applied factored axial load, kips.
1230 PCR	Maximum strength of an axially loaded compression member, kips (i.e. P_{cr} of equation (N4-2), reference 14).
8240 PE	Euler buckling load, P_e, kips.
1340 PY	Plastic axial load, P_y, kips.
.85 CM	Coefficient applied to bending term, C_m.
USER NEXT	Display output functions.
M2	$M = 0.857\ M_m$, kip ft.
M3	$M = 0.916\ M_p$, kip ft.

A.4.4: STPL: COLUMN STIFFENER PLATES

PROGRAM LISTING:

KEYSTROKES	COMMENTARY
MEMORY HOME USER STEEL	Open parent directory, STEEL.
' STPL MEMORY CRDIR	Create subdirectory, STPL.
USER STPL	Open current directory, STPL.
≪ ' (PBF−FYC * T * (TBF+5 * K))/ FYST ' EVAL ENTER	Define the required area of column-web stiffeners (i.e. A_{st} of equation (K1-9) reference 15).
' AST STO	Store AST
≪ ' 4100 * TCW^3 * √FYC/PBF ' EVAL ENTER	Define the column-web clear depth, DC, which, when exceeded, requires stiffeners opposite the compression flange (i.e. d_c of equation (K1-8) reference 15).
' DC STO	Store DC.
≪ ' .4 * √(1.8 * PBF/FYC) ' EVAL ENTER	Define the column flange thickness, TCF, below which stiffeners are required opposite the tension flange (i.e. t_f of equation (K1-1) reference 15, as modified by reference 16 for special moment-resisting frames in seismic zones 3 and 4).
' TCF STO	Store TCF.
≪ {STO PBF FYC T TBF K FYST TCW} MENU ENTER	Set up an input menu for data entry.
' DAT STO	Store the data menu.

EXAMPLE

MEMORY HOME USER STEEL STPL	Recall program STPL.
NEXT DAT	Prepare data entry.
189 PBF	Force in beam flange, P_{bf}, kips.
36 FYC	Column yield stress, F_{yc}, ksi.
.91 T	Column web thickness + doubler plate thickness, in.
.585 TBF	Beam flange thickness, t_b, in.
1.69 K	Column flange thickness plus depth of fillet, k, in.
36 FYST	Stiffener yield stress.
NEXT	
.66 TCW	Column web thickness, t, in.
USER NEXT	Display output functions.
AST	A_{st} = -2.98 sq in...stiffeners are not required.
DC	d_c = 37.42 in...stiffeners are required opposite the compression flange if the column web clear depth exceeds 37.42 in.
TCF	t_f = 1.23 in...stiffeners are required opposite the tension flange if the column flange thickness is less than 1.23 in.

SECTION A.5: PROPERTIES OF SECTIONS

A.5.1: IXX: MOMENT OF INERTIA OF COMPOSITE SECTIONS

PROGRAM LISTING:

KEYSTROKES	COMMENTARY
MEMORY HOME ' PROPS CRDIR	Create parent directory, PROPS.
USER PROPS ' IXX MEMORY CRDIR	Create subdirectory, IXX.
USER IXX	Open current directory, IXX.
≪ 0 ' ΣI ' STO 0 ' ΣA ' STO 0 ' ΣAY ' STO 0 ' ΣAYY ' STO SUBR ENTER	Set to zero the values of ΣI, ΣA, Σ(A * Y) and Σ(A * Y^2) and recall the subroutine, SUBR.
' STRT STO	Store the above routine in user memory under the label, STRT.
≪ {STO A Y I} MENU HALT I ' ΣI ' STO+ A DUP ' ΣA ' STO+ Y * DUP ' ΣAY ' STO+ Y * ' ΣAYY ' STO+ SUBR ENTER	Set up on input menu for data entry and, for each element of the cross section in turn, input the relevant data whilst the program operation is suspended. Then, key CONT to increment the values of ΣI, ΣA, Σ(A * Y) and Σ(A * Y^2).
' SUBR STO	Store the subroutine SUBR.
≪ ' ΣAY/ΣA ' EVAL ENTER	Define the height of the centroid, YBAR.
' YBAR STO	Store YBAR.
≪ ' ΣI+ΣAYY-ΣA * YBAR^2 ' EVAL ENTER	Define the moment of inertia, IXX.
' IXX STO	Store IXX.
≪ ' IXX/YBAR ' EVAL ENTER	Define the bottom fibre section modulus, SBOT.
' SBOT STO	Store SBOT
≪ ' √(IXX/ΣA) ' EVAL ENTER	Define the radius of gyration, RGYR.
' RGYR STO	Store RGYR.

EXAMPLE

MEMORY HOME USER PROPS IXX	Recall program IXX.
NEXT NEXT STRT	Initialize all storage registers to zero and prepare for data entry for element no.1.
34.2 A	Area of element no.1, sq.in.
25.56 Y	Height of centroid of element no.1, in.
26 I	Moment of inertia of element no.1, in^4.
CONT	Increment the values of ΣA, ΣI, Σ(A * Y) and Σ(A * Y^2) and suspend program operation to allow data entry for element no.2.
16.7 A	Area of element no.2, sq.in.
10.58 Y	Height of centroid of element no.2, in.
1170 I	Moment of inertia of element no.2, in^4.
CONT	Increment the values of ΣA, ΣI, Σ(A * Y) and Σ(A * Y^2).
USER	Display output functions
ΣA	Total area = 50.9 sq. in.
ΣAY	Σ(A * Y) = 1,051 in^3.
ΣAYY	Σ(A * Y^2) = 24,212 in^4.
NEXT	
ΣI	ΣI =1,196 in^4
YBAR	Height of centroid of composite section = 20.65 in.
IXX	Moment of inertia of composite section = 3,714 in^4.
SBOT	Bottom fibre section modulus = 180 in^3.
RGYR	Radius of gyration = 8.54 in.

A.5.2: RIG: RIGIDITY OF WALLS

PROGRAM LISTING:

KEYSTROKES	COMMENTARY
MEMORY HOME USER PROPS	Open parent directory, PROPS.
' RIG MEMORY CRDIR	Create subdirectory, RIG.
USER RIG	Open current directory, RIG.

≪ ' 3 * H/L ' EVAL ENTER	Define the deflection due to shear deformation in a wall with an applied unit load, SHR.
' SHR STO	Store SHR.
≪ IF WTYP C SAME THEN ' 4 * (H/L)^3 ' EVAL ELSE ' (H/L)^3 ' EVAL END	Define the deflection due to flexural deformation, flex, in either a cantilever wall or a fixed ended wall with an applied unit load.
' FLEX STO	Store FLEX.
≪ ' SHR+FLEX ' EVAL ENTER	Define the wall flexibility, DEFL.
' DEFL STO	Store DEFL
≪ ' 1/DEFL ' EVAL ENTER	Define the wall rigidity, RIG.
' RIG STO	Store RIG
≪ {STO H L WTYP} MENU ENTER	Set up an input menu for data entry.
' DAT STO	Store the data menu.

EXAMPLE

MEMORY HOME USER PROPS RIG	Recall program RIG.
DAT	Prepare for data entry.
20 H	Height of wall, ft.
35 L	Length of wall, ft.
F WTYP	The wall is fixed ended.
USER NEXT	Display output functions.
SHR	Shear deflection = 1.7143/Et in/kip where, E = Youngs modulus of wall, ksi. t = thickness of wall, ins.
FLEX	Flexural deflection = 0.1866/Et in/kip.
NEXT	
DEFL	Wall flexibility = 1.9009/Et in/kip
RIG	Wall rigidity = 0.5261 * Et kip/in.

REFERENCES

1. American Concrete Institute, *Commentary on Building Code Requirements for Reinforced Concrete (ACI 318-83),* Section 10.3, Detroit, MI, 1985

2. International Conference of Building Officials, *Uniform Building Code - 1994,* Section 1910.2, Whittier, CA 1994.

3. American Concrete Institute, *Commentary on Building Code Requirements for Reinforced Concrete (ACI 318-83),* Table 10.3.2, Detroit, MI, 1985.

4. International Conference of Building Officials, *Uniform Building Code - 1994,* Section 1910.5, Whittier, CA, 1994.

5. International Conference of Building Officials, *Uniform Building Code - 1994,* Equations (11-1), (11-2), (11-3), Whittier, CA 1994.

6. International Conference of Building Officials, *Uniform Building Code - 1994,* Equation (11-14), Whittier, CA, 1994.

7. International Conference of Building Officials, *Uniform Building Code - 1994,* Section 2101.4, Whittier, CA, 1994.

8. International Conference of Building Officials, *Uniform Building Code - 1994,* Section 2107.2, Whittier, CA, 1994.

9. Schneider R.R. & Dickey W.L. *Reinforced Masonry Design,* p158, Prentice-Hall, Inc., Englewood, Cliffs, NJ, 1987.

10. American Forest and Paper Association. *National Design Specification for Wood Construction, Eleventh Edition (ANSI/NFoPA NDS-1991).* Washington, 1991.

11. American Forest and Paper Association. *Commentary on the National Design Specification for Wood Construction, Eleventh Edition (ANSI/NFoPA NDS-1991).* Washington, 1991.

12. International Conference of Building Officials, *Uniform Building Code - 1994,* Section 2251 F1.3, Whittier, CA, 1994.

13. International Conference of Building Officials, *Uniform Building Code - 1994,* Section 2251 H1, Whittier, CA, 1994.

14. International Conference of Building Officials, *Uniform Building Code - 1994,* Section 2251 N4, Whittier, CA, 1994.

15. International Conference of Building Officials, *Uniform Building Code - 1994,* Section 2251 K1, Whittier, CA 1994.

16. International Conference of Building Officials, *Uniform Building Code - 1994,* Section 2211.7.4, Whittier, CA, 1994.

This page left blank intentionally.

INDEX

Exam Files

Professors around the country have opened their exam files and revealed their examination problems and solutions. These are actual exam problems with the complete solutions prepared by the same professors who wrote the problems. Exam Files are currently available for these topics:

Calculus I
Calculus II
Calculus III
Circuit Analysis
College Algebra
Differential Equations
Dynamics
Engineering Economic Analysis
Fluid Mechanics
Linear Algebra
Materials Science
Mechanics of Materials
Organic Chemistry
Physics I Mechanics
Physics III Electricity and Magnetism
Probability and Statistics
Statics
Thermodynamics

For a description of all available **Exam Files**, or to order them, ask at your college or technical bookstore, or call **1-800-800-1651** or write to:

Engineering Press
P.O. Box 1
San Jose, CA 95103-0001